Stereology and Stochastic Geometry

Computational Imaging and Vision

Volume 28

Stereology and Stochastic Geometry

by

John E. Hilliard †

formerly Northwestern University, Evanston, IL, U.S.A.

and

Lawrence R. Lawson

Allegheny Institute of Natural History, University of Pittsburgh, Bradford, PA, U.S.A.

SPRINGER-SCIENCE+BUSINESS MEDIA, B.V.

A C.I.P. Catalogue record for this book is available from the Library of Congress.

ISBN 978-90-481-6455-4 ISBN 978-94-017-0163-1 (eBook)
DOI 10.1007/978-94-017-0163-1

Printed on acid-free paper

Originally published by Kluwer Academic Publishers in 2003
Softcover reprint of the hardcover 1st edition 2003

This book is dedicated to

John Cahn

and

Gwendolen Hilliard

U.S. Copyright History of *Stereology and Stochastic Geometry*

Contents

PREFACE

Somebody had to do it. The Chinese speak of deep water wells called "grandfather wells" because they take three generations of diggers to complete. Imagine the thought of such a well being abandoned incomplete by the third generation. What a loss! This book is like a grandfather well except that it has taken only two generations, John Hilliard's and mine, to finish. When I saw his manuscript lying in a heap, I decided that I must spend the time to put it and his notes into a publishable form. Now, it is done.

This book is mostly about performing spatial measurements through the statistical sampling of images; it is a text on classical stereology as John Hilliard saw it. His vision of the subject was broad. Consequently, its title is broad too. It presents this subject and some of its modern extensions from the classical perspective of the one of the founders of the field, and my first advisor at Northwestern University, John Hilliard. There is nothing new in this book but much that may have been lost over time. It rediscovers many useful discussions about such subjects as the variances of stereological measurements, anisotropy etc. It recovers some of the dialogues between John Hilliard and his students on such topics as fractals and Monte Carlo simulations. It recaptures a little of John Hilliard's unique and subtle wit. Best of all, it goes through all derivations in a step by step manner without the difficult personalized notation which characterizes much of what is now called "modern".

There are two kinds of forces at work in applied mathematics. One bends and one straightens. They tend to separate into two schools. In stereology, what has been called the "East German" school and their allies took over the lead in the field from Hilliard's application-oriented generation in the late

1970s. They have contributed order and rigor to the subject through set and measure theory. Their greatest accomplishment has perhaps been the breaking of the ties between stereology and dimensionality. Stereology now treats all dimensionalities including those of hyperspace at once. In this book, Chapter 9 presents Hilliard's digest of Miles's treatment of these hyperspatial generalizations. But, along with the elimination of inconsistencies, some useful work has been pushed from the foreground into the background because the strict mathematical apparatus to support it does not yet exist. Thus the work of El-Soudani on surfaces and the Hilliard-Cahn variance estimators seem to have faded. Both are worth reexamining. El-Soudani's ideas about surfaces are discussed in Chapter 7. The Hilliard-Cahn estimators of the variances of stereological measurements are discussed in Chapter 8.

This book is intended for materials scientists, anatomists, histologists and similar workers who want to understand the nuts and bolts of the subject rather than just do the applications by rote. These applications have procedural protocols. John Hilliard had hoped in his lifetime to provide a discussion of all the protocols as well. Although he worked several decades on this book, he died before he could do so. In reviewing the state of the art, I concluded that I could not live long enough myself. This book is conceived as a textbook, but it should not be a bitter pill for students to swallow. It is to be enjoyed. Its perspective is delicious and a little irreverent.

Stereology and Stochastic Geometry descended from an earlier work, *Stereology, Principles and Practices.* Although having blocked-off sections marked "to be completed" in the future, the work was privately published in several editions and used as the principal stereology text at Northwestern University from the 1970s into the 1990s -- even after John Hilliard's death. The strength of that work was that it presented the derivations of stereology formulas in a compact organized step-by-step manner totally accessible to students having little or no knowledge of set theoretic analysis and measure theory. Since the derivations were given step-by-step, the intermediate results were immediately available. I found this helpful in my own work on Monte Carlo Markov chain (MCMC) modeling of crack coalescence. For example, the geometric problem of crack coalescence is closely related to Buffon's needle problem (discussed in this book) except that isotropy cannot be relied upon. Hilliard's original text (as does this one) uniquely included an intermediate integral that did not require isotropy and could be applied to solve the coalescence problem.

After John Hilliard's death my then current advisor, Morris E. Fine, asked me for a copy of the manuscript to circulate among senior faculty. He hoped to find someone willing to put it in completed publishable form. The manuscript circulated for six years without anyone wanting to take on the task. At that point I asked John Hilliard's sister-in-law, Gwendolen Hilliard, for the rights to the book, which she gave me.

The scope of *Stereology, Principles and Practices* was simply too large to manage. Accordingly, I decided to omit the discussion of specific practices and rename the work *Stereology and Stochastic Geometry*, since that was descriptive of the work as it stood in 1995. Another limitation of the original work was that it was often too laconic. It lacked the oral tradition given in John Hilliard's lectures that both explained the content and brought it up to date. These explanations I have tried to add from the lecture notes and handouts used the last time John Hilliard taught the course.

Although a textbook on classical stereology, modern topics such as fractals, numerical methods, information theoretic and mathematical morphology approaches were under discussion and became class projects at the time I was involved with John Hilliard. I have tried to capture the essence of some of these discussions hoping that they might entice the reader to further work and study in these areas. I have added discussions of the "new stereology" including star volume (a nucleator or "local" method) and the disector. This book is intended to be usable as a text for a one-semester course on stereology. As John Hilliard taught the course, it began with a brief review of Chapter 2 and then turned to the subject itself, Chapters 4 through 6. The material pertaining to transmission electron microscopy from Chapter 7 was presented in lecture, since many students were electron microscopists. Chapter 9 was also discussed in lecture but few assignments were made. The "non-problems" given at the end of Chapter 10 are problems taken from handouts called "non-assignments", which were not required to be turned in.

This is an age in which evolution is often measured by changes in notation and semantics. Even so, I have retained the original notation of *Stereology, Principles and Practices* (which conforms the "Bureau of Standards" notation as promulgated at the Fourth International Congress for Stereology) as much as practicable. One change I have made is in the area of the treatment of anisotropy, one of John Hilliard's major contributions to the field. The original treatment of anisotropy in the plane is reworked to favor a discrete Fourier transform notation with complex coefficients, since this change makes it possible to do anisotropy analyses using the FFT algorithm. The treatment of anisotropy in 3-space is substantially revised. The notation for spherical harmonics has been replaced by that defined in such popular quantum mechanics texts as Richard Liboff, *Introductory Quantum Mechanics* and implemented in calculating packages such as Mathematica© and MAPLE©.

John Hilliard was born in England on May 14, 1926. He was educated at the University of Liverpool. and emigrated from England to America. He became a graduate student in Materials Science at MIT. At MIT, he met John Cahn who became his lifelong collaborator. Together they worked for General Electric Corporate Research doing seminal research in stereology and in materials physics. John Cahn taught at MIT and later joined the National Bureau of Standards (He continues there now as a Fellow under its new title, the National Institute of Standards and Technology.) John Hilliard was recruited as faculty by Northwestern University, and they continued their collaboration until John Hilliard's death on April 21, 1987. John Hilliard never married. When I knew him, he lived alone in an apartment near to the campus. I used to see him from time to time having a thoughtful lunch by himself at a Greek cafe near the Technological Institute.

I would like to thank those who have helped make this project possible. First I want to thank Gwendolen Hilliard. Then, I want to thank those "famous men" in the field who helped through editorial assistance and whose names alone will identify them. In alphabetical order they are: John Cahn, Jiri Janacek, Pertti Mattila, Ivan Saxl and Martina Zaehle; I hope they will be pleased with this result. Then, I want to thank those not yet quite as famous in this field. I want to thank Gabriel D. Carroll of the Berkeley Math Circle for a result from combinatorial number theory used in a Chapter 8 discussion and Richard Shorell, the photographer/mathematician, who taught me that nature conveniently repeats her patterns on widely differing length scales. Next, I want to thank the librarians who helped with the reconstruction of John Hilliard's lost bibliography, specifically Marietta Frank and Nancy Cook of the University of Pittsburgh at Bradford, Godfrey Waller of Cambridge University and also those at SUNY University at Buffalo and at Cornell University who were invaluable to this project. Again and again experience teaches that the generosity of librarians is under-appreciated. Finally, I want to thank those who helped with the copy: Ms. Julie Carr for transcribing a great deal of the equations and Prof. Elyce Helford whose group at Middle Tennessee State University's English Department performed the initial scanning of John Hilliard's manuscript into digital format.

In addition to the individuals who helped with this project, I wish to acknowledge the organizations which sponsored the research from which some of the photographic examples were taken: These include the IBM Corporation, The National Science Foundation through the Northwestern University Materials Research Center under Grant. No. DMR 8821571 and the National Institute of Standards and Technology under Contract No. 70NNANB9H0916.

Finally, I notice that some of the references cited herein begin with disclaimers. Should I be different? I have spent over 6 years checking and verifying the contents of this book. But, there has never been an instance that in rereading these chapters, I do not see something that could have been done better. I cannot guarantee that every formula is error-free or that every phrase is unambiguous. Should the reader find an error, please let me know at *Cal_Mant@yahoo.com.*

-Larry Lawson, Bradford, Pennsylvania, April 2003

Chapter 1

IMAGES AND PATTERNS

1. INTRODUCTION: WHAT IS STEREOLOGY ALL ABOUT?

Stereology studies spatial statistics as seen through images. It particularly relates to problems of measuring quantities such as volume densities from photographs. A UK website in 2002 listed 21 separate definitions for stereology garnered from the literature. Some of these definitions are very broad; others are narrow. The problem of defining stereology is existential: since existence precedes essence, we will not know exactly what stereology is until stereology is dead; then we will know what stereology was. Is stereology mathematics or physics? This book views stereology as physics. Accordingly, the mathematical terminology used should be interpreted in its associated physical contexts.

Stereology began with the examination of rocks. For instance, gold has long been a preoccupation of mankind. Every speck of it is precious. Gold is widely distributed in the earth's crust. Almost every pound of earth has some gold in it. But, gold is very scarce. Most pounds of earth have only a few atoms of gold and these are invisible. When a prospector looks at a rock, he can see only its surface. If he can see some gold or gold ore in the rock, he asks, "How much gold is there in it?" The prospector sometimes carries his piece of rock to the assayer. The assayer weighs it, grinds it to powder and does some form of chemical or gravimetric analysis on the powder to measure the amount of gold. Grinding the rock to powder presumably exposes all the gold -- but, only if the powder is fine enough. The assayer may then scorify the sample, dissolving it so that all atoms are in a sense exposed. We can concern ourselves with how fine a powder is meaningful. Do we need to determine gold if it is only on an atomic scale? What if the rock is too big to pound into dust? Well, the assayer can at least analyze a piece of it. If it has a vein of gold, he can analyze material from the vein. For comparison, he can analyze the matter outside the vein, but he is now confronted with the question of how much of the rock is "vein" and how much is other stuff. This is just the same problem restated. On a larger scale a mining engineer may look at a rock formation or a geographical region and ask, "How much gold is there in it?" It is the same question again on a still bigger scale. Except for small pieces that can be analyzed by reducing them to molecules, as by dissolving them into a solution, the problem of determining the amount of something not directly observable has long persisted. Answering this problem as it pertains to rocks through geometric inference was the beginning of what we now call "stereology".

From a geometric viewpoint, the prospector's problem arises because he can only see things in two dimensions but the quantity he wants to measure, the amount of gold, is a three dimensional one. As humans, distinct from Superman, we can see the surface of a rock, but not the inside of it. Even if we were to x-ray the rock, we might be hard pressed to trace out the amount of gold, even with computerized tomography. Computerized tomography has a limited resolution compared to surface observations with a microscope. (X-ray diffraction is also confined to spaces near surfaces. Samples can be pulverized to overcome this difficulty, but this is much the same as dissolving the specimen.)

Geometric inference may give a more accurate answer with less work than tomography, diffraction or chemistry. In medicine tomography and stereology are often used together. Determining the composition of rocks is only one example of the type of geometric inference this book examines. Historically, it was the first. As a result, stereology is often thought of as the study of how to make three-dimensional measurements from two-dimensional data. This is by far its most significant use. In its broader sense it deals with any kind of measurement problem dependent on spatial sampling especially where the dimensionalities do not agree in some way, e.g. the dimensionality of the sample is smaller than that of the specimen. How does stereology solve these problems? It does so through the analysis of patterns. It uses highly general auxiliary information, such as that a three dimensional pattern which is not wholly observable may be statistically "isotropic". It then applies this external isotropy condition to the data available -- typically photographs -- through statistical analysis.

This book deals with the quantitative interpretation of patterns. Quantitative interpretation of patterns is a very broad definition, since the study of patterns in nature has been used as a definition for all of physical science. So, some restriction needs to be applied. Here we will be studying only spatial patterns and only through passively measuring and characterizing them. Some would say we are studying only their "kinematics" because we ask nothing about how these patterns came about. (The term "kinematic" has special meaning to stereologists. We won't find out what it is until Chapter 10.) We won't inquire into causes nor infer determinism. Instead we will develop a set of statistical tools, which we will call "stereology". Being both statistical and geometric in their nature, these tools are part of what is called "stochastic geometry". Loosely speaking, if stochastic geometry is the forward problem analogous to probability, stereology in general terms is the inverse problem analogous to statistics. Although geometric randomness is a key concept in stereology as randomness is in ordinary statistics, it would be an oversimplification to say that all stereological measurements are randomized or made on random patterns. Some simply compile "statistical" data.

When stereology first appeared, its definition was very narrow. Since stereology has evolved, its definition has broadened. The student may seek its definition's "centroid" in order to better understand the provenance of the subject. History is often a good way of locating the center of any discipline. Perhaps the earliest definition of stereology is that given by Hans Elias.

"Stereology, sensu stricto, deals with a body of methods for the exploration of three-dimensional space, when only two-dimensional sections through solid bodies or their projections on a surface are available. Thus stereology could also be called extrapolation from two-to three dimensional space."

Stereology therefore is largely concerned with deriving three-dimensional data from two-dimensional pictures. These two-dimensional pictures can come about in basically three ways. Most often, they are obtained by slicing a three-dimensional object with some instrument such as a microtome and viewing the resulting surface. A second way is simply to photograph the natural surface of a three-dimensional object. The third way is to project the silhouette of a three-dimensional object or (using penetrating radiation) of the features inside a three-dimensional object. Both of these latter two methods involve explicit projection.

Historically, stereology has been preoccupied with slicing things and viewing the slices with microscopes. Early books on stereology talked about it as "quantitative microscopy" and this term is still used. Quantitative interpretations of the second type, surface photographs, of data -- usually macroscopic -- have been somewhat neglected by stereologists and left to photogrammeters. Photogrammeters however are mostly concerned with making maps and less concerned with the geometric properties of what is being mapped. Photogrammetric microscopy is in its infancy. The earliest work of this kind is only about 20 years old. The narrow field of view of most microscopes has hindered its development. The first stereological approaches to the third type of image, through projection, evolved from the interpretation of transmission electron micrographs (TEM). In these, the features tend to overlap leading to difficulties in determining such quantities as their area or volume.

Consider an example of a stereological problem. Let us be contrary and build this example about the second way of making two dimensional pictures from three dimensional data because, although it is least used by stereologists, it is the most familiar to the ordinary reader: simply taking a photograph. Recently, interpretation of space probe telescope photos has raised questions like "are the craters on Mars (or Phobos) randomly distributed?" Answering this sort of question is certainly within the province of stochastic geometry. The surface of Mars is a two-dimensional manifold in three dimensions. We wish to examine the spatial distribution of features called "craters" using photographs, two-dimensional projections. We might probe these projections with an array of random points and see which intersect a crater. Point probes of this type are basic stereological tools. (Again: stereology is a sub-discipline of stochastic geometry. Quantitative microscopy is a sub-discipline of stereology.) We might like to find say the average number (or probability) of craters of a certain size per square mile. Because in doing this we would need to account for the fact that Mars is not flat, we need to map the topography. It follows logically that photogrammetry becomes connected to stereology in problems requiring the interrogation of curved surfaces. But, most of the surfaces studied by stereologists are deliberately made flat. If we did not correct for curvature, we would find that most craters would appear to be nearer the edge of the visible surface.

Stereology is not limited to just two-dimensional representations of three-dimensional data. In Chapter 9 we look at ourselves as flat-landers and explore stereology formulas for 4, 5 and any arbitrary number of dimensions. If as our three-dimensional selves we were like the two-dimensional flat-landers in the story “Flatland”, unwittingly in a three-dimensional world, we could pose the question, "How many dimensions are we in"; and, answer it stereologically. It is remotely possible that such questions might be relevant to physics on very small or very large scales.

Since stereology evaluates geometric quantities that often cannot be directly perceived, examples of the kinds of quantities measured gives a feeling for what the subject is about. The amount of interfacial area per unit volume of an opaque solid having two constituent phases is a typical example. There is no direct way of determining such a quantity. Even if you could view the solid with x-ray tomography, you would be frustrated by the fact that you could still only see plane images. With computer solid modeling you might be able to see a bug's-eye view of the interface between phases. But, if that interface were tortuous, this visualization would not do much to help you measure its area. The closest means to a direct measurement would have to be made using a computer to identify the interfaces and then integrate the surface in three dimensions. Anyone with experience in trying to do this in two dimensions will know how easy it is to disagree with a computer about where an interface actually lies. This is because real interfaces often appear to be discontinuous. We could oversee such a measurement in two dimensions and order the computer to act according to what our eyes would tell us. This would result in a decision, which would be comfortable if not always correct.

Finding the area of a human lung with all its tubes-within tubes would be intractable from the standpoint of combining x-ray tomography with computer surface tracing. Here is another case where statistical inference is needed and stereology supplies it. But, behind statistical inference lurk statistical assumptions in the form of explicit or often implicit models. Mathematical continuity is a major assumption implicit in many models. Reckoning with the conflict between our idealized notion of continuity and the discontinuities of reality is a subject which stereology needs to explore further. The mathematical tools to do it are now being made. Classical stereology, the best place to begin studying the subject and the primary topic of this book, assumes a modest amount of smoothness and continuity. The absence of these in a particular sample or image has been traditionally identified as the result of "mistakes".

Another example of a stereological measurement is the determination of the amount of volume of one constituent of material per unit volume of another, the gold problem just discussed. For measuring the amount of "rat dirt" in a pound of flour, a chemical measurement might suffice. For measuring the amount of pearlite, a so-called constituent of steel, that is in fact only a pattern of precipitation, chemical analysis would be of no avail since it would tell nothing about the *pattern* of precipitation. Pearlite forms in microscopic "colonies". Consequently some means is required to extract the information from micrographs. Micrographs are two-dimensional pictures and here again, the desired information is three-dimensional. This illustrates the first part of the stereological perspective: the dimensions of the entity we want to measure are usually higher than those of our vision. Also, steel is decidedly opaque. If it were transparent, we could maybe look at each colony of pearlite from several angles and match its shape to some mensuration formula and thereby measure its volume. If we added all these volumes together for say a nice easily measured cube of a sample, we would have determined the volume fraction. But, we cannot see inside the steel. Even if we could, the colonies would still overlap and obscure each other.

Any time we slice the steel and look at the face of the cut, we have made that cut unknowing of what was inside. Slicing and looking is a random procedure. (A great deal has been said about the rightness and wrongness of random cutting and slicing. This is a practical side of stereology, which this text will leave to others.) Part of the stereological perspective is that the material we are examining is largely unknown. There is an element of probability which enters when we use a two-dimensional picture to represent a three dimensional object. Probability has two interpretations: "don't know" and "don't care". We are now facing the "don't know" part in that we cannot see into an opaque sample. The "don't care" part arises when we consider the amount of computational effort required to solve a problem and arises subtly when we will talk about *regionalized variables*. Even if we could see and measure each pearlite colony, we probably wouldn't choose to because we know that a statistical sample can tell us what we need to know with much less effort. For both these reasons, probability makes itself an integral part of the stereological perspective.

Another class of stereological problems deals with surfaces. We may wish to know the ratio of area of tile to that of grout in a mosaic. Again we could divide the surface into patches and apply mensuration formulae, but it is simpler to draw a line on the surface and determine the relative areas from the proportions of the line that fall in grout or on tile using a stereology formula. Or we could choose a set of random points on the surface and count the relative fraction of these that fall on tile or grout and use another stereology formula to determine the ratio of areas. These two geometrical statistical sampling techniques are further examples of what stereologists term "probes". When the surface is not flat, new problems arise. Finding the length of river per square mile of land or the relative amounts of farmland planted in corn to that planted in soybeans are typical examples of this latter type. Unlike flat tiles, land is usually curved distorting the proportions of lengths and areas as viewed in photographs. The amount of formal systematization of the solution of this type of problem for practical purposes seems to have fallen short of that devised by microscopists for dealing with sections of specimens. One reason is that cartographers often use ready approximations rather than stereological probes. *Orthophotography* is a photogrammetric technique that corrects for the distortion that results from the point-projection construction of photographs [Wolf, 1983], which is used for comparing surface areas between say pine tree forest and hardwood forest. It does not, however, correct for the contour of the land. Numerical integration after fitting the contour with polynomials is another approach. The question of what is the best polynomial is not beyond the reach of stereology since it is a spatial sampling question. While you can see a surface in a picture even if it is mountainous and there is nothing hidden as in the opaque sample, this does not mean that geometric and statistical inference are not needed to obtain the best result.

There is one aspect of cartographic measurement problems that is so profound that it has given rise to what amounts to at least a whole new branch of mathematics. It is the measurement of the length of a coastline. The length of a coastline is not fixed; it depends on how finely you measure it. The "science of fractals" resulted from consideration of this measurement scale issue. Retrospectively, we see that this unexpected difficulty in measuring coastline may also be related to many of the boundary discontinuity "mistakes" mentioned earlier. Measuring coastline is a one-dimensional task. Surface areas are two-dimensional and are usually met embedded in three-dimensional space. Measurement scale problems are certainly not confined to one dimension and are likely to be a part of any real surface area measurement problem. These are new areas for stereology to explore.

Stereology often determines quantitative measurements of geometric properties that take the form of ratios. Volume fraction, areal fraction, number per unit volume, length per volume, area per volume, total length, total integrated curvature and length of curve per unit angle per unit area are examples of the types of ratios stereologists measure. Stereological measurements are always statistically sampled measurements.

2. VARIABLES AND INVARIANCES

There are three kinds of variables important to stereology, continuous variables, random variables and regionalized variables. But, having so said, there is even at this outset, a problem. Real-life variables are none of these. These are just three conventions we apply to variables in order to characterize them. The concept of the continuous variable is a product of analysis: continuous variables are *analytic* (usually). For a variable to be analytic, it must be differentiable at all points. In order for this to happen there has to be some region of space within which some measurable property takes definite values at all points. The values of that measurable property are said to be a function of their positions within that space. Commonly the "spaces" are *Borel sets* (see Appendix and Chapter 3 for more details.) They represent what is called the *domain* of the function. Examples are the points on a line segment or the points within some bounded region of what we often call the "x-y plane". If the function is continuous and analytic, viz. "well-behaved", then around each point we must be able to construct a neighborhood of size δ. Within that neighborhood the absolute value of the difference between the value of the function at the center of the neighborhood and at any other point within that neighborhood must always be less than some positive number ε. As δ gets smaller, ε must likewise diminish in some uniform way so that ε and δ approach zero simultaneously. Continuity implies infinite divisibility. This is a useful construct, and the development of calculus depended on it. The Heisenberg uncertainty principle is one example of the limitations of this construct; it sets a resolution limit on spatial detail. The presence of noise in all real-life images is a more immediate example of the limitation of continuity; noise directly limits the measurement accuracy of variables. This limits their differentiability.

Random variables are not expected to be continuous although they can be. Random variables have an average value called the *mean.* Because they are not well behaved they have a dispersion that behaves like the ε for the continuous variable but rarely does it ever go to zero. Many random variables are not bounded by any finite ε. Most of the random variables of interest to us are random functions. If the mean of a random function is not affected by translation, the function is termed *stationary*. Thus, were the average price of stocks to remain the same every year, the price of stocks would be stationary -- which it is not. If the average concentration of gold in a lump of ore does not depend on where we sample it, the average gold concentration is stationary. Obviously, some kind of scale is implied over which this stationarity applies since on a small enough scale there must be random fluctuations. It follows that for any finite body of material, any time-independent quantity can be regarded as stationary since the average of it over the finite sample is unchanging. For many random variables there is a sort of continuity in the ε-δ sense. For example, because solids are subject to diffusion of their molecules, concentrations of constituents such as gold often do not change too abruptly. Random functions which behave this way may be said to have a *correlation distance* over which changes take place. If this correlation distance (or more technically its *covariance function*) does not vary with translation and the mean also does not vary with translation, a random function of position is said to be stationary in "wide sense". These are two examples of invariances.

Similarly we can specify the same invariances but include a variable function of angle. Or, contrarily we might have invariance with respect to rotation. We might allow the mean to vary as a function of distance from a point but only as a function of distance. In that case the function is said to be *isotropic* with respect to the center of that symmetry. We could also define isotropy in the "wide sense" if the covariances vary with only relative displacement but not with rotation. Continuing this discussion of definitions, consider a function in polar coordinates. Let that function's mean vary with distance along the polar axis but be invariant with angle. Such a function is said to be *orthotropic.* In Chapter 2 we will begin our discussion of probability through a general revue of the parts of the subject we will need. This will better clarify these concepts. In most problems we will encounter, we will tacitly assume that whatever parameter we are interested in can be represented as the sum of a random isotropic and a random *anisotropic* function. In most cases, the anisotropic component will be regarded as part of the error of estimation.

We have now discussed continuous variables and random variables. The third type of variable we will encounter is called the regionalized variable. It has its origin in geostatistics [Matheron, 1963]. Regionalized variables are a mathematical construct to represent the grade of an ore. In a sense they are in between a continuous and a random variable. Their purpose is to find a way of dealing with the problems of limited spatial resolution, limited differentiability and noise. A regionalized variable does not take values everywhere but only on its *geometrical field* a term that replaces or augments the idea of domain. The *support* of a regionalized function representing the grade of an ore may be the ore sample itself including its geometrical shape. For example, we might define grade in terms of samples of specified dimensions taken with a core drill. The term "support" implies some sort of averaging of the grade over the size of the sample. When the term support is used, there is no assumption that says infinitesimal continuity is a meaningful concept. Rather the support contributes what may be thought of as an innate and irreducible δ which is intimately connected with sample size. As a hint toward understanding its meaning, the term support is often used in the context of fractal and multifractal functions. The support is what forms the container or holder for the variable defined by the function. The support and the regionalized variable are often inseparable. In the same rock formation changing the geometrical support will generally result in the formation of a new regionalized variable. Like random variables, regionalized variables are regarded as having qualities which are unknowable in detail. But, unlike truly random variables, this unknowability results from the complexity of the variable not from any fundamental uncertainty. Anywhere in the rock formation, whatever ore there is has a grade. In principle that grade could be known with absolute certainty. From a practical point of view all we can do is take samples of a relatively tiny amount of the rock. To cope with unknowability, the concept of the regionalized variable has been applied to weather-related and pollution-related variables as well. Both are difficult to predict. But, especially in hindsight, pollution and weather are always present. As the old adage says"...we will have weather, whether or not."

Regionalized variables have three sets of attributes. The first refers to the relation of the variable to its geometric support. If we are looking for alluvial gold and we could have "point" or *punctual* support: the function can take only two values, e.g 1 if there is an atom of gold at the point and 0 if there isn't. If the regionalized variable describes the proportion of gold in 10-Kg core samples it will not be the same as one which describes the proportion of gold in 10 ton blasts.

Secondly, the regionalized variable may be idealized as having continuity and variation. This continuity is only assured when one speaks of continuity "in the average". Continuity in the average allows regionalized variables to be analytical. This is their advantage. As with random variables, regionalized variables are describable as being sums of continuous variables and step functions. In the nomenclature of geostatistics these step functions are called the *nugget effect*. Namely, in plotting the concentration of gold in the ore there are jumps in local concentration due the presence of pure gold nuggets. The term is also applied to autocovariances, the covariance of the concentration at a point with that at another point some distance away. Normally one expects this covariance to start dropping immediately as distance increases. If it remains at a high value for some particular distance and then begins dropping, then that particular distance is the size of the nugget -- as would happen if the gold were not uniformly dispersed but concentrated into nuggets with non-gold in between. In practice, this spacing might be smaller than the smallest practicable spacing between samples.

Thirdly, the regionalized variables may exhibit anisotropies. With ores there may be a stratification effect: the ore lies only in certain geological strata. These anisotropies are often patterned and are given names in the geological literature. We will discuss the quantification of geometric anisotropy in Chapter 5 and again in Chapter 10. In Chapter 10 we will briefly discuss the geostatistical method called *kriging* that allows one to treat cases where the quantity being measured is not uniform and isotropic. In practice all the major results of classical stereology are based on the assumptions of uniformity and isotropy of the measured attribute within the sample—either because it truly is or because the attribute is sampled in a way that induces uniformity and isotropy in the measured variable. (This is termed “random sampling” in spite of the fact that it implies the opposite.) In deriving “random sampling” results, directional variables are integrated out. One advantage of this book over others is that it presents much of the intermediate results. In situations where variables do not integrate out as they are "supposed to", the intermediate results may still apply and the reader can do her own integration for the special case at hand.

Classical stereology is founded on the subject called integral geometry. Many of the major results are over 100 years old and their derivations assume strict continuity. The use of the Lebesgue integral and the concept of continuity in the average allow these results to be applicable in realistic settings. In what follows, variables will be assumed to be at least piecewise continuous. Lines and surfaces will be assumed to be rectifiable, that is, having definite length or area. This is not always true. Some quantities will be found to depend on magnification in a way suggesting that the are unbounded. A brief practical discussion of this will be given in Chapter 10 although the subject of the stereology of true fractals is beyond the scope of this book. Regionalized variables, because the support is a part of the variable, get around the fractal problem. Most of the quantities in this book may be thought of as regionalized variables without introducing any conflict.

3. QUANTITATIVE ATTRIBUTES OF PATTERNS

Figure 1.1 shows a picture of a section through a piece of metal, in this case an alloy of lead with a small amount of tin. It is used as a solder in microelectronics.

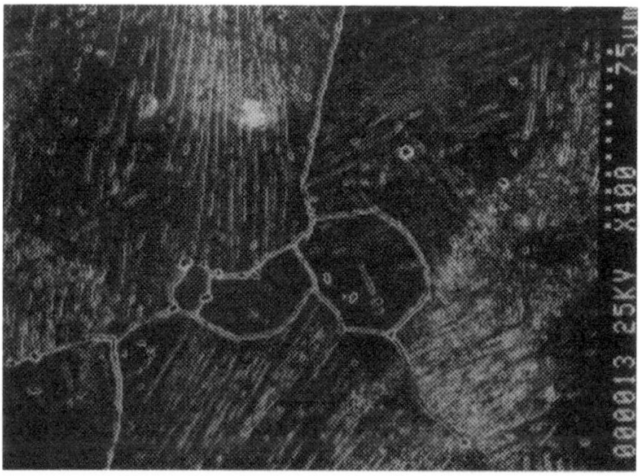

Figure 1.1:. Polished solder surface

The magnification of the picture is large. The picture is made with an electron microscope. Connecting lines resembling roads divide the surface into separate sections. In the middle there is a rotary with an island in the center. Within most sections are fields of parallel lines. These lines have some width but do not connect to each other. A metallographer could tell us that the parallel lines are strips of tin while the balance of the surface is nearly pure lead. The roads are dividing lines between crystals of different orientation. In this picture the fact that what we are seeing as roads are the interfacial surfaces of three-dimensional crystals is not evident. Having apparent width, the "roads" are like the grout between the tiles. Each tile has a different orientation -- we just cannot see it. The roads are called *grain boundaries* because the individual crystals are called *grains*.

Knowing the magnification, the properties which we could directly measure in Fig. 1.1 include the total length of the grain boundaries, the apparent area subtended by the grain boundaries and the total length and area of the tin strips. At higher magnification, we could measure the spacing between the tin strips. For the most part the grain boundaries appear straight but the "rotary" has curvature. We could measure its curvature at various points. At a lower magnification we could measure the areas of the grains. These are examples of some of the measureables on a plane surface. Because the surface is flat, we know that all the lengths and areas will be in the relative proportion in which they appear. If the surface were not flat, we would have to account for the distortion. Metallographers and geologists have developed methods for making flat surfaces, which go back to the quarrying and polishing of building stones in ancient Egypt. The assumption of flatness is so ingrained that stereologists often refer to the two dimensional surface being observed as the *plane of polish*.

We have just enumerated some of the properties we could measure on a photograph. These properties belong to the photograph and would be different on a different photograph. None of them are properties of the object photographed. In order to begin to relate these measurements to the object itself we can make ratios out of them. We can divide by the area and find grain boundary length per unit area on this photograph. We could calculate the areal fraction of the tin plates. We could calculate an areal fraction of grain boundary. We could find a grain boundary curvature per unit area. The advantage of working with ratios is that they are capable of expressing quantities that at least do not formally depend on the size of the sample. If we could know that these ratios do not depend on how the sample is prepared, which way we look at the sample or where on the sample we look, then we are still closer to finding properties of the sample rather than of the picture of the sample. It turns out that the apparent area of the grain boundaries depends on the kind of polish we use. But, their length does not. The area of the tin strips also depends a little on the kind of polishing. For the grain boundaries, the ideal polish reduces their width to atomic dimensions. Hence, we can ignore their width unless we are looking at atomic resolution electron micrographs. On the other hand, the width of the tin strips remains about the same with any good polish. We could find a bad polish that corrodes the edges and makes them appear larger. We could also find a bad polish that doesn't really bring out the edges or obscures them altogether. We could find an etch that would dissolve some of the tin and leave the lead making measurable slots with sharp edges at the sides. This is in fact the treatment given to the specimen in Fig. 1.1, as a higher magnification photograph would show. These then are issues of specimen preparation not stereology per se. So, apart from specimen preparation, the issues of specimen orientation and where on the specimen we look remain.

Here we formally introduce the working assumptions upon which classical stereology is based: 1) There exist mean values which are properties of the material. 2) These mean values are *stationary*; they do not vary with translation --- moving the specimen under the microscope. 3) These mean values do not vary with specimen orientation, i.e. rotating the specimen. In the language of mathematics, we say that these mean values are invariant under *affine transformations*, that is transformations which are non-distorting, namely translations and rotations.

Now that we have defined some properties of the surface of the specimen, we need to see what sorts of things we can infer about the bulk of the specimen, that is properties of the three-dimensional structure of the specimen. In order to do this we must be able to make similar assumptions about the bulk of the specimen as we made about its surface. Namely that there exists a mean value for the quantity we want to measure and that its covariances are either isotropic or at least stationary with respect to both observation and slicing or in projection for that matter, since are the methods we use to reduce three-dimensional structures down to measurable two-dimensional ones.

For instance, from the standpoint of gross anatomy, enough slicing would eventually reveal that the strips of tin in Fig.1.1 are in fact thin tin plates, lamellae. A random slice through a plate gives a strip. The width of this strip depends on the angle at which the tin plate is sliced. But, this width will have a mean value. And this mean value can be related back to the thickness of the plate in the ordinary sense. We will see that quantities like the volume fraction of say gold or tin can be related to the areal fraction of gold or tin averaged over slices taken in many different directions. Similarly the length per unit area of grain boundary on an averaged slice can be related back to the surface area of grains per unit volume of material.

4. IMAGES AS SLICES AND PROJECTIONS: ESTIMATING GEOMETRIC PROPERTIES

Sectioning and projection form the two basic techniques by which stereologists explore three-dimensional matter by reducing it to two-dimensional images. This is not all there is to the subject of stereology. But, it is a large part of it.

4.1 Flat Surfaces: Sectioning Methods

First lets consider something like a sausage roll. It is an elongated irregular lump of dough filled with sausage. Figure 1.2 illustrates a sausage roll.

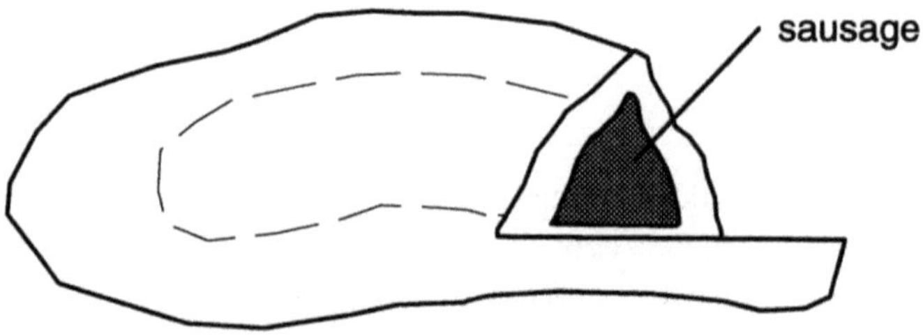

Figure 1.2: Sausage roll cut open

In Fig. 1.2, the end of the sausage roll has been cut open so we can see the sausage. The dotted line indicates where the sausage is. Let us say we would like to know the volume of the sausage and the volume of the entire sausage roll. One approach would be to slice the sausage roll into thin slices of thickness t like a loaf of sliced bread. We could measure the areas of sausage and roll for each slice with a planimeter. Then we could sum these areas and multiply by t to obtain the total volume of each. Doing this is simply an application of a Riemann sum in a geometric context. It is called *Cavalieri's principle*. We can write

$$V = t\sum a_i \ , \tag{1.1}$$

where a_i is the area of the i-th slice. If we define the set of sausage to be a subset of the roll then we can find the volume fraction of sausage as

$$V_V = \frac{\sum\limits_{(sausage)} a_i(sausage)}{\sum\limits_{(roll)} a_i(roll)}, \tag{1.2}$$

since t divides out. We could also write this in terms of the averages. If there are n slices,

$$V_V = \frac{n\bar{a}(sausage)}{n\bar{a}(roll)} = \frac{\bar{a}(sausage)}{\bar{a}(roll)} . \tag{1.3}$$

We see that when we take ratios, variables like t and n divide out. Maybe we could use a sampling method and not have to slice up the entire sausage roll to find the volume fraction of sausage. We could look at the areal fraction of sausage to roll (where, again, "roll" includes both sausage and pastry). From Eqn.(1.3) we might assume that the average areal fraction A_A equals the volume fraction. This is called the *Principle of Delesse* in stereology. On inspection of Eqn.(1.3) we see a dark side to this. Were we to expand both averages back into the series from which they were derived, we could not "legally" re-arrange the terms into a sum of ratios as would be required unless the individual $a_i(roll)$ were all equal. Were we to weight the areal fraction of each slice by the slice's own area, than the re-arrangement of terms becomes legal:

$$V_V = \sum_{(i)} A_A(sausage)_i \cdot \frac{a_i(roll)}{\sum\limits_{(i)} a_i(roll)} = \frac{\sum\limits_{(i)} a_i(sausage)}{\sum\limits_{(i)} a_i(roll)} \tag{1.4}$$

In the language of stereology, averaging the areal fraction of a constituent, e.g. sausage, using the area of the section as a weight, as in Eqn.(1.4), is termed the use of *A-weighted sections*. To summarize, when appropriate averaging is used,

$$V_V = A_A . \tag{1.5}$$

Area can be difficult to measure unless the image is in a digital form. There is another stereological principle based on line length fraction L_L. It is called the *Principle of Rosiwal*. Simply stated,

$$V_V = L_L \ . \tag{1.6}$$

The detailed interpretation of this principle will be discussed in Chapter 5. In practice, we would still take slices as before. But, instead of measuring the areas, we would draw random lines on photographs of the slices and find L_L by appropriate averaging. If we average over slices of different sizes, we would still need to use A-weighted sections. So, in practice Eqn.(1.6) reduces to Eqn.(1.5) and the assumption that,

$$A_A = L_L \ . \tag{1.7}$$

Perhaps the simplest way of establishing volume fraction from areal fraction is to take a point fraction. Oddly, this is the most recent method--dating from well into the 20th Century. By that time, estimating volume fractions was "old hat". So, the use of point fractions was not named after any of its originators. For the record, using appropriate averaging to be discussed in Chapter 5, the point fraction P_P equals the areal fraction and by Eqn.(1.5),

$$P_P = V_V \ . \tag{1.8}$$

We will explore the concept of the appropriate averaging for points shortly. The reader has now been introduced to the basic stereological probes: the *point probe,* the *linear probe* and the *areal probe.*

In the remainder of this book we will not be calling the principles of Cavalieri, Delesse or Rosiwal by name very often. But, there is yet another ancient named principle, the *Principle of Buffon*. This is one of the most useful stereological results when it comes to applications other than interpreting micrographs. It concerns the probability of intersection of a line segment, randomly placed and randomly oriented, with a pattern of lines occupying the same surface space. It says that the probability per unit length of the random line segment intersecting the pattern of lines is proportional to the length of line pattern per unit area L_A. Specifically,

$$\overline{P}_L = \frac{2}{\pi} L_A \ . \tag{1.9}$$

where $\overline{P}_L$ is the average number of intersections per unit length of the probe line segment with the line pattern. Assuming that ants crawl randomly and that sidewalks have lines on them (transverse lines, "cracks", every three or four feet) Eqn.(1.9) gives you a good estimator of the chances of an ant crossing a line on the sidewalk. Since, ants are typically smaller than the width of the lines in a concrete sidewalk, this allows you to calculate the probability of an ant "stepping on a crack and breaking its "mother's back", if you are superstitiously inclined. The derivation of Eqn.(1.9) is given in Chapter 5. This equation also allows us to estimate the total length of a linear (or lineal) array using a linear probe. In Fig.1.1 we saw a plane polished surface of a piece of metal. We learned that the length of the grain boundaries visible in the picture was related to the amount of grain surface area per volume of metal. How can we quickly estimate the length of grain boundary? One answer is to draw a "random" line and see how many grain boundaries it intersects. Then using Eqn.(1.9) we can estimate the total length of these grain boundaries.

4.2 Projection Methods and TEM

As we already mentioned, there are two categories of projection method. In one, parallel or *orthogonal* projection, we cast a shadow with parallel rays. In the other, point projection, we take a picture as with a pinhole camera. Unlike the case of sectioning methods, there is no hallowed body of named formulas for projections. There is merely a named nuisance called the Holmes effect, which we will discuss in Chapter 7. The impetus to develop parallel projection methods arose from transmission electron microscopy (TEM). In TEM a slice of something is magnified and projected onto a plane. The slice has thickness. If we use x-y-z Cartesian coordinates and z is the thickness direction, all the objects contained in the slice retain their x-y locations but the z-axis information is lost. The same applies to conventional x-rays; the organs of the patient appear to overlap. Of course the actual projection does not use truly parallel rays. But, in TEM, the image usually looks as if parallel rays were used and then the magnification was applied separately. Parallel projection is thus a convenient simplification. The theory of parallel projection methods depend a lot on the theory of random point processes. The simplest of these, the Poisson process, will be discussed in Chapter 7. Until we get there, we still need an example of a projection method to give us a feeling for the subject. Let us discuss the craters on Mars problem mentioned earlier in terms of point projection.

We are given some pictures of Mars. These pictures show craters. We are asked, "Are the craters on Mars randomly located?" In Chapter 4 we will look at what the meaning of "random" is in a geometric context. It turns out the question is almost meaningless without some supplementary definition. We could re-phrase the problem: Is the area of craters per unit area of Mars constant over its surface? This re-phrasing simplifies the problem of calculation a little but does not solve our problem of what do we mean by "random". We can see that crater areal fraction (we can call it density) is one of those regionalized variables which depends on its support. With point support there either is or is not a crater. This is not going to be a constant over the surface of Mars. Alternatively, we could use all of Mars as the support. Since Mars is just so big, there can only be so much crater; the amount of crater is bounded. Consequently, there is a mean for the planet. That mean is a constant since there is only one planet Mars. But, this is no help. Following the same line of reasoning, we might divide the planet into sections and look at the variance in the crater density as a function of the size of the sections. To be sure we would have to average the variance in density over all possible choices of section boundaries since a boundary falling in the center of a large crater could make the distribution appear more uniform than it is. Having done this, we still need to decide whether the distribution of craters in the sections is uniformly uniform. The statistically inclined might use a chi-squared distribution (see Section 2.4.13) to compare the amount of crater in each section with the mean for the planet. (Note that the sections are the support for the density function.) If these statisticians were lucky, the probability of uniformity associated with the chi-squared estimate would not vary much with the size of the support. Other statisticians might agree with the choice of the chi-squared distribution, but in so doing, our problem has been quietly pushed under the rug by a tacit definition of randomness. Furthermore, we have not yet studied the chi-squared distribution; computationally our example has become too difficult to use.

One part of the problem that we might be able to tackle at this point is that of determining the crater density for the planet as a whole. Say we start with a picture of the planet, a round ball with craters on it. We probe the surface with a grid of points. We lay the grid of points on the picture and count how many points fall in craters. Would the ratio of the number of points in craters to the total number of points be a good estimate of the crater density?

If the surface were flat, we would expect that an array of randomly located points would give a good estimate of areal fraction. So long as there was no periodicity in the arrangement of craters (Intuitively periodic craters would not be "random" but would not be detected with the chi-squared test just mentioned, since periodicity is a different kind of non-randomness.) a regular grid (e.g. rectangular or hexagonal) of test points would give an even better estimation of the areal fraction. Unfortunately, the surface is not flat.

We saw that in the case of slicing the sausage roll we should weight our slices by the area of the slice if we wanted to find the correct areal average. On a flat surface, a regular array of points has a constant fixed area associated with each point. Accordingly the sampling probability per unit area is a constant anywhere on the surface. The sampling is therefore uniform. If we consider Mars to be a sphere, we can analyze the effect of sampling its plane projection with a uniform grid of points. Figure 1.3 illustrates this.

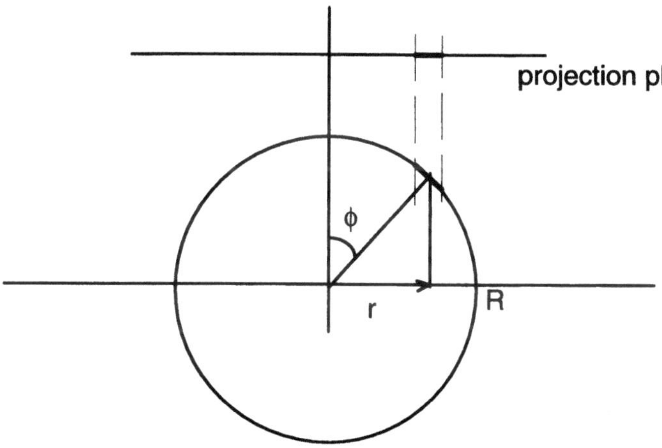

Figure 1.3: Projection of a plane onto a sphere

Being a point projection, our picture of Mars is similar to that which would be obtained by a parallel ray projection except for a scale factor, so long as our camera is on a telescope very far away. So, for the purpose of visualization, we can use parallel rays (Fig. 1.3). Phi is the co-latitude of a point on the sphere of radius R representing Mars. It corresponds to a point at a distance r from the center of the projection. We see that as we move away from the center, a given area on the projection corresponds to an increasing area on the sphere. Consequently a uniform sampling on the projection plane samples the sphere densely near the pole and far less densely near the equator. We have already said that we could associate an area with each sample point. Let that area be δ^2. For a small δ^2, dense sampling, the corresponding area on the surface of the sphere is $\delta^2 / \cos \vartheta$ where ϑ is the angle between the normal to the surface of the sphere at the sampled point and the normal to the projection plane. Let w be the ratio of the per-point area on the sphere to the per point area on the projection plane. From the construction of Fig. 1.3 we can deduce,

$$w = \frac{1}{\sqrt{1-\left(\frac{r}{R}\right)^2}} \quad . \tag{1.10}$$

In order to make our sample of the surface of Mars weighted in proportion to the area, we could weight the value obtained at a sample point (1 or 0) with the value of w for that point. This would give an A-weighted sample in the sense of Eqn.(1.4) and in principle the correct average. However, the polar region would still be more heavily sampled than the equator. We might expect that this would contribute to a higher amount of error in our estimate even though the estimate would be *unbiased* in the sense that as much positive error would be expected as negative error. In the spirit of Eqn.(1.4) we could make the density of points on our sampling grid proportional to w. This would have the effect of uniform sampling and each point would then be equally weighted. Apart from the fact the Eqn.(1.10) shows that the density at the equator would be infinite, a difficulty arises if we do not really have a sphere. Instead of estimating the average surface fraction of craters on Mars, we might want to estimate the total area devoted to crops on a hillside. There would be no a priori formula to occupy the role of Eqn.(1.10).

Apart from fitting the curve of the hillside with some sort of polynomial and numerically fitting a w function for locations on the hillside -- which may in fact the most economical approach given the availability of computer solid modeling -- we could use a grid with too many points and then delete some. This avoids the problem of having an infinite density at places like the equator in Fig. 1.3. Here is an example of a deletion scheme. Let $w_{\max}$ be the largest value of w in our particular set of probe points. If we set our probability of deletion equal to

$$P(\text{deletion}) = 1 - \frac{w}{w_{\max}} \quad , \tag{1.11}$$

we have no chance of deleting our highest weighted point and still retain an approximately uniform sampling point density on the surface being sampled.

If we do not have a continuous function supplying us with w, we may still approximate w locally, at least in principle. Consider that around our sample point for which we want to determine w we draw an equilateral triangle having sides of length a on the projection plane. We can estimate $1/w$ as being the ratio of the area of the triangle to that formed by projecting the equilateral triangle's vertices onto the surface in question. Let the coordinates of the triangle's vertices after projection onto the curved surface be (x_1,y_1,z_1), (x_2,y_2,z_2) and (x_3,y_3,z_3). Then, if the triangle is small relative the curvature of the surface,

$$w = \frac{2}{a^2\sqrt{3}}\sqrt{\begin{vmatrix} y_1 & z_1 & 1 \\ y_2 & z_2 & 1 \\ y_3 & z_3 & 1 \end{vmatrix}^2 + \begin{vmatrix} z_1 & x_1 & 1 \\ z_2 & x_2 & 1 \\ z_3 & x_3 & 1 \end{vmatrix}^2 + \begin{vmatrix} x_1 & y_1 & 1 \\ x_2 & y_2 & 1 \\ x_3 & y_3 & 1 \end{vmatrix}^2} \quad . \tag{1.12}$$

Note that in Eqn.(1.12) a can be made to divide out. The purpose of this exercise has been primarily to introduce the reader to thinking about the various geometric concepts important to stereology. Before completing this "training circuit" we will give an answer to how the point coordinates in Eqn.(1.12) might be obtained.

We have discussed the use of sampling grids of points as applied to images and their projection onto the usually curved surfaces of real objects. We have tacitly assumed that small dots might be substituted for points. We know that small dots can be the support of a regionalized variable. While the quantities we have used in our examples are binary quantities at true points, we realize that this limitation is not fundamental. In fact the application of regular grids of small dots as test probes forms the basis of the theory of digitized images [Serra 1982].

In the system of image analysis called *mathematical morphology* the construction of digital images is viewed as a transformation of the actual image by probing the image with an array of small dots. The small dots themselves are referred to as *structuring elements*. The process of transforming the image through taking intersections of the structuring elements with the features of the image is referred to as a *hit or miss transformation*. Hit or miss transformations of various sorts can be used to implement stereological measurements in the context of digital imaging and represent an alternative way of expressing stereological probing.

4.3 Curved Surfaces: Stereography and Photogrammetry

While being of little importance to mainstream stereology, photogrammetry formulas will be used in Chapter 10 in connection with "star volume" and are geometrically interesting in themselves. The reader who is not interested in this specialized subject may skip this section. Stereography is the making of stereo photographs. At the turn of the 19th Century these 3-D images were "all the rage". Cameras for making them were still popularly sold into the 1960s. A few are still manufactured now, but the novelty has worn off. Until the advent of the hologram, all three-dimensional imaging was based on this technique of having two separate pictures, one for each eye. While stereopticons are no longer to be found in every parlor, the subject is undergoing a resurgence of interest due to its application to machine vision, remote sensing by satellites, "virtual reality" and computer terminals having three-dimensional imaging.

Photogrammetry is a quantitative technique based on stereography. It is the technique of making three-dimensional maps from two-dimensional projections. Because it involves reconstructing something in three dimensions from lower-dimensional data, it has a kinship with stereology. But, there is also another link between the two. Most stereological formulae based on slicing apply to surfaces of any contour. However, points must be points on the surface, lines must be lines on the surface and areas must be areas on the surface rather than on a projection. When only a plane projection of the contoured surface is available for study, explicit knowledge of the surface contour is needed. For example, the estimation of the area of a forest that is devoted to fir trees might be attempted by point counting as in the previous section. Since aerial photographs are point projections, the density of points would be effectively higher for areas of the photograph physically nearer to the camera. Similar bias problems would affect other stereological probes such as test lines or area patches. For test lines, the length of the line's projection on the terrain would be the appropriate length to insert into a stereological formula. But, knowing this length also requires knowing the contour of the terrain. Here is where photogrammetry becomes helpful. It allows a quantitative determination of surface contours. It can be applied to any case where stereology is attempted on a curved surface or section. Take another example. The volume fraction of certain pebbles in an aggregate is desired. A core section is drilled from an aggregate and polished. The diameter of the core is measured and a stereo pair of photographs are made. All the needed information is now present since the projected (as they appear in the photos) areas of the pebble sections can be corrected to their actual surface areas in order to reduce bias. What would amount to bias? Were a large pebble in the center of the photographs, it would distort the apparent statistical sample in favor of itself unless corrections were made. The application of photogrammetry can be made whether the scale is geological or microscopic.

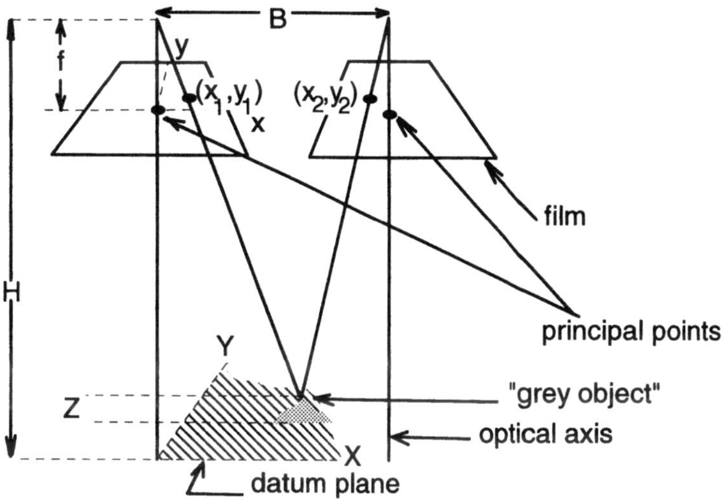

Figure 1.4: Photogrammeteric geometry

Photogrammetry is a geometric interpretation of ordinary binocular depth perception. Its simplest representation is in the *truly vertical photograph* pair at a constant *flying height*. The terminology, of course, comes from aerial photography. A truly vertical photograph is one taken looking exactly straight down on the land, so that the film plane is parallel with sea level of the earth's surface. This condition gives rise to a simplifying geometry shown in Fig. 1.4.

Figure 1.4 shows two overlapping vertical photographs as they are being taken. The tilted rectangle on the left represents a piece of film in the aerial camera as does the tilted rectangle on the right. The letter *H* denotes the *flying height*, the distance between the focal point of the camera and the *datum plane*, a plane normal to the film plane. The distance *B* is called the *air base* and extends along the X-direction. In fact, it defines the X-direction. It is the distance the airplane has flown between photographs. The letter *f* denotes the focal length and the astute reader will notice that these pieces of film are drawn in their mirror image, as if projected in front of the camera, for convenience -- they are emulsion side up and represent the 1:1 positive prints which would be used for calculating the coordinates of objects on the ground. The datum plane is a hypothetical reference plane usually taken as being somewhere slightly below the surface of the earth. It contains the origin of an XYZ coordinate system, the one shown. In this example, a gray object on the ground has coordinates (X,Y,Z) in the object field. This object field coordinate system needs to be matched with the two-dimensional coordinate systems of each photograph.

Each photograph has what is called a *principal point* located at the intersection of the film and the optical axis of the camera. This is ideally in the exact center of the picture. The principal points are the origins of these film coordinates. On the film the x-axis must be chosen such that it is parallel to the X-axis on the ground (datum plane). Similarly, the y-axis on the film is parallel to the Y-axis on the ground. In Fig. 1.4, the optical axis of the left hand photograph's optical system forms the Z-axis when it intersects the datum plane (at the origin). On each film itself there is a local coordinate system, each having its origin at the principal point for that film. These film-based coordinates are used to locate objects seen on the ground. For the left hand film, the image of the gray object is at (x_1, y_1) with the positive directions as shown. For the right hand film, the image of the gray object is at (x_2, y_2). As a result of the motion of the plane between photographs, *parallaxes* occur in the X-direction, but not in the Y-direction. The parallax of a given object is defined as the difference between its two x-coordinate values in the two photographs; for the gray object, the parallax $p = x_1 - x_2$. Given these definitions, Z is the elevation, B is the air base and

$$Z = H - \frac{Bf}{p}. \tag{1.13}$$

$$X = \frac{x_1}{f}(H - Z) = \frac{x_1}{p}B. \tag{1.14}$$

$$Y = \frac{y_1}{f}(H - Z) = \frac{y_1}{p}B. \tag{1.15}$$

These three equations, called the *parallax equations*, are readily derived from a similar triangles construction [Wolf, 1983]. They represent the heart of photogrammetry. Rarely can so much seemingly sophisticated service be had from such simple expressions. Notice that for the X and Y (termed the "planimetric") coordinates, the left hand film is used as a reference since its principal point is the image of the origin of the coordinate system. The balance of the subject of photogrammetry is largely concerned with transforming these equations into other coordinate systems and correcting for errors. These latter tasks have made photogrammetry a complex subject. Coordinate transformations are needed whenever the datum plane becomes tilted with respect to the terrain; the optical axes of the photographs are not exactly parallel etc. Actual cartography requires multiple photographs. Least squares methods are often used to obtain coordinate estimates from a tiling of multiple photographs. Note that as Eqns.(1.14 and 1.15) show, the scale is a function of the elevation because the photographs are point projections. Consequently, photographs can only tessellate, that is be matchable together at their boundaries like tiles, at one elevation and will be mismatched at others!

In practical applications the first step in photogrammetry is to align the stereo pair beginning with identifying the principal points. Their location is usually approximated using fiducial marks placed by the aerial camera on the borders of the negative. The next step is to position the films such that the air base is parallel to the X-axis. Since you may want to actually do this as a part of a stereological procedure, I will explain to you how it is done manually. The digital implementation simply generalizes the manual procedure. Once the principal points are located, the x-axes must be identified. Parallax occurs only in the direction parallel to the direction of flight. Accordingly, by aligning the photos with the principal points and their *conjugates* all in a line, not only is the x-axis identified on each photo, but the x-axes of the two photos become aligned with the X-axis. The *conjugate principal point* is simply the image of the object corresponding to the left hand film's principal point as it appears in the right hand film (and visa versa). After identifying and possibly marking the principal point in each film, the corresponding objects, say a tree-top in the left film and a smokestack in the right film are each located and possibly marked on both films. Each film then has two identified points: a principal point and a conjugate principal point. Drawing a line through each pair locates the x-axis for that film. Placing these two axes on a common line and the principal points a known distance apart completes the alignment. Usually the separation of the principal points is chosen to accommodate viewing in a stereoscope, but this is not needed. Once the photos are aligned, the parallax p can be measured with a type of micrometer called a *parallax bar* by positioning each of two fiducial marks in the micrometer in contact with the images of some object in question on the photos. The same measurement can be automated using various computerized image scanners. By projecting a movable artifact with a known parallax onto a stereo pair viewed through a stereoscope (which is simply a more precise version of a stereopticon) a virtual *floating mark* may be created. This floating mark can be positioned onto any point in the image and that object's elevation and coordinates can be read or plotted on a map. Extremely elaborate mechanical plotting systems based on this principle have been devised for cartography.

For stereological purposes, the above equations can be used to find the XYZ coordinates of points which may in turn fitted with say cubic splines to approximate the contours of any curved surface. From the fitted curves, which may form a net, the ratio of photo area to actual area (or photo length to actual length) may be estimated at any point on the "landscape".

5. BIBLIOGRAPHY

G. Matheron, "Principles of Geostatistics", *Economic Geology*, **58** (1963) 1246-1266.

J. Serra, *Image Analysis and Mathematical Morphology* (Academic Press, New York, 1982) Chapter 2.

P.R.Wolf, *Elements of Photogrammetry* (McGraw-Hill, New York, 1983).

Chapter 2

MATHEMATICAL PREMIMINARIES

This chapter comprises a list of the various definitions and formulas in mathematics and probability theory that will be required in later Chapters. In most cases the results are given without derivations, but exceptions are made where the reader may have difficulty in locating a proof elsewhere. Readers unfamiliar with the theory of integral transforms and some other matters in this chapter are to be assured that we will be making only occasional use of these techniques; most of the derivations in this book involve nothing more than elementary calculus. Part of the function of this chapter and the one which follows is to set down the conventions for the use of symbols. A symbol table although helpful is rarely adequate. A symbol table is included as part of the Index.

1. A REVIEW OF SOME INTEGRAL FUNCTIONS

1.1 Dirac Delta Function

The Dirac delta function, $\delta(x)$, is formally defined by:

$$\int_{-\infty}^{\infty} g(t)\delta(t-y)dt = g(t=y)\,, \tag{2.1}$$

where $g(t)$ is any arbitrary continuous function. More loosely, one may regard $\delta(x)$ as being everywhere zero except at $x = 0$. The trouble is that it does not have a defined value at zero either. It is therefore not really a function at all. Initially the theory of "distributions" developed to explain delta functions might have been flawed slightly in a strict mathematical sense. It has evolved to a point where there are no apparent inconsistencies [Strichartz, 1993]. The delta "function" is currently best described as a "one-form" since it combines with a function to give a single number. Note that substituting the derivative of the delta function into Eqn.(2.1) gives the derivative of g at $t = y$ as the result. Delta functions are obtained from gaussians and similar functions by taking the limit as the parent function is rescaled to fit into a smaller and smaller interval. At the limit the integral of the result remains unity, but the domain over which this "function" is not zero becomes infinitesimal.

1.2 The Unit Step Function

The symmetric unit step function, $U(x)$, (Fig. 2.1) has the following properties:

$$U(x) = \begin{cases} 0 \; for \; x < 0 \\ \frac{1}{2} \; for \; x = 0 \\ 1 \; for \; x > 0 \end{cases} \tag{2.2}$$

and is "formally" related to $\delta(x)$ by:

$$\delta(x) = \frac{d}{dx} U(x) \tag{2.3}$$

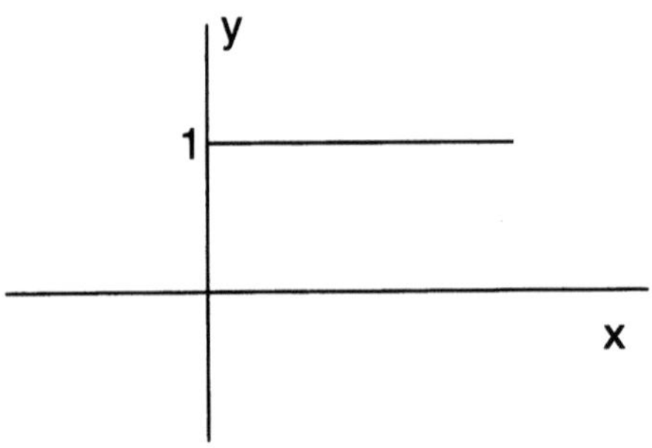

Figure 2.1: Unit Step Function

It is sometimes called the "Heaviside Step Function" after its use by Oliver Heaviside in the development of his Operational Calculus. For that reason, the letter H is used for it in some texts.

1.3 The Complete and Incomplete Gamma Function

The gamma function, $\Gamma(z)$, is a generalization of the factorial that we will be encountering frequently. It is one of those functions defined by an integral and it is known that it does not satisfy any differential equation with rational coefficients. It is one of the most interesting transcendental functions and is sometimes called the Eulerian integral of the second kind. It is defined by:

$$\Gamma(z) = \int_0^\infty t^{z-1} \exp(-t) dt \; t > 0 \tag{2.4}$$

The variable, *t*, is what is called a dummy variable in that it is entirely contained in the integration process and has no external definition. Note that the exponent of *t* is *z*-1. If *z* is less than unity, the integral becomes improper since the integrand tends to infinity at the origin. If *z* is less than or equal to zero, the integral will fail to converge at all. We will be primarily concerned with arguments that are integers or multiples of 1/2. For these cases there is a recursion relation:

$$\Gamma(z+1) = z\Gamma(z) \quad \forall z > 0 \text{ (for all } z{>}0) \tag{2.5}$$

and

$$\Gamma(n+½) = (2n)! \frac{\pi^{½}}{4^{\mathrm{n}} n!} , \tag{2.6}$$

which follows from Eqn.(2.5) and that $\Gamma(½) = \sqrt{\pi}$. In particular,

$$\Gamma\left(\frac{1}{2}\right) = \pi^{\frac{1}{2}}, \Gamma\left(\frac{3}{2}\right) = \frac{1}{2}\pi^{\frac{1}{2}}, \Gamma\left(\frac{5}{2}\right) = \frac{3}{4}\pi^{\frac{1}{2}}$$

and

$$\Gamma\left(\frac{7}{2}\right) = \frac{15}{8}\pi^{\frac{1}{2}} .$$

From Eqn.(2.5) comes the highly useful formula,

$$\Gamma(n+1) = n!, (n = 0,1,2,...) . \tag{2.7}$$

Another formula, not quite so useful, is the duplication formula,

$$\Gamma(2z)=(2\pi)^{-\frac{1}{2}}2^{2z-\frac{1}{2}}\Gamma(z)\Gamma(z+½) \tag{2.8}$$

that is sometimes helpful in simplifying expressions involving products of gamma functions.

The incomplete gamma function is defined by,

$$\gamma(p,x)=\int_0^x t^{p-1}\exp(-t)dt \tag{2.9}$$

Its values are given in tables and it is used to simplify similar integrals.

1.4 The Complete and Incomplete Beta Functions

The beta function is sometimes called the Eulerian Integral of the first kind. But since it can be reduced to gamma functions, the gamma function was introduced first. The (complete) beta function is defined by

$$B(\alpha,\beta)=\int_0^1 t^{\alpha-1}(1-t)^{\beta-1}dt \;\; \beta>0 \tag{2.10}$$

or by

$$B(\alpha,\beta)=\frac{\Gamma(\alpha)\Gamma(\beta)}{\Gamma(\alpha+\beta)}. \tag{2.11}$$

This expresses the symmetry of the beta function in its two parameters. The incomplete beta function is defined by,

$$B_x(\alpha,\beta,)=\int_0^x t^{\alpha-1}(1-t)^{\beta-1}dt. \tag{2.12}$$

1.5 Error Function

The error function is another defined by an integral and often arises in solutions to parabolic partial differential equations. It plays a prominent role in the theories of probability, diffusion, heat conduction, refraction of light etc. It is perhaps best known as the integral of the normal curve of statistics. The error function is defined here by,

$$erf(x) = -erf(-x) = \frac{2}{\sqrt{\pi}} \int_0^x \exp(-t^2)dt . \qquad (2.13)$$

Because of its use in divers contexts, caution should be exercised when consulting tables of this function, since it is tabulated under several different forms. In mathematical contexts, the factor of $2/\sqrt{\pi}$ is often absent. The bell-shaped integrand is doubly symmetric and has a maximum value of unity when t = 0. It is asymptotic to the t-axis in both positive and negative directions. The variable, t, is a dummy. As x increases to infinity, Eqn.(2.13) approaches unity. Using this definition, the complementary error function, *erfc*(), equals 1-*erf*().

2. INTEGRAL TRANSFORMS

2.1 Fourier Transform

The Fourier transform is used to effect a change of variable by means of an integral, the Fourier integral that, like the error function, has several variant forms. This change is easily visualized in one dimension as being the alternate descriptions of the sound of a musical note having a beginning and an end: The sound may be thought of as an air pressure that is a function of time or as a spectrum of tones given as a function of frequency (that is nearly the same as pitch). In three dimensions, the most common application of the Fourier transform is x-ray crystallography. For crystallographers visualizing the Fourier transform in terms of a crystal, we have, untransformed, the crystal itself mapped out in ordinary spatial coordinates. As in the case of the musical note, the crystal must have a beginning and an end for reasons of convergence of the Fourier integral. Ignoring this size issue, the Fourier transform of a very very large crystal's spatial distribution is it's x-ray structure factor where the Miller indices, imagined in this case to form a continuum, are the frequencies but scaled by their conjugate crystallographic cellular dimensions. Infinitely periodic functions or functions on a finite interval may be reduced to Fourier series and through distribution theory the Fourier series and the integral representations can be unified. The exponential Fourier transform, F[$g(x)$], of the function, $g(x)$ is here defined by,

$$\mathrm{F}[g(x)] = \int_{-\infty}^{\infty} g(x)\exp(ixy)dx . \qquad (2.14)$$

The inverse transform is defined by,

$$g(x) = \frac{1}{2\pi} \int_{-\infty}^{\infty} F(y) \exp(-ixy) dy . \tag{2.15}$$

There are several definitions of the transform that are in use which differ in the sign of the exponent and factors of pi in the exponent and pre-exponential term. We have selected this particular form so as to agree with the definition of the characteristic function [section 2.3.3], thus avoiding the need to introduce a separate symbol for the latter.

Among the useful properties of the Fourier (and other) integral transforms is their ability to simplify expressions involving integral relationships, for example, the convolution.

Convolution integrals take the form,

$$h(x) = k \otimes g = \int_{-\infty}^{\infty} k(x-z) \cdot g(z) dz .$$

The symbol used for convolution and that used for the Cartesian product are very similar although usually clear in a given context. Here the convolution symbol has been made boldface. The transform of the convolution integral is the product of the transforms of the two functions,

$$\mathrm{F}[k \otimes g] = \mathrm{F}[k(x)] \mathrm{F}[g(z)] . \tag{2.16}$$

It follows from Eqn. (2.14) that the transform of the n-th derivative of a function is

$$\mathrm{F}[g^{n}(x)] = (iy)^{n} \mathrm{F}[g(x)] , \tag{2.17}$$

and y is used as indicated in Eqn.(2.15). Conversely, if $I_n(x)$ is the n-th integral of $g(x)$,

$$\mathrm{F}[I_n(x)] = \left(\frac{1}{iy}\right)^{n} \mathrm{F}[g(x)] . \tag{2.18}$$

As is the case with other integral transforms, there are rules for change of scale and shifting of the origin. Finally, by the change of scale and shift theorems,

$$\mathrm{F}\left[g\left(\frac{x}{a} + b\right)\right] = a \exp(-iaby) \cdot F[ay] , \tag{2.19}$$

where $F[y] = \mathrm{F}[g(x)]$.

2.2 Laplace Transform

The raison d'etre of the Laplace transforms is that, being one-sided, it need not have the convergence problems associated with the Fourier transform. The secret of this lies in the exponential factor of the integrand. The variable, p, is a complex number with a positive real component. This allows for the positive real component to perform the duty of creating an exponential convergence factor. For example, a Fourier integral of a sine function does not converge at infinity since the integrand is a periodic oscillating function. The Laplace integral does converge since the integrand contains the factor $\sin(ax)/\exp(bx)$ which tends to zero at large x. Because of the extensive use of the Laplace transform in vibration and electricity, extensive tables of Laplace transforms have been constructed. Since we will be mainly concerned with variables that can take only positive values, it will be advantageous to work with Laplace rather than Fourier transforms.

The Laplace transform of $g(x)$ is ,

$$\mathrm{L}[g(x)] = \int_0^\infty g(x)\exp(-px)dx . \tag{2.20}$$

The Laplace transform of a convolution is,

$$\mathrm{L}[k \otimes g] = \mathrm{L}[k(x)]\mathrm{L}[g(x)] . \tag{2.21}$$

The Laplace transform of the derivative of a function is,

$$\mathrm{L}[g'(x)] = p\mathrm{L}[g(x)] - g(+0) , \tag{2.22}$$

where $g(+0)$ is the limit of g at zero approached from the positive side. If I is the integral of the function $g()$ from zero to x, then

$$\mathrm{L}[I(x)] = \frac{1}{p}\mathrm{L}[g(x)] . \tag{2.23}$$

For a change in scale and shift of the variable,

$$\mathrm{L}\left[g\left(\frac{x}{a} + b\right)\right] = a\exp(-abp) \cdot L[ap] , \tag{2.24}$$

where $L[p] = \mathrm{L}[g(x)]$.

It is frequently necessary to determine the Laplace transform of variable raised to an arbitrary power. For any value of $\upsilon > -1$,

$$\mathrm{L}\left[x^{\upsilon}\right] = \Gamma(\upsilon + 1) p^{-(\upsilon+1)}. \tag{2.25}$$

2.3 Mellin Transform

The Mellin transform of a function, $g(x)$ is,

$$\mathrm{M}[g(x)] = \int_0^{\infty} g(x) x^{s-1} dx \tag{2.26}$$

where, in general, s is a complex variable. However, the transform also yields the (s-1) moment about the origin of a probability distribution $g(x)$, section [section 2.3.2]. We will be encountering instances where it will be easier to derive the moments of the distribution rather than the distribution function itself. In such cases it is frequently possible to recover the distribution function from a table of inverse Mellin transforms. The following relationship,

$$\mathrm{M}\left[x^{\alpha} g\left(a x^{\beta}\right)\right] = \frac{1}{\beta} a^{-\frac{s+\alpha}{\beta}} M\left[\frac{s+\alpha}{\beta}\right], \tag{2.27}$$

where $M[s] = \mathrm{M}[g(x)]$ is often useful in reducing a transform to a tabulated form. The change of shape of the letter "M" denotes only whether the argument is the function to be transformed or the transform variable, s.

2.4 Relationships Between Transforms

Using the definitions given above and two additional ones,

$$F[y] = F[g(x)] \tag{2.28a-d}$$

$$L[p] = L[g(x)]$$

$$L_{-}[p] = L[g(-x)]$$

$$M[s] = \mathrm{M}[g(\log(x))],$$

then the three transforms we have considered are interrelated as follows,

$$F[y] = L[-iy] + L_{-}[iy] \tag{2.29}$$

and

$$F[y] = M[iy] \ . \tag{2.30}$$

These relationships can be used to produce further transforms from tabulated ones. Extensive tables of Fourier, Laplace and Mellin transforms and their inverses are to be found in Ref. 5 [Erdelyi, 1954] in the bibliography at the end of this chapter.

3. PROBABILITY THEORY

3.1 Distribution Functions

The distribution function, $f(x)$, is defined such that $f(x)dx$ is the probability that the continuous random variable x has a value in the range x $\pm\, dx/2$. It follows from this definition that the probability, $p(a \le x \le b)$, of x having values between a and b is given by:

$$p(a \le x \le b) = \int_a^b f(x)dx \tag{2.31}$$

In particular:

$$\int_{-\infty}^{\infty} f(x)dx = 1 \quad . \tag{2.32}$$

Other names that have been assigned to this function include "frequency", "probability density function", PDF and "density". The last is the one currently favored by most authors although PDF is more common in reliability engineering. It should also be noted that some authors, for example, Feller [1971], use the name "distribution" for what we will be terming "cumulative distribution", or CDF. (In this and the integrals that follow we will adopt the convention of assuming $f(x) = 0$ outside the domain over which x can exist.) It should be noted that $f(x)$ is of dimension x^{-1} and is not in itself a probability.

The usual procedure, at least in the context of this book, for determining a distribution function is to select a variable that is uniformly distributed on an interval and then, if necessary, to use a variable transformation (section 2.3.5) to obtain the distribution of the variable of interest. As an example, consider the distribution of the distance, x, of a point from the end of a line of length ℓ (Fig. 2.2).

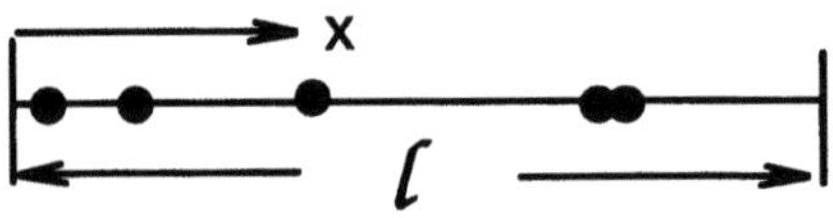

Figure 2.2: Random Points on a Line

If the point is randomly distributed then all values of x are equally probable; consequently, $f(x)$ = constant. Substituting in Eqn.(2.32) and integrating between 0 and ℓ, the constant is found to be $1/\ell$. Thus:

$$f(x) = 1/l \quad ,[0 \le x \le l\ ] \ .$$

The cumulative distribution function, often abbreviated CDF, is defined by:

$$\mathrm{F}(x) = \int_{\infty}^{x} \mathrm{f}(t)\,dt\,, \qquad (2.33)$$

from which it follows that F(-∞) = 0 and F(∞) = 1. (It will be noted that $F(x)$ is a probability.)

The density function $N(x)$ is defined such that $N(x)$dx is the number of times that the random variable x has a value in the range $x \pm$ (dx/2). It is obviously related to $f(x)$ by:

$$N(x) = Nf(x)\,,$$

where N is the total number of occurrences. Also, as expected,

$$\int_{\infty}^{\infty} N(x)dx = N \ . \qquad (2.34)$$

3.2 Moments of a Distribution and Expectation

The moments of a distribution are the averaged or, in more precise mathematical language, the *expected* values of the variate raised to some, usually integer, power. This is not the same as obtaining the expected value, which is the mean, and then raising that mean to a power. Rather, each variate that occurs, as in some process such as the numbers found on numbered cards drawn from an urn, is first raised to a power and then subsequently averaged. Expectation is an important enough concept to deserve a pause for contemplation. Paraphrasing Gnedenko [1967], we may imagine that a certain gun is fired at a particular target. Suppose that, given this gun and target, one shot will hit it with probability, p_1, two shots will hit it with probability, p_2 , three shots will hit it with probability, p_3 and so on. How many shots are needed to hit the target on the average? Surely, no matter how poorly we aim -- disallowing that we deliberately avoid the target -- eventually we must hit it. Thus summing over all values of p must yield unity as in Eqn.(2.32). And, supposing a very large number of shots are fired, we may apply Bernoulli's probability theorem and say that p_1 is the relative frequency that the target is hit in one shot. The relative number hits (not firings) taking two shots is p_2. One hit in one shot takes one shot. One hit in two shots takes two shots. Thus, on the average,

$$p_1 + 2p_2 + 3p_3 + \sum_{i=4}^{\infty} ip_i$$

shots are required to get a hit. When the probability function is continuous, summation may be replaced by an integral and any function for which the expected value is desired may be substituted for the integers, *i*, in the above. So, the *n*-th moment about the origin is defined by:

$$\mu_n'(x) = \int_{-\infty}^{\infty} x^n \, f(x) d\,x, \tag{2.35}$$

and is also known as the absolute *n*-th moment. As previously noted, the *n*-th moment is equal to the Mellin transform of the distribution evaluated at $s = n+1$. Moments about the origin are also called *absolute moments*. The first moment yields the average (or mean, expected value), $\bar{x}$ of x. We can also define moments about the average:

$$\mu_n(x) = \int_{-\infty}^{\infty} (x - \bar{x})^n \, f(x)\, dx. \;, \tag{2.36}$$

which are also known as *central moments*. An important special case is $\mu_2(x)$ which is termed the variance, $\sigma^2(x)$, of x. The square root, $\sigma(x)$, is the *standard deviation* of x and is a commonly used measure of the dispersion of the distribution. The ratio $\sigma(x)/\overline{x}$ (expressed as a percentage) is known as the *coefficient of variation*. It is also to be noted that:

$$\mu_0'(\mathrm{x}) = \mu_0(\mathrm{x}) = 1, \tag{2.37}$$

$$\mu_1(x) = 0 \tag{2.38}$$

and

$$\mu_{-1}'(x) = 1/\mathrm{H}(x), \tag{2.39}$$

where H(x) is the *harmonic mean* of x. The variance is related to the second moment about the origin by:

$$\sigma^2(\mathrm{x}) = \mu_2(x) = \mu_2'(\mathrm{x}) - \overline{\mathrm{x}}^2 \ . \tag{2.40}$$

This is a special case of a general relationship between the two sets of moments.

There exists a simple relationship between absolute and central moments,

$$\mu_n = \sum_{k=0}^{n} \binom{n}{k} \left(-\mu_1'\right)^{n-k} \mu_k' . \tag{2.41}$$

For the case of the third moment this becomes,

$$\mu_3 = \mu_3' - 3\mu_2'\overline{x} + 2\overline{x}^3 \tag{2.42}$$

It is also to be noted that not all moments of a distribution may exist. An extreme example often cited is the Cauchy distribution for which all moments (excluding the zeroth but including the average) are infinite. The existence of $\mu'_k(x)$ or $\mu_k(x)$ implies the existence of all moments of order $O < n \leq k$ and, conversely, if $\mu'_k(\mathrm{x})$ or $\mu_k(x)$ are infinite, all moments of order $O \geq k$ are also infinite.

The value of a random variable raised to a power is only one example of expectation. Any function of the variable can be similarly integrated to find the *expected value* of that function, denoted $E(y)$, where y is the value of the function of the random variable. Under fairly general assumptions, the expected value for the sum of random variables is the sum of the expected values and the expected value for the product of independent uniform random variables is the product of the expected values. The ratios of random variables are more difficult to handle. They are discussed in Section 5.4.1.7.

3.3 Characteristic Function

The characteristic function of a distribution $f(x)$ is

$$\phi(y) = \int_{-\infty}^{\infty} f(x)\exp(ixy)dx \ , \tag{2.43}$$

and is therefore equal to the Fourier transform $F[f(x)]$ as we have defined it in Eqn.(2.14). Moments about the origin can be generated by repeated differentiation of the transform:

$$\mu_n' = (-i)^n \left. \frac{d^n\phi}{dy^n} \right|_{y=0} . \tag{2.44}$$

However, it is usually easier to derive the moments from Eqn.(2.35). An important application of characteristic functions is the determination of the distribution of the sum of independent random variables. If

$$X = x_1 + x_2 + \ldots x_n \ ,$$

then:

$$\phi(Y) = \mathrm{F}[\mathrm{f}(X)] = \mathrm{F}[f_1(x_1)] \cdot \mathrm{F}[f_2(x_2)] \cdot \ldots \cdot \mathrm{F}[f_n(x_n)] . \tag{2.45}$$

If the random variable can take only non-negative values, the characteristic function can be replaced by the Laplace transform, which has similar properties. But, care must be taken as to the signs of the variables etc. See problem 2.1 for a test of the characteristic function.

3.4 Bivariate Distribution Functions

The bivariate (or joint) distribution function, $f(x_1, x_2)$, is defined such that $f(x_1, x_2)dx_1dx_2$ is the probability that the variables simultaneously have values in the range $x_1 \pm (dx_1/2)$ and $x_2 \pm (dx_2/2)$. It therefore follows that

$$\int_{-\infty}^{\infty}\int_{-\infty}^{\infty} f(x_1, x_2)dx_1\, dx_2 = 1. \tag{2.46}$$

Integration with respect to one variable yields the *marginal distribution*:

$$f(x_2) = \int_{-\infty}^{\infty} f(x_1, x_2)\, dx_1 \ . \tag{2.47}$$

This process is sometimes termed "integrating out" a variable. A distribution function is a probability and a joint distribution function is the probability of two events simultaneously. Accordingly we can write $f(x_1, x_2) = f(x_2|\, x_1)f(x_1)$ where $f(x_2\,|\, x_1)$ is the *conditional distribution function*[1]. If the two variables are independent, then $f(x_2|\, x_1) = f(x_2)$ and

$$f(x_1, x_2) = f(x_1)f(x_2). \tag{2.48}$$

Moments about the origin are defined by

$$\mu'_{m,n}(x_1, x_2) = \int_{-\infty}^{\infty}\int_{-\infty}^{\infty} x_1^m x_2^n f(x_1, x_2)dx_1 dx_2, \tag{2.49}$$

and about the mean by

$$\mu_{m.n}(x_1, x_2) = \int_{-\infty}^{\infty}\int_{-\infty}^{\infty} (x_1 - \bar{x}_1)^m (x_2 - \bar{x}_2)^n f(x_1, x_2)dx_1 dx_2 \tag{2.50}$$

The first-order moment about the mean, $\mu_{11}(x_1, x_2)$ is known as the *covariance* of x_1 and x_2. It follows from Eqns.(2.48) and (2.50) that the covariance is zero if the variables are independent. The reader can easily extend the foregoing results to distribution functions containing more than two variables.

3.5 Transformation of Variable

We will frequently be faced with the problem of determining the distribution function of a variable given the distribution function of another variable and the relationship between the two variables. Let $f(x)$ be the known distribution, $f^T(y)$ the distribution that we are seeking, and $y = g(x)$ the relationship between the two variables. We further require that the events, which result from the variable being transformed, belong to a

---

[1]see B. V. Gnedenko, *The Theory of Probability*, Chelsea, N.Y.,1967 p. 167.

suitably defined sample space (here, an intuitive term to be defined later). This is unlikely to pose a problem if $g(x)$ is continuous and bounded. In addition, for reasons that will be apparent $g(x)$ must be monotonic and the inverse relationship, $x = g^{-1}(y)$, must exist. Let x_1 be a particular value of x; and let y_1, which equals $g(x_1)$, be the corresponding value of y. By definition, $f(x_1)dx$ is the probability that x lies in the interval x_1 to $x_1 + dx$. But this must be equal to the probability that y is in the interval y_1 to $y_1 + dy$. However, $dy = [dg(x)/dx]\, dx$. Therefore:

$$f^{\mathrm{T}}(y_1)dx\frac{dg(x)}{dx} = f(x_1)dx\,,$$

or, dropping the subscript:

$$f^{\mathrm{T}}(y) = f(g^{-1}(y)\left|dx/dy\right|. \tag{2.51}$$

The absolute signs have been added because negative values of $f\,(y)$ are not allowable and because the derivative is used here as ratio of the absolute measures of two small increments of the domains over which a probability is defined rather than as a slope of something. This is apparent when the two cases, $g(x)$ is monotonic increasing and $g(x)$ is monotonic decreasing are worked out separately. The restriction that $g(x)$ be monotonic is required because, otherwise, there will be more than one value of x corresponding to a given value of y. No general rule can be given for non-monotonic functions. However, all the non-monotonic ones we will be encountering are symmetric about $x = 0$ and are monotonic on each side. In this case we need only consider $x \geq 0$.

As an example of a variable transformation consider the estimation of $f(y)$ given that $y = x^2$ and $f(x) = 1/\ell$ for $0 \leq x \leq \ell$. [This corresponds to finding the distribution of the square of the distance of a random point from the end of a line of length ℓ cf. example in section (2.3.1), finding the distribution of the distance to a random point from the end of the line.]

For $y = x^2$:

$$dy/dx = 2x = 2y^{1/2}$$

Substituting in Eqn.(2.51) we find:

$$f^{\mathrm{T}}(y) = (2l\, y^{\frac{1}{2}})^{-1}\,,\ [0 \leq y \leq \ell^2]$$

The reader can easily verify that this satisfies Eqn. (2.32). It is important to note that when changing a variable all references to the old variable must be eliminated. Thus, in the example just considered it would have been improper to write the result as $f^{T}(y)=1/(2|x)$.

It will be noted that, apart from the matter of sign, the rule for changing the variable of a distribution function is identical with that for a change of variable in an integral. The reason for this, of course, is that integration yields a probability that, for corresponding limits, has to be the same for the two distributions.

It is also possible to use the characteristic function to change the variable without any restriction on the form of $g(x)$. However, the work involved is usually greater than that in using Eqn.(2.51). Equation (2.51) can be extended to multivariable distributions.

If:

$$y_1 = g_1(x_1,x_2, \ldots x_n) \text{ and } x_1 = h_1(y_1,y_2, \ldots y_n)$$

$$y_2 = g_2(x_1,x2, \ldots x_n) \text{ and } x_2 = h_2(y_1,y_2, \ldots y_n)$$

$$y_n = g_n(x_1,x_2, \ldots x_n) \text{ and } x_n = h_n(y_1,y_2, \ldots y_n);$$

are continuous random variables related by a reciprocal one-to-one (non-singular) transformation, then their respective distribution functions are related by:[2]

$$f^{T}(y_1, y_2, \ldots y_n) = f(x_1, x_2, \ldots x_n) \cdot |J| \tag{2.52}$$

where J is the Jacobian, i.e., the determinant:

[2]A derivation (which is fairly difficult) for the general case is to be found in: C. A. Stewart, *Advanced Calculus*, Methuen Co., Ltd., London, 1940, p. 289. A simple derivation for two variables has been given by T. C. Fry, Probability and Its Engineering Uses, D. Van Nostrand Co., Inc., New York, 1928, p. 155. A less difficult discussion showing that this follows from the change of variable formula for multidimensional integrals is given in T. Apostol, *Calculus*, vol. 2, John Wiley and Son, 1969, p.271.

$$J = \begin{vmatrix} \frac{\partial x_1}{\partial y_1} & \frac{\partial x_1}{\partial y_2} & \cdots & \frac{\partial x_1}{\partial y_n} \\ \frac{\partial x_2}{\partial y_1} & \frac{\partial x_2}{\partial y_2} & \cdots & \frac{\partial x_2}{\partial y_n} \\ \cdots & \cdots & \cdots & \cdots \\ \frac{\partial x_n}{\partial y_1} & \frac{\partial x_n}{\partial y_2} & \cdots & \frac{\partial x_n}{\partial y_n} \end{vmatrix}$$

provided that the Jacobian exists and is continuous. The x's on the r.h. side of Eqn.(2.52) are, as before, written in terms of y rather than x. Jacobians will be discussed again in section 3.2.4.

3.6 Discrete Random Variables

If a random variable X can only take discrete values $x_1, x_2, \ldots x_k$ then the corresponding probabilities $p(x_1)$, $p(x_2)$, $p(x_k)$ define a discrete distribution. Apart from changing the variable, all the results listed for a continuous distribution are directly applicable to the discrete case if the integrals are replaced by the corresponding summations over the limits 1 to k. For example, the n-th moment about the origin is:

$$\mu_n'(x) = \sum_{i=1}^{k} x_i^n \, p(x_i) \tag{2.53}$$

An alternative approach is to use a "continuous" distribution consisting of a sum of Dirac delta functions (section 2.1.1). Thus:

$$f(x) = p(x)\ \delta(x\text{-}x_1) + p(x)\ \delta(x\text{-}x_2) + \ldots + p(x)\ \delta(x\text{-}x_k) \tag{2.54}$$

corresponds to the discrete distribution $p(x_1)$, $p(x_2)$, ... $p(x_k)$ so long as $f(x)$ appears in an integral. The delta functions work as selectors. Since $p(x)$ can be factored out, the remaining sum of delta functions is sometimes called a "comb" when they are evenly spaced.

Because of the similarity between the handling of discrete and continuous distributions, a commonly used generalization, which eliminates the need for knowing in advance whether the distribution is discrete or continuous, is applied to probability integrals. This is called Stieltjes integration and the result is a Stieltjes integral. The idea is that were you to graph the cumulative distribution function, CDF, Eqn.(2.33), for a discrete distribution, the graph would look like a staircase, a sequence of steps. As each $p(x)$ was encountered with increasing x, the CDF would advance by one step in height. But, while differentiation of the CDF for a continuous distribution merely recovers the probability density function, differentiation of the CDF for a discrete distribution results in an expression of the form of Eqn.(2.54) by way of Eqn.(2.3). The Stieltjes integral is written $\int ...dF$ where F designates the cumulative distribution function of f. Thus the n-th moment about the origin as given by Eqn.(2.53) becomes,

$$\mu_n'(x) = \int_{-\infty}^{\infty} x_i^n \, dF \, . \tag{2.55}$$

Evaluation of Eqn.(2.55) is done by applying the differentiation chain rule to dF.

3.7 The Central Limit Theorem

The distribution of the sum of n independent random variables having the same distribution function and a common finite average and a common finite variance[3], approaches a normal distribution as $n \rightarrow \infty$. This is called the central limit theorem. Examples of its application include the approach of the binomial to the normal distribution as the number of trials increases and the square root dependence of accumulated roundoff errors. But also, under some relatively weak assumptions, which include the Lindberg-Feller condition, the distribution of the sum of a large number of independent random variables still approaches a normal distribution even if neither the distribution functions, the means nor the variances are common. The Lindberg-Feller theorem is an augmentation of the central limit theorem.

The result of a stereological analysis is usually the sum of a large number of individual observations. Thus, from the central limit theorem, we know that the results will be normally distributed (assuming that the measurement has a finite variance, which is not always the case). The importance of this is that we can ascribe a probabilistic interpretation to the standard deviation of the analysis. However, we must take caution in using the presence of a normal distribution as an indicator of procedural quality. The Lindberg-Feller theorem tells us that the results will be normally distributed even if the analysis is faulty.

[3] The theorem is also valid under certain less restrictive conditions; see Feller [1971].

4. SOME THEORETICAL DISTRIBUTIONS

The following is a summary of the properties of certain theoretical distributions that we will be encountering. In some cases expressions for $\mu'(x)$ or the characteristic function, $F[f(x)] = F(p)$, Sect. (2.3.3), have been omitted when they are too complicated to be of practical use. The first two distributions are discrete, the remainder continuous. The variable p is substituted instead of y to minimize confusion.

4.1.1 2.4.1 Binomial Distribution

The binomial distribution is a discrete probability distribution. It is usually the first distribution taught in statistics courses. In engineering importance it ranks perhaps second to the Poisson distribution, that for rare events. Like the normal distribution, the Poisson distribution is also a limiting case of the binomial distribution. The binomial distribution arises whenever there is a constant probability of some event happening in a sequence of unrelated or *independent* trials. The typical illustration is that of dice. Taking a simpler case first, what is the probability of getting 3 sixes is a row? The answer is $(1/6)^3$, since the probability of getting a six is 1/6. We make the problem a little harder. What is the probability of getting 2 sixes in 3 tosses? The probability of getting two sixes is $(1/6)^2$. The set of these outcomes consists of those where a six is gotten twice and something else is gotten once. The probability of getting something else, viz. not getting a six, is 1-1/6 or 5/6. The result which is not 1/6 may occur either on the first, second or third toss. There are thus three ways this result can occur. Hence the total probability is 3·(1/6)(5/6). The prefactor, 3, is the number of combinations of n things, the three tosses, in r ways, the number of pairs of sixes. The reader will recall that this is the same as the binomial coefficient for, in this case, the second term of a third degree polynomial. Hence arrives the name, "binomial distribution". The prefactor is given by $\binom{3}{2}$, the binomial coefficient. Binomial coefficients can be constructed through Pascal's triangle or the formula, $\binom{n}{k} = \frac{n!}{(n-k)!k!}$. Thus, $p(k)$ may be considered to be the probability of k successes in n independent trials when the probability of success in a single trial is p. As $n \to \infty$, the binomial distribution approaches a normal distribution with a mean of np and a variance of $np(1 - p)$.

$$p(k \mid n) = \binom{n}{k} p^k (1-p)^{n-k} \quad . \tag{2.56}$$

$$P(k \mid n) = \sum_{m=0}^{k} p(m \mid n). \tag{2.57}$$

Interval: $k = 0, 1,2,3, \ldots n$.

Mean: $\mu = np$. (2.58)

Variance: $\mu_2 = np(1-p)$. (2.59)

$$\mu_3'(k) = n(n-1)(n-2)p^3 + 3n(n-1)p^2 + np.^{4} \tag{2.60}$$

$$\mathrm{F}[f(x)] = [1 - p + p \exp(iy)]^n. \tag{2.61}$$

Equation (2.61) for the characteristic reminds us that the binomial probabilities are of the same form as the terms of a binomial expansion to the power, n. The vertical bar in Eqn.(2.57) denotes a conditional probability. Here the left had side of Eqn.(2.57) is to be read, "the cumulative probability of up to *k* positive outcomes given *n* trials", where *k* is of course less than or equal to *n*. Bernoulli is credited with early work on the binomial distribution; his name is associated with it especially when the concept of *independent* trials is emphasized. A Bernoulli random variable is the number of successes observed in a single independent trial.

4.2 Poisson Distribution

The Poisson distribution is a discrete distribution like the binomial. It results from taking the limit of the binomial distribution as $p \rightarrow 0$ and the number of trials, $n \rightarrow \infty$, while the product, *np*, remains constant. The Poisson distribution underlies the Poisson point process which is perhaps the simplest point process and is used in stereological sampling. The number of fine inclusions of some specific type found in a sequence of equal-area samples is apt to be Poisson distributed. It is also often used to represent noise processes in imaging.

---

[4]The third moment about the origin is cannot be a function of the variable *k* but is meant to be understood as taken over *k*. This notation is a tradition of sorts.

$$p(n) = \frac{\mu^n}{n!}\exp(-\mu), \ [n > 0]. \tag{2.62}$$

$$P(n) = \sum_{k=0}^{n} p(k). \tag{2.63}$$

Interval: $n = 0, 1, 2, \ldots \infty$.

Mean: $\mu = \mu$ (2.64)

Variance: $\mu_2 = \mu$. (2.65)

$$\mu_3'(\mathrm{n}) = \mu(\mu^2 + 3\mu + 1). \tag{2.66}$$

$$\mu_4'(n) = \mu(\mu^3 + 6\mu^2 + 7\mu + 1). \tag{2.67}$$

$$\mathrm{F}[p(n)] = \exp(\mu \exp(iy) - \mu). \tag{2.68}$$

We will frequently be encountering the Poisson distribution in models for the random dispersion of features in space. It is important to note that it holds only if the "events" are independent of one another. For example, we might expect that the number of collisions of automobiles with lamp posts during a period when the traffic density remains constant would be Poisson distributed, because the number of trials is large but the probability of success is small. However, we would not expect that the number of people injured in such collisions would be Poisson distributed, because the average occupancy of a vehicle is greater than one.

4.3 Standard Normal Distribution

The normal distribution is a continuous distribution. Continuous distributions apply where the space of possible outcomes is not denumerable or where their number is so large that it is not worth counting. The normal distribution becomes a limiting case for the binomial distribution when the number of trials approaches infinity. When it becomes too cumbersome to ask, "how many trials...?", because the number is so large, it is easier to ask, "what fraction of trials?". In that case, the discrete variable, k, of the binomial distribution is replaced by a continuous variable that is expressed as a fraction (which may even be irrational) of *n*, the number of variates. Here the continuous variable is represented by the letter *t*. In books on statistics, it is frequently designated by the letter, *z*, called the *z-score*. This distribution is usually regarded as a special case of the normal distribution and most authors make no distinction in name.

$$f(t) = (2\pi)^{-1/2} \exp(-t^2/2). \tag{2.69}$$

$$F(t) = (1/2)[1 + \mathrm{erf}(t/\sqrt{2})] \quad . \tag{2.70}$$

Interval: $-\infty < t < \infty$.

Mean: $\mu = 0$. (2.71)

Variance: $\mu_2 = 1$. (2.72)

$$\mu_3' = \mu_{2m+1}' = 0 . \tag{2.73a}$$

$$\mu_4' = 3 . \tag{2.73b}$$

$$\mu_6' = 15 . \tag{2.73c}$$

$$\mu_8' = 105 . \tag{2.73d}$$

$$\mathrm{F}[f(t)] = \exp(-y^2/2) . \tag{2.74}$$

4.4 Normal Distribution

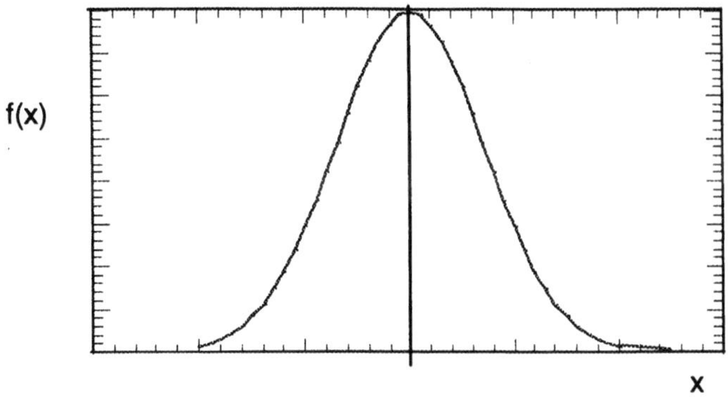

Figure 2.3: The Normal Distribution.

This distribution follows from the standard normal distribution with the substitution $t = (x - \bar{x})/\sigma$ and the use of Eqn.(2.51).

$$f(x) = (2\sigma^2\pi)^{-1/2}\exp\left[-\frac{1}{2}\left(\frac{x-\bar{x}}{\sigma}\right)^2\right]. \tag{2.75}$$

$$F(\mathrm{x}) = \frac{1}{2}\left[1 + \mathrm{erf}\left(\frac{x-\bar{x}}{\sigma\sqrt{2}}\right)\right]. \tag{2.76}$$

Interval: $-\infty < x < \infty$.

Mean: $\mu = \bar{x}$. (2.77)

Variance: $\mu_2 = \sigma^2$. (2.78)

$$\mu_3'(x) = \bar{x}\left(\bar{x}^2 + 3\sigma^2\right). \tag{2.79a}$$

$$\mu_4'(x) = \bar{x}^4 + 6\bar{x}^2\sigma^2 + 3\sigma^4. \tag{2.79b}$$

$$\mathrm{F}[f(x)] = \exp(\iota\bar{x}y - \sigma^2 y^2/2). \tag{2.80}$$

4.5 Log-Normal Distribution

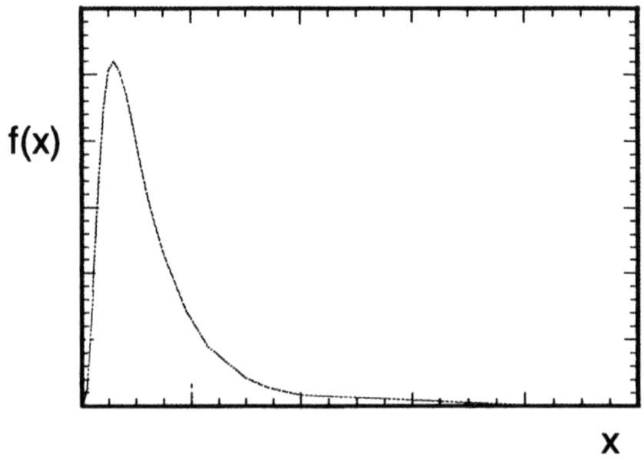

Figure 2.4: The Log-Normal Distribution

This smallest value in this distribution must always be greater than zero. It is frequently used for inclusion sizes, microcrack lengths and the like. It follows from the standard normal distribution with the substitution $t = (\ln(x)-\alpha)/\beta$ and the use of Eqn.(2.51). This distribution (which is skewed to the left) is often encountered in particle sizing and has the property that x^n follows the same distribution (but with different values of α and β). Thus, if the particle diameters are log-normally distributed, then so will be their surface areas and volumes. (Readers are to be warned that the factor of x^{-1} in the pre-exponential term is sometimes erroneously omitted.)

$$f(x)=\frac{1}{\beta x\sqrt{2\pi}}\exp\left\{-\left[\frac{\ln(x)-\alpha}{\beta\sqrt{2}}\right]^2\right\}. \tag{2.81}$$

$$F(x)=\frac{1}{2}\left[1+\operatorname{erf}\left(\frac{\ln(x)-\alpha}{\beta}\right)\right]. \tag{2.82}$$

Interval: $0 < x < \infty$.

Mean: $\mu = \exp(\alpha + \beta^2/2)$. (2.83)

Variance: $\mu_2 = \exp(2\alpha + 2\beta^2) - \exp(2\alpha + \beta^2)$. (2.84)

$$\mu_n'(x) = \exp(\alpha n + \beta^2 n^2/2) \quad . \tag{2.85}$$

Note that the median of this distribution is located at $x = e^{\alpha}$ which is not the same as the maximum (or mode). The characteristic function is omitted from the list since it is relatively complicated. Equation (2.85) is derived by first finding the *moment generating function*, $\int_0^{\infty} e^{tx} f(x)dx$ for ln(x). The moment generating function is similar to the characteristic function except that no imaginary numbers are involved; and, in the formula for the moment (where $t \to 0$), the factor involving negative one to a power is eliminated. Characteristic functions always exist for every distribution. The moment generating function integral may diverge.

4.6 Rectangular Distribution

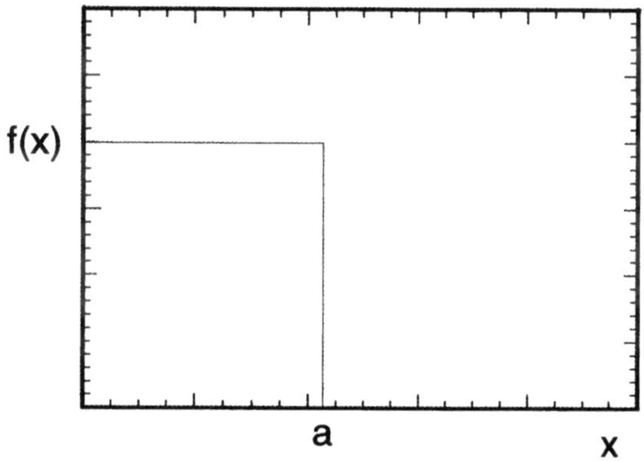

Figure 2.5: The Rectangular Distribution

This is often referred to as the uniform distribution. The width of the rectangle is a.

$$f(x)=1/a. \tag{2.86}$$

$$F(x)=x/a. \tag{2.87}$$

Interval: $0 < x < a$.

Mean: $\mu = a/2$. (2.88)

Variance: $\mu_2 = a^2/12$. (2.89)

$$\mu_n'(x) = a^n/(n+1),\ [n > 0]\ . \tag{2.90}$$

$$\mathrm{F}[f(x)] = [\exp(iay) - 1]/iay\ . \tag{2.91}$$

4.7 Parabolic Distribution

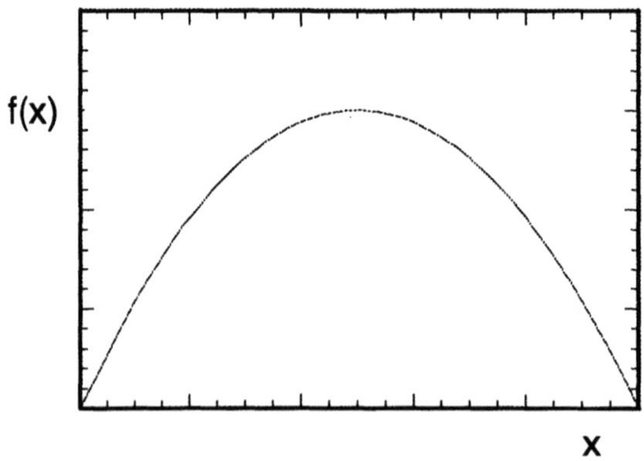

Figure 2.6: The Parabolic Distribution

$$f(x) = 6x(1-x)\ . \tag{2.92}$$

$$F(\mathrm{x}) = 3x^2 - 2x^3 \tag{2.93}$$

Interval: $0 < x < 1$.

Mean: $\mu = 1/2$. (2.94)

Variance: $\mu_2 = 1/20$. (2.95)

$$\mu_n'(x) = \frac{6}{(n+2)(n+3)}, \quad [n > -1]. \tag{2.96}$$

$$\mathrm{F}[f(x)] = 6(iy)^{-3}[iy + 2 + (iy-2)\exp(iy)]\ . \tag{2.97}$$

Because it is simple to handle analytically, the parabolic distribution is often been used for simulating particle-size distribution data when testing computational procedures.

4.8 Exponential Distribution

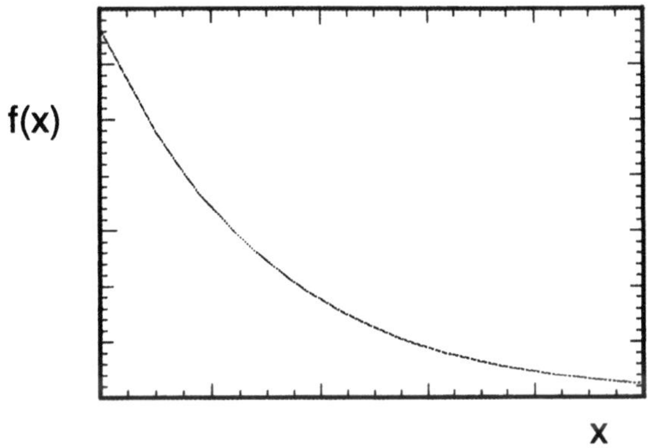

Figure 2.7: The Exponential Distribution

The exponential distribution applies to the intervals (in time or space) between random events. It tends to arise in situations analogous to the following. Suppose one filled one's car with very cheap tires and drove around the world. If all tires had some constant probability of blowing out per mile (and never lasted long enough to wear out), the total distance traveled by the truck before it ran out of tires would be exponentially distributed.

$$f(x) = (1/\alpha)\exp(-x/\alpha), \quad [\alpha > 0] \ . \tag{2.98}$$

$$F(x) = 1 - \exp(-x/\alpha) \ . \tag{2.99}$$

Interval: $0 \leq x \leq \infty$.

Mean: $\mu = \alpha$. (2.100)

Variance: $\mu_2 = \alpha^2$ (2.101)

$\mu_n' = \Gamma(n+1)\alpha^n, \quad n > 0$. (2.102)

$$F[f(x)] = \frac{1}{1 - i\alpha y} . \tag{2.103}$$

In the example above, when one tire is changed, the car behaves the same as if all tires had been changed as far as the probability of breakdown in the next mile is concerned. This is because the tires don't last long enough to wear out and the probability of any one of them blowing out is independent of how long they have been on the car or how far they have gone. This property, a characteristic of the exponential distribution, is termed "lack of memory".

4.9 Beta Distribution

The beta distribution may be regarded as a continuous generalization of the binomial distribution and has one of the more flexible distribution functions in that it can be fitted to both skewed and non-skewed data. When n independent variates are located randomly in the interval (0,1), the probability density for the location of the k-th point from the left is given by the beta distribution with parameters $\alpha = k$, $\beta = n\text{-}k\text{+}1$. The uniform distribution itself is given by the beta distribution with $\alpha = \beta = 1$.

$$f(x) = \frac{\Gamma(\alpha+\beta)}{\Gamma(\alpha)\Gamma(\beta)} x^{\alpha-1}(1-x)^{\beta-1}, \qquad \alpha > 0, \beta > 0 . \tag{2.104}$$

$$F(x) = \frac{\Gamma(\alpha+\beta)}{\Gamma(\alpha)\Gamma(\beta)} B_x(\alpha, \beta) \quad . \tag{2.105}$$

Interval: $0 \le x \le 1$.

Mean: $\mu = \dfrac{\alpha}{\alpha + \beta}$. (2.106)

Variance: $\mu_2 = \dfrac{\alpha\beta}{(\alpha+\beta)^2(\alpha+\beta+1)}$. (2.107)

$$\mu_n' = \frac{\Gamma(\alpha+n)\Gamma(\alpha+\beta)}{\Gamma(\alpha)\Gamma(\alpha+\beta+n)} . \tag{2.108}$$

F[$f(x)$]: (complicated)

$B_x(\ )$ is the incomplete beta function, Eqn.(2.12). The characteristic function for the beta distribution is known as Kummer's confluent hypergeometric function for these values of the arguments, $M(\alpha, \alpha+\beta, iy)$. For certain values, the latter function can be expressed in terms of elementary functions. A list of these values is to be found on p. 509 of Ref. (2) [Abramowitz and Stegun, 1964]in the bibliography for this chapter.

4.10 Gamma Distribution

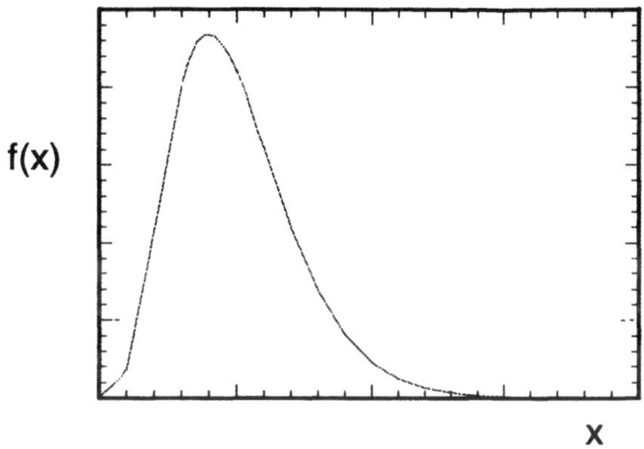

Figure 2.8: The Gamma Distribution

The gamma distribution looks similar to the beta distribution except that the beta distribution is more variable in shape. Like the beta distribution the gamma distribution involves the gamma function and may be written with two parameters designated α and β. In this form it subsumes the chi-squared distribution as a special case. The one-parameter form is given below which is the same as the two-parameter form where β is set to unity. In this one parameter form, p is the parameter.

$$f(x) = \frac{x^{(p-1)}}{\Gamma(p)} \exp(-x) \ . \tag{2.109}$$

$$F(x) = \gamma(p, x)/\Gamma(p) \ . \tag{2.110}$$

Interval: $0 \leq x < \infty$

Mean: $\mu = p$. (2.111)

Variance: $\mu_2 = p$. (2.112)

$$\mu_n'(x) = \Gamma(p+n)/\Gamma(p). \tag{2.113}$$

$$F[f(x)] = (1-iy)^{-p} . \tag{2.114}$$

$\gamma(p,x)$ is the incomplete gamma function, Eqn.(2.9). For a Poisson distribution with a mean and variance of unity, the gamma distribution gives the distribution of distance of the p-th event. For example, if in a linear scan the average distance between intersections with voids is x, than the distribution of distances measured in x-units from the origin to the p-th void is given by Eqn.(2.109).

4.11 Delta Distribution

This is the distribution for which "distribution theory" is most known. It looks like a single spike of zero width and infinite height.

$$f(x) = \delta(x-\alpha). \tag{2.115}$$

$$F(x) = U(x-\alpha) . \tag{2.116}$$

Interval: $-\infty < x < \infty$

Mean: $\mu_2 = \alpha$. (2.117)

Variance: $\mu_2 = 0$. (2.118)

$$\mu_n'(x) = \alpha^n . \tag{2.119}$$

$$\mathrm{F}[f(x)] = \exp(i\alpha\, y) . \tag{2.120}$$

Comment: $\delta(x)$ and $U(x)$ are defined in section (2.1).

4.12 p - Distribution

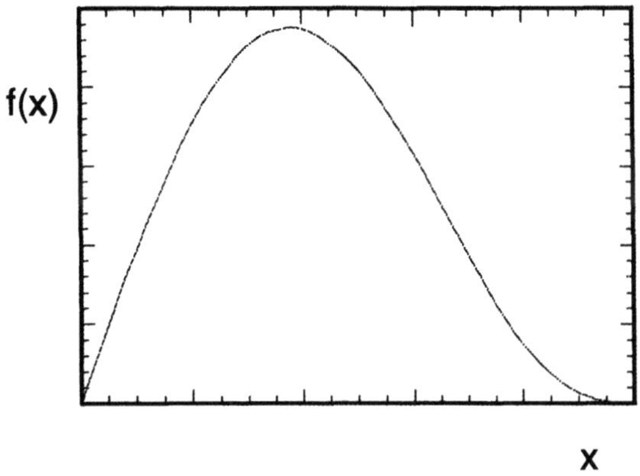

Figure 2.9: The p-Distribution

$$f(x) = 2(p+1)x(1-x^2)^p, \quad [p \geq -1] . \tag{2.121}$$

$$F(x) = 1-(1-x^2)^{p+1} \quad . \tag{2.122}$$

Interval: $0 \leq x \leq 1$.

Mean: $\mu = \dfrac{\sqrt{\pi}\,\Gamma(p+2)}{2\Gamma\left(p+\frac{5}{2}\right)}$. (2.123)

Variance: $\mu_2 = \dfrac{1}{p+2} - \mu^2$. (2.124)

$$\mu_n^{'}(x) = \frac{\Gamma\left(\frac{n}{2}+1\right)\Gamma(p+2)}{\Gamma\left(p+2+\frac{n}{2}\right)} . \tag{2.125}$$

F[$f(x)$]: (complicated)

Comment: Since we may be encountering this distribution in several different contexts, it is convenient to give it a name (although the reader should be cautioned that the name is not an "official" one). The function is an interesting one; as $p \to -1$ it approaches a delta function at $x = 1$. For $p < 1.5$ the distribution is skewed to the right and for $p > 1.5$ it is skewed to the left (although the skewness is more apparent than real, since the distribution remains approximately symmetric about its maximum). Finally, as $p \to \infty$ the distribution again becomes a delta function this time centered at $x = 0$. With $p = (N-3)/2$, $(1/D)f(x = L/D)$ is the distribution of intercept lengths in an N-dimensional sphere of diameter D, and $(W/L^2)F(x = W/L)$ is the distribution of intercept lengths in an N-dimensional plate of thickness W. With $p = (N-5)/2$, $(1/D)f(x = d)$ yields the distribution of profile diameters d in a three-dimensional section through a collection of N-dimensional spheres of constant diameter D in N-dimensional space. The same distribution also arises in a theoretical treatment [Woodhead (1968)] for the size distribution of particles when growth is diffusion controlled. In this case p is equal to the exponent in the expression for the time dependence of the nucleation rate.

4.13 Chi-Square Distribution

The chi-square distribution is another that resembles the gamma distribution in form. It is in fact an example of the two-parameter gamma distribution.

$$f(x) = \frac{x^{\frac{\nu}{2}-1} \exp\left(\frac{-x}{2}\right)}{2^{\frac{\nu}{2}} \Gamma\left(\frac{\nu}{2}\right)}. \tag{2.126}$$

$$F(x) = 1 - \frac{\gamma\left(\frac{\nu}{2}, \frac{x}{2}\right)}{\Gamma\left(\frac{\nu}{2}\right)}. \tag{2.127}$$

Interval: $0 \leq x < \infty$.

Mean: $\mu = \nu$. (2.128)

Variance: $\mu_2 = 2\nu$. (2.129)

$$\mu_n' = \frac{2^n \Gamma\left(\frac{\nu}{2} + n\right)}{\Gamma\left(\frac{\nu}{2}\right)}. \tag{2.130}$$

$$F[f(x)] = (1 - 2iy)^{-\frac{\nu}{2}} \quad . \tag{2.131}$$

Comment: This is actually the gamma distribution [Eqn. (2.109)] with $p = \nu/2$ and the substitution of $t = x/2$. The parameter ν is known as the "degrees of freedom". The chi-square distribution is used extensively in making statistical comparisons involving several events. The chi-square statistic (the x in the above formulas) is a discrete variable and is defined as $X^2 = \sum_{i=1}^{k} \frac{(o_i - e_i)^2}{e_i}$ where o_i is the observed frequency and e_i is the expected frequency of the k variables. The parameter, ν, is k minus the number of relations between the variables. One use of the chi-square distribution in stereology is the comparison of observed distributions with theoretical ones. For instance, it is often convenient to assume that the sizes of flaws such as pores, inclusions or microcracks follow a Weibull distribution. Precipitate sizes are often assumed to be lognormally distributed. These are not always good assumptions but are easily tested using the chi-square distribution.

4.14 Parthenogenic Distribution

Here is another distribution that resembles the gamma distribution in form.

$$f(x) = \frac{x}{\alpha^2} \exp\left(-\frac{x^2}{2\alpha^2}\right) \quad . \tag{2.132}$$

$$F(x) = 1 - \exp\left(-\frac{x^2}{2\alpha^2}\right) \quad . \tag{2.133}$$

Interval: $0 \le x < \infty$.

Mean: $\mu = \alpha\sqrt{\frac{\pi}{2}}$. (2.134)

Variance: $\mu_2 = \alpha^2\left(2 - \frac{\pi}{2}\right)$. (2.135)

$$\mu'_n(x) = \alpha^n 2^{\frac{n}{2}} \Gamma\left(\frac{n+2}{2}\right) \qquad [n > -1]. \tag{2.136}$$

F[$f(x)$]: (complicated)

Comment: This is an interesting distribution since it is self-reproducing under the following circumstances: if the diameters of an aggregate of spherical particles obey this distribution, then the diameters of the circular profiles on a planar section through the particles have an identical distribution. The same is also true for the distribution of intercept lengths on a linear probe through the particles. The distribution is also of some practical interest in that it is skewed to the left, which is typical of many observed particle-size distributions. However, its use to represent such distributions is limited by the fact that it contains only a single parameter, and thus the spread of the distribution cannot be varied independently of the average.

4.15 An Example of a Bimodal Distribution

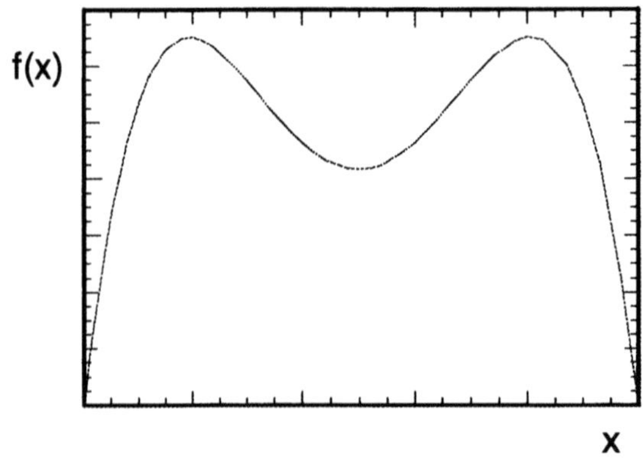

Figure 2.10: A Bimodal Distribution

Strictly speaking a bimodal distribution must have two modes. In common parlance any distribution having twin peaks is called "bimodal". Bimodal distributions are usually, but not always, sums of two unrelated distributions. Scores on mathematics examinations are notorious for being bimodal. Defenders of this situation say that there are two populations: "those who know it" and "those who don't". Opponents say that there is an arbitrary gradient-related process that splits an initially uniform distribution; the teacher teaches to the faster learners in proportion to their speed of learning.

$$f(x) = 10\left[\frac{1}{12} + \frac{(2x-1)^2}{4} - \frac{(2x-1)^4}{3}\right]. \qquad (2.137)$$

$$F(x)=10\left[\frac{1}{120}+\frac{x}{12}+\frac{(2x-1)^3}{24}-\frac{(2x-1)^5}{30}\right]. \tag{2.138}$$

Interval: $0 \le x \le 1$.

Mean: $\mu = 1/2$. (2.139)

Variance: $\mu_2 = 19/252$. (2.140)

$$\mu_n'(x)=10\left(\frac{-16}{3(n+5)}+\frac{32}{3(n+4)}-\frac{7}{n+3}+\frac{5}{3(n+2)}\right). \tag{2.141}$$

F[*f*(x)]: (complicated)

Comment: This is included as an example of a bimodal distribution that has been used in a simulation of particle-size unfolding [Anderssen and Jakeman (1975)]. It is symmetric about x = 1/2 and has maxima at x = 1/2 ± $(3/32)^{1/2}$.

4.16 Weibull Distribution

The Weibull distribution is yet another variation on the gamma distribution and looks similar to it. It exists in both three-parameter and two-parameter forms. The three-parameter form was the first to be introduced. The two-parameter form is given below.

$$f(x)=\frac{m}{\theta}\left(\frac{x}{\theta}\right)^{m-1}\exp\left(-\left[\frac{x}{\theta}\right]^m\right). \tag{2.142}$$

$$F(x)=1-\exp\left(-\left[\frac{x}{\theta}\right]^m\right). \tag{2.143}$$

Interval: $0 \le x < \infty$.

Mean: $\mu = \theta\Gamma\left(1+\frac{1}{m}\right)$. (2.144)

Variance: $\mu_2 = \theta^2\left\{\Gamma\left(1+\frac{2}{m}\right)-\left[\Gamma\left(1+\frac{1}{m}\right)\right]^2\right\}$. (2.145)

$$\mu_n'(x) = \theta^n\,\Gamma\left(1+\frac{n}{m}\right). \tag{2.146}$$

F[$f(x)$]: (complicated)

Comment: The Weibull distribution was introduced by Waloddi Weibull circa 1951 as a "statistical distribution of wide applicability". It is not a consequence of any particular process; nothing generates inherently Weibully-distributed variates. Nevertheless, the Weibull distribution has become the most popular distribution for representing flaws in materials. Pores, microcracks, inclusions, weak fibers and similar entities are among those that have been fitted with Weibull distributions. Microscopists are thus often called upon to fit their results to the Weibull distribution function. The reason for its popularity is that is the simplest mathematical expression to express a weakest series link property. This property is simply that the probability of a chain of m independent and identically-distributed links holding at a load, x, is $[1-F(x)]^m$ where $F(x)$ is the probability that the failure load of an individual link is less than x. Since $F(x)<1$, $1-F(x)$ is also less than one. It follows from the requirement that the chain must hold some load even when m becomes infinite that $F(x)=1-e^{-q(x)}$ where q is some simple function such as x to a power. Eqn.(2.143) is the result. The extremel property that a group of samples has a smallest something is analogous. What is the smallest failure load in a group of m links? It is the same as that of a chain made from them. In this way the Weibull distribution finds application as an *extreme value distribution*. Similarly Wiebull distributions are used to represent the distribution of the largest of things. For example, the distribution of the size of the largest tree on an acre might be conveniently represented by a Weibull distribution.

5. PROPERTIES OF A HYPERSPHERE (AND HYPERBALL)

Extension of stereological results to higher dimensions often requires the "N-1 content" or "surface area", $S|b|$ and the "N-content" or "volume", $V|b|$, of an N-dimensional sphere, b_N. Taking volume first, we need a definition for it on which we may easily agree and which is not dependent on the number of dimensions of the space in which we are working. The key to the geometry of hyperspace is that it is consistent with vector algebra and analytic geometry. Since volumes can be obtained as vector products and vector spaces have an arbitrary dimensions, it is not surprising to find that the volume of an n-dimensional cube is a^n, where a is the length of an edge. What is "length"? It is the distance between two points however you define it. We explore this question in the next chapter and in the Appendix. The definition we use here is the ordinary Euclidean one consistent with the square of the length of the hypotenuse of a right triangle being equal to the sums of the squares of the lengths of the other two sides. Not all possible definitions of length have that property. This approach is consistent with defining the N-content of an N-dimensional sphere (or point set) as its "volume"; volume is the measure that has the same number of dimensions as the space in which it is taken. Given that we may propose an integral as a definition of volume for a sphere,

$$V|b|= \int\limits_{\left(\sqrt{\sum\limits_{(N)} x_n^2} \le R\right)}^{(N)} dx_n \ . \tag{2.147}$$

We are saying that we are taking the N-th integral of unity subject to the condition that all the points in the integral summation lie within a specified radius defined by $\sqrt{\sum\limits_{(N)} x_n^2}$ which is the distance between a point and the origin in ordinary or Euclidean space. Each x_n in this expression is a component of an N-vector to the particular point being summed in the integration. Distance formulas such as these are called, "metrics" and this distance formula is the Euclidean metric. More will be said of this in Section 3.3.1. This formula is more easily stated than it is evaluated. But by evaluating it we learn some techniques for wrestling with hyperspace.

To find the volume of a sphere in 3-space, we customarily sweep a disk through a dimension not contained in the disk. If done correctly, the locus of the disk is a sphere. The sweeping is done in two ways. Either, in spherical coordinates, a disk is rotated through π radians to form a solid of revolution; or, in rectilinear coordinates, a disk of varying diameter is swept along a line perpendicular to it. The diameter of the disk is reduced as the disk moves away from the center of the sphere to keep the radius from the center of the sphere to the edge of the disk constant. We are not yet on sure footing with the concept of an n-dimensional sphere; so, we will not attempt to use n-

dimensional spherical coordinates. Rather, we will define one-, two-, three and 4-dimensional spheres in rectilinear coordinates. By extending this series, we will find the volume and surface area of an n-dimensional hyperball. The volume of a one-dimensional ball is simply,

$$V_1 = 2\int_0^{R_1} dx = 2R_1 . \tag{2.148}$$

It is the length of a line extending in both directions from the origin out to a distance, R_1. The volume of a two-dimensional ball is similarly found, Fig. 2.11, taking a little liberty with the way of expressing the order of integration,

$$V_2 = 2\int_0^{R_2} 2\sqrt{R_2^2 - x_2^2}\, dx_2 = 2^2 \int_0^{R_2} \int_0^{R_1=\sqrt{R_2^2-x_2^2}} dx_2\, dx_1 . \tag{2.149}$$

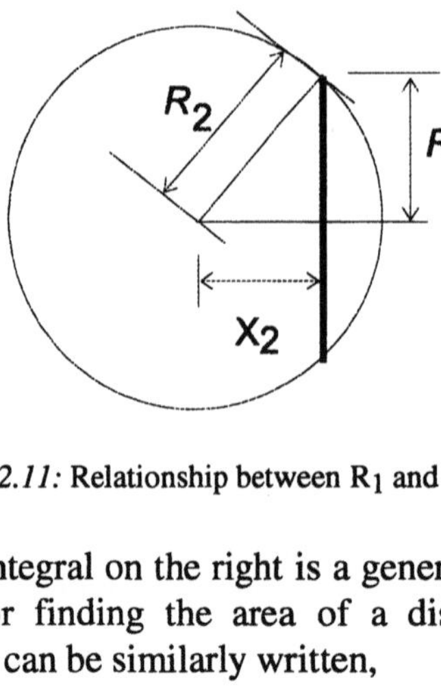

Figure 2.11: **Relationship between R_1 and R_2**

The double integral on the right is a generalization of that on the left, the common one for finding the area of a disk. The volume of the three-dimensional ball can be similarly written,

$$V_3 = 2^3 \int_0^{R_3} \frac{\pi}{4}\left(R_3^2 - x_3^2\right) dx_2 = 2^3 \int_0^{R_3} \int_0^{\sqrt{R_3^2-x_3^2}} \int_0^{\sqrt{R_2^2-x_2^2}} dx_1\, dx_2\, dx_3 . \tag{2.150}$$

Extending this, we may find the volume of a four-dimensional ball as,

$$V_4 = 2^4 \int_0^{R_4} \int_0^{\sqrt{R_4^2 - x_4^2}} \int_0^{\sqrt{R_3^2 - x_3^2}} \int_0^{\sqrt{R_2^2 - x_2^2}} dx_1\, dx_2\, dx_3\, dx_4 \ . \tag{2.151}$$

Direct evaluation of this integral is straightforward but cumbersome. It is evident that R_4 is the only independent variable and that the others are dummies. This process of sweeping and integrating can be extended algebraically to any number of dimensions. Consequently, by hypothesis, it can be extended geometrically as well.

In order to evaluate Eqn.(2.151) generally, we re-write it as a series of convolution integrals. We begin by making the substitutions, $\rho_i = x_i^2$ and $P_i = R_i^2$. Since $dx = \frac{1}{2}\rho^{-\frac{1}{2}} d\rho$,

$$V_4 = 2^4 \int_0^{P_4} \left(2\sqrt{\rho_4}\right)^{-1} \int_0^{P_3} \left(2\sqrt{\rho_3}\right)^{-1} \int_0^{P_2} \left(2\sqrt{\rho_2}\right)^{-1} \int_0^{P_1} \left(2\sqrt{\rho_1}\right)^{-1} d\rho_1 d\rho_2 d\rho_3 d\rho_4 \, . \tag{2.152}$$

Noting that the ρ_i are dummies and all the P_i are equal, integral (2.152) can then be broken down into:

$$\begin{aligned} V_4 &= 2^4 \int_0^{P} \left(2\sqrt{\rho}\right)^{-1} F_3(P-\rho)\, d\rho \\ F_3 &= \int_0^{P} \left(2\sqrt{\rho}\right)^{-1} F_2(P-\rho)\, d\rho \\ F_2 &= \int_0^{P} \left(2\sqrt{\rho}\right)^{-1} F_1(P-\rho)\, d\rho \\ F_1 &= \int_0^{P} \left(2\sqrt{\rho}\right)^{-1} d\rho \end{aligned} \tag{2.153 a-d}$$

from which, $V_4 = 2^4 g \otimes g \otimes g \otimes \int g\, d\rho$ follows by abbreviation of notation. Using equations (2.20, 2.21, 2.23) this form is easily transformed by the LaPlace transform into one *in which the number of dimensions of the space is a parameter, n,*

$$L[V_n]=\frac{2^n}{p}L[g]^n\,, \tag{2.154}$$

where $g=\left(2\sqrt{\rho}\right)^{-1}$, and its transform is $\dfrac{\Gamma\left(\frac{1}{2}\right)}{2p^{\frac{1}{2}}}$. Inverting Eqn. (2.154) gives

$$V|b_N|=\frac{\pi^{\frac{n}{2}}R^n}{\Gamma\left(\frac{n}{2}+1\right)}\,. \tag{2.155}$$

R is the radius of the ball in the chosen space.

Finding the surface "area" of this ball is done by making an analogy using the idea that the surface area of a shell times its thickness equals its volume. The surface thus has n-1 dimensions. For the sake of consistency, then, the surface is,

$$S|b_N|=\frac{dV_n}{dR}=\frac{\pi^{\frac{n}{2}}R^{n-1}}{\Gamma\left(\frac{n}{2}\right)}\,. \tag{2.156}$$

More details on the transformation can be found in [Sneddon, 1972] p. 173.

6. BIBLIOGRAPHY

Books suggested as good references by J.H. are numbered and separated. Non-numbered references are citations.

R.S. Anderssen and A.J. Jakeman, "Computational methods of solution and the random spheres approximation", *Journal of Microscopy*, **102** pt. 2 (1975) 135-153.

(1) G. A. Korn and T. M. Korn, *Mathematical Handbook for Scientists and Engineers*, McGraw-Hill, 1968.

As its name implies, this is a compendium of definitions, theorems and formulas. Although no derivations are given, there is an extensive bibliography of where they can be found. It is a very valuable reference work, which any stereologist doing theoretical work should have on the shelf.

(2) M. Abramowitz and J. A. Stegun, (Eds.), *Handbook of Mathematical Functions* (National Bureau of Standards, Washington, D C, 1964).

The increasing availability of small computers and programmable calculators has rendered the tables in this volume less necessary than when it was first published. However, it is still extremely useful for its comprehensive treatment of the properties of the higher functions and the relationships between them.

B.V. Gnedenko, *The Theory of Probability*, B.D. Seckler, trans. (Chelsea, Publications Co., New York, 1967).

Stereological relationships frequently involve the evaluation of integrals. Short tables of integrals are to be found in most handbooks. A more extensive collection is:

(3) W. Gröbner and N. Hofreiter, Integraltafel Vol. 1; *Bestimmte Integrale, Vol. 2; Unbestimmte Integrale* (Springer-Verlag, Wien, 1961).

Another very comprehensive source is:

(4) I. S. Gradshteyn and I. M. Ryzhik, *Table of Integrals,Series and Products* (Academic Press, New York, 1965).

Both of the foregoing are considerably superior to the older tables of D. Bierens de Haan.

One of the problems of using integral tables is that, by substitution of the variable, any given integral may be written in many different forms and it is often difficult to judge under what form it will be listed. In those cases where a definite integral can be written as one of the standard integral transforms it can often be located more rapidly in a table of transforms. Such tables therefore serve as integral tables in addition to their original purpose. A collection of transforms that I (J.H.) have found very useful is:

(5) A. Erdelyi (Ed.) *Tables of Integral Transforms*, Vols. 1 and 2, (McGraw-Hill, New York, 1954).

These tables include Fourier sine, cosine and exponential transforms; Laplace, Mellin, (and their inverse) transforms in addition to ten less well-known varieties.

Derivations of most of the results in probability and statistics that we will be using are to be found in any introductory text. For example:

(6) P. G. Hoel, *Introduction to Mathematical Statistics* (John Wiley and Sons, Inc., New York, 1962).

An extremely valuable source of general information about statistics is

S. Kotz and N.L. Johnson, editors-in-chief, *Encyclopedia of Statistical Sciences* (John Wiley and Sons, New York, 1981).

See also,

S. Kotz, N. Balakrishnan and N.L. Johnson, *Continuous Multivariate Distributions*, Volume 1, 2nd. Edition (John Wiley & Sons, New York, 2000).

More advanced texts relevant to this work include:

(7) W. Feller, *An Introduction to Probability Theory and Its Applications* (John Wiley and Sons, Inc., New York, Vol. 1: 1968, Vol. 2: 1971).

(8) T. C. Fry, *Probability and Its Engineering Uses* (D. Van Nostrand, Princeton, 1928).

(9) M. G. Kendall and A. Stuart, *The Advanced Theory of Statistics* (Charles Griffin and Co., London, Vol. 1: 1969, Vol. 2: 1973).

Hyperspace and its geometry are becoming popular and introductory works such as published by Scientific American are available. These can help intuition but do little for ones ability to calculate. Some traditional works oriented toward calculations include:

(10) M.G. Kendall, *A Course in the Geometry of n Dimensions* (Hafner, New York, 1961).

J.H. Sneddon, *The Use of Integral Transforms*, (McGraw-Hill, New York, 1972)

(11) D.M.Y. Sommerville, *An Introduction to the Geometry of N Dimensions* (Dover, New York, 1958).

R.S. Strichartz, *A Guide to Distribution Theory and Fourier Transforms* (CRC Press, Boca Raton, 1993).

G. Wimmer and G. Altmann, *Thesaurus of univariate discrete probability distributions* (Stamm, Essen, 1999).

J.L. Woodhead, "Stabilized zirconia particles by sol-gel processes", *Science of Ceramics*, **4** (1968) 105-111.

7. PROBLEMS

1) It follows from the definitions of $\mu_n'(x)$ and the characteristic function F[*f*(*x*)] that $\mu_0'(x)=1$. And therefore, F[*f*(*x*)] evaluated at $y = 0$ also equals unity. Verify that this is true for the properties of the distributions listed in section 2.4.

2) Verify that Eqns.(2.155) and (2.156) yield the correct results for a circle and a sphere.

3) A point is placed at random in a circle of diameter *d*. (We define a random location of a point as being that for which the probability of a point falling within a given domain inside the circle is equal to the ratio of the area of the domain to that of the circle, and is independent of the location of the domain.)

Show that for such a placement the distribution of the distance, *q*, of the point from the center of the circle is given by:

$$f(q) = 8q/d^2 \quad (2.157P)$$

4) In Chapter 6 it is shown that the distribution of profile diameters, *f*(*d*), generated by the random interception of a plane on a sphere of diameter, *D*, is:

$$f(d)=\frac{d}{D\left(D^2-d^2\right)^{\frac{1}{2}}} .$$

Prove that the cumulative distribution is given by:

$$F(d)=1-\frac{1}{D}\left(D^2-d^2\right)^{\frac{1}{2}} . \quad (2.158P)$$

Chapter 3

DEFINITIONS, SETS AND MEASURES

The purpose of this chapter and of the Appendix is to acquaint the student with some of the elements of set theory that are needed to understand stereology in the larger framework of stochastic geometry and its literature. This is only a beginning, and yet we expect that the reader will have encountered many of these things before in Calculus. Understanding this chapter is not essential to understanding most of the material in this book, since the derivations given in later chapters are kept close to their applications. They are thus intentionally "old-fashioned", so that the reader does not have to be an expert on set and measure theory to understand them. Like the material presented in the previous chapter, the definitions presented in this chapter and in the Appendix are written to be more intuitive than rigorous. They are to refresh the memory or give the reader a clue before the reader looks up the fuller more rigorous definitions elsewhere. We want the beginner understand at least some of the terminology in the literature. We encourage the interested reader to study these matters further. Because there can be subtle changes in terminology as it evolves, we should inform the reader from where the definitions given here and in the Appendix arose. The primary source is a "classic", Kolmogorov and Fomin's *Introductory Real Analysis* [1970]. While not particularly oriented to the geometer, its definitions are extensive and seem fundamental to all disciplines. Some of the measure theory of geometry is adapted from the excellent and readable *Geometry of Sets and Measures* by Pertti Mattila [1995]. Additional references on measure theory include Stoyan and Stoyan's *Fractals Random Shapes and Point Sets* [1994] and Stoyan, Kendall and Mecke's *Stochastic Geometry and Its Applications* [1995]. This chapter presents those aspects of the subject that are most likely to be helpful to the reader of this book.

The Appendix contains many descriptions of terms that may come to the reader's mind in the course of reading this and related material.

When Elias first defined *stereology*, the term specifically meant the study of three-dimensional objects through two-dimensional sections. The second author (L.L.) once asked John Hilliard whether or not the interaction of an eddy current with a metal (for the purpose of measuring its distributed discontinuity) could be termed stereology. He said that indeed it was and that the statistics of interaction of any sort of probe no matter how general with any sort of structure was the province of stereology. With appropriate phrasing and some added definitions, this broad definition could encompass most all of applied stochastic geometry and a good number of other things as well. Accordingly, we prepare ourselves.

1. SETS, PROPERTIES AND MAPPINGS

We have already used the word "set". Although the language of set theory is ubiquitous, at this point we review some of the foundations of the subject.

1.1 Definitions

"Sets" are made up of elements of various kinds e.g. the set of sides of a square or the points that comprise a line. We will generally use capital roman letters to designate sets. For emphasis or to enclose the elements of a set we will use curly braces, $\{\}$. Usually we will use small roman letters to indicate the elements of a set. We will say $a \in A$ to indicate that a belongs to the set A, as we have already. Similarly, $b \notin A$, indicates that b does not belong to A. $A \subset B$ means "the set A is contained in the set B" or is a subset of B. If $A \neq B$ in this case, A is a *proper* subset of B. Some sets have no elements. The set containing no elements, the *null set*, is designated here by $\varnothing$.

Some sets of particular interest include the real line and the n-dimensional real space. In order to define these we reaffirm the common notation for intervals. An open interval is one that does not include its bounds. An open interval on a line is designated with parentheses thus (a,b) where a and b are the bounds. A closed interval includes its bounds and is designated with square brackets, $[a,b]$. The bounds a and $b \in [a,b]$ but not to (a,b). Intervals may be half closed and half open: $(a,b]$ and $[a,b)$. The real line is denoted by $\mathbf{R} = (-\infty, \infty)$. The real space, $\mathbf{R}^n$, is composed of n-tuples of real numbers each of which lies on the real line. For convenience the null set, $\varnothing$, is defined as open. There is no such thing as a set of sets. Defining sets of sets leads to paradoxes such as that of the definition which defines a set that contains a specific set only if it does not contain that set. The history of mathematics is fun. We refer to *families* of sets in order of evade such paradoxes.

1.2 Operations on Sets

The sum or union of two sets is designated $A \cup B$. In terms of a Venn-type diagram (frequently used to describe probability spaces) it is shown this way.

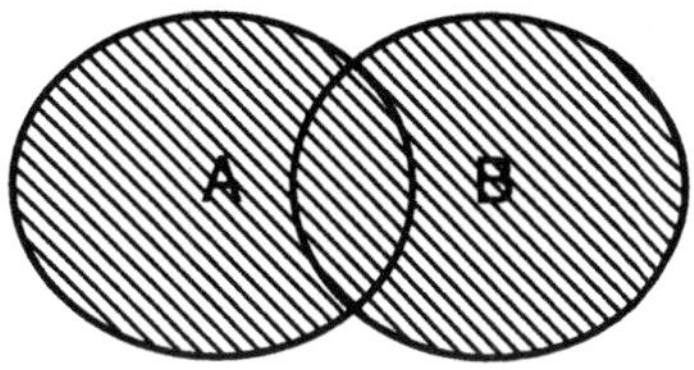

Figure 3.1: The union of two sets.

In the region of overlap, Fig. 3.1, the elements are not duplicated. Rather, the union is the set of elements belonging to at least one of the parent sets. The intersection of two sets, $A \cap B$ is the set of elements common to both parent sets. In terms of a Venn diagram it is shown
this way, Fig.3.2.

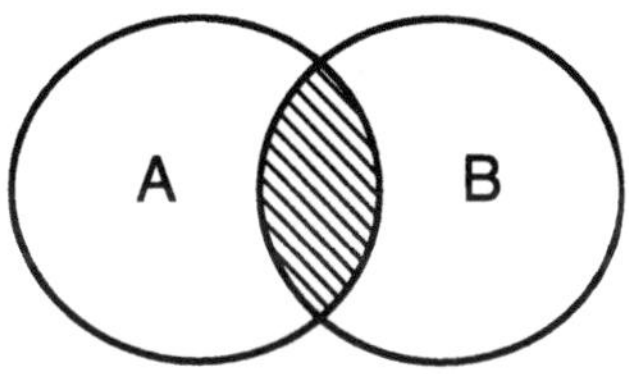

Figure 3.2: The intersection of two sets.

The union of an arbitrary number of sets is written $\bigcup_{\alpha} A_\alpha$ where the union is taken over the indexing parameter, α. Similarly the intersection of an arbitrary number of sets is designated $\bigcap_{\alpha} A_\alpha$. Two sets, A, B, are *disjoint* if $A \cap B = \varnothing$. Unions and intersections are commutative and obey the distributive laws as described in the Appendix for cuts. (Cuts were important when the authors took their high school math courses. They are not much discussed today but make a good introduction to the theory of Borel sets.) The difference between two sets, $A - B$ is the set A with any common elements of B subtracted from it. The difference is most often written $A \setminus B$. Many geometers use the notation $\varphi - x$ to mean the set φ shifted to the left by the amount or vector x.

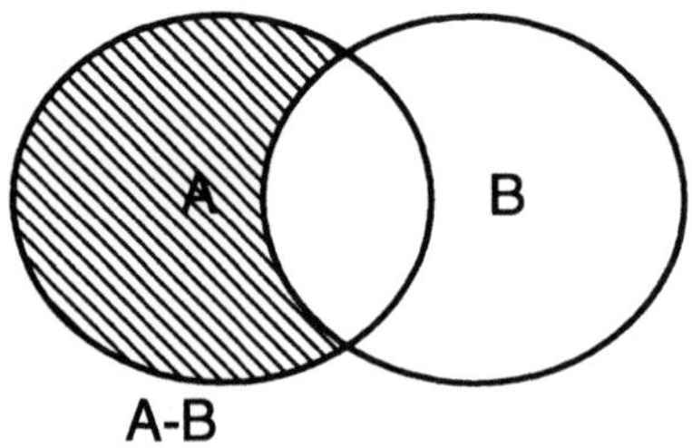

Figure 3.3: The difference of two sets.

The hatched portion of the diagram, Fig. 3.3, at the left is the difference. As in ordinary arithmetic, the difference is not commutative. Another type of difference is called the symmetric difference. It is commutative and in Boolean terms is an exclusive OR. It is designated as
$A \Delta B$.

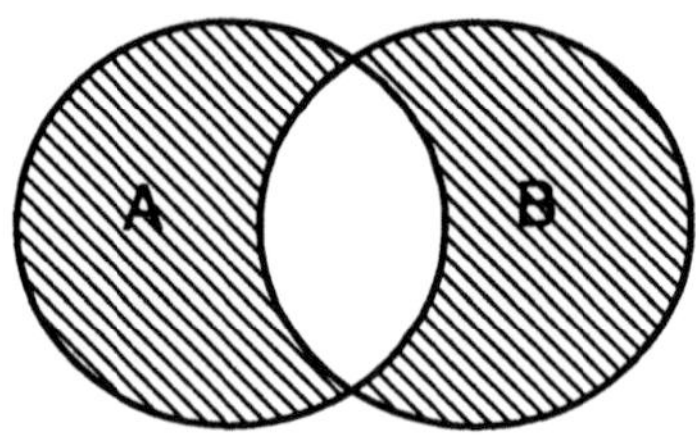

Figure 3.4: The symmetric difference of two sets.

The crosshatched area of Fig. 3.4 represents $A\Delta B$. $A\Delta B = (A\text{-}B) \cup (B\text{-}A)$. This is much like the exclusive OR of Boolean logic. We will often be concerned with subsets of some basic set. If the basic set is R then the *complement* of $A \subset R$ is $R\text{-}A$. This complement is usually written A^c or $R \setminus A$.

A useful rule of set theory is the "duality principle". It is written

$$R - \bigcup_{\alpha} A_{\alpha} = \bigcap_{\alpha} (R - A_{\alpha}) \text{ and } R - \bigcap_{\alpha} A_{\alpha} = \bigcup_{\alpha} (R - A_{\alpha}) \qquad (3.1\text{a,b})$$

In other words, the complement of the union equals the intersection of the complements and visa versa. (These relations, Eqns.(3.1), are referred to as the DeMorgan formulas.) It follows that any theorem involving a family of subsets can be converted into another by replacing all subsets by their complements, all unions by intersections and all intersections by unions.

1.3 Mappings

A map from a space M to a space N is a rule for associating a unique element of M with a unique element of N. The example with which the reader will most likely be most familiar is an ordinary function on $\mathbf{R}^1$, which is the same as $\mathbf{R}$. For every x-value, in this case, there is also a unique y-value. Since both x and y belong to $\mathbf{R}^1$, we see that M and N may be the same set. There does not have to be a unique x-value for every y-value however. A mapping from M to N under a function f is written f: $M \to N$. When there are several x-values for one y-value, the mapping is said to be "*many to one*". The set of values in N corresponding with a given set, S, of values in M according to some mapping is called the *image* of S in N. If there is also a unique $x \in S$ for every point in the image of S, here called T, then S is the *inverse image* of T under f^{-1} the inverse of f. When both f and f^{-1} exist, then the mapping is one-to-one.

A geographical map is a 1-1 (one-to-one) map of a geographical part of the Earth's surface onto a piece of paper. We note that it is a 1-1 map that maps something *onto* something else if say it is a map of Georgia and Georgia is the entire set M and the map is only of Georgia. If the paper map includes Alabama, then the mapping is an "into" mapping. In an "onto" mapping $f(M) = N$. In an "into" mapping $f(M) \subset N$. Every "onto" mapping is also an "into" mapping. There is a notation for the elements in a mapping. Let f: $M \to N$ and let $x \in M$ and $y \in N$. Then $f : x \mapsto y$. Another symbolic notation arises in the description of a sequence of mappings. Let f map $M \to N$ and g maps $N \to P$. Then $g \circ f$ maps $M \to P$. Note the reversal of order. The map acting first is the one on the right. "Onto" mappings are called *bijections* when they are also one-to-one. "Onto" mappings which are not one-to-one are called *surjections*.

In stereology we will be concerned with those mappings, isometries, which preserve the distances between points.

2. RINGS OF SETS

A *system of sets* is another way of referring to a "set of sets". A *ring of sets*, $\mathcal{R}$, is a system of sets closed under the operations of taking unions, intersections, differences and symmetric differences. The formal definition is limited to symmetric differences and intersections. The other operations follow as consequences because

$$A \cup B = (A \Delta B) \Delta (A \cap B)$$
$$A / B = A \Delta (A \cap B)$$

Rings are called rings because they are closed. Since $A \backslash A = \varnothing$ for all sets, A, a ring of sets must contain $\varnothing$. Note well that if a ring of sets contains a *unit*, E, such that $E \cap A = A$, it is called an *algebra of sets*. This is an important concept in what follows.

The concept of a ring of sets has proven too restrictive for some applications. To meet this need, the concept of the semiring has evolved.

A *semiring* of sets is a system of sets, $\mathcal{S}$, meeting these requirements:

(1) $\mathcal{S}$ contains $\varnothing$.

(2) $A \cap B$ belongs to $\mathcal{S}$ if $A \in \mathcal{S}$ and $B \in \mathcal{S}$.

(3) For all $A \in \mathcal{S}$, there exists a finite expansion in terms of pairwise disjoint sets of $\mathcal{S}$,

$$A = \bigcup_{k=1}^{n} A_k .$$

Requirement (3) does not imply that a semiring is closed under the operation of taking finite unions. This is discussed further in the Appendix. The concept of a semiring is useful to us because the finite union expansion in important to the concept of measure. A ring of sets is called a *σ-ring* if it contains the union, $\bigcup_{n=1}^{\infty} A_n$, whenever it contains A_1, A_2, ... etc. A *σ-algebra* is a semiring of sets belonging to some ground set containing all the elements of the ground set and their complements. In a *σ-algebra* all (infinite) unions of elements are included as in a *σ-ring*.

Extremely important to the foundations of stereology is the concept of a Borel algebra of Borel sets, which are subsets of $\mathbf{R}^n$. A Borel algebra is the smallest *σ-algebra* on $\mathbf{R}^n$ containing all the open subsets of $\mathbf{R}^n$.

3. MEASURE AND RELATED TOPICS

A measure is monotonic, subcover-additive and vanishes for an empty set. By subcover-additive, we mean, Eqn.(3.2),

$$\mu\left(\bigcup_{i=1}^{\infty} A_i\right) \leq \sum_{i=1}^{\infty} \mu(A_i)\,, \tag{3.2}$$

whenever $A_i \in X$ where X is the ground set.

Measures are used in the context of volumes and surface areas but for the sake of consistency must apply to distances as well. For our purposes measures may be assumed non-negative (under most circumstances). Measures are identified with integrals. A measure is associated with a metric; what it measures is the content of a metric space.

3.1 The R^n Space and the Euclidean Metric

The real line is designated by R. The Cartesian product of R with itself taken n times is the space R^n. The three dimensional space of ordinary acquaintance is R^3. When we measure it (by conventional means e.g a ruler), it becomes a metric space. Generally a *metric space* is the combination of a set and the means of measuring distance called a metric. Similarly a *measurable space* is the combination of a family of measurable sets and a measure. A good metric is one that assigns a non-negative real number to any pair of elements in the associated set. The metric, ρ, must also have three properties:

(1) $\rho(x,y) = 0$ iff $x = y$.

(2) Symmetry: $\rho(x,y) = \rho(y,x)$.

(3) Triangle inequality: $\rho(x,z) \leq \rho(x,y)+\rho(y,z)$.

This leaves a lot of room for choosing the metric. However, the Euclidean metric is sufficient for stereology. Proving that the Euclidean metric meets the triangle inequality in an n-dimensional space is usually accomplished through the use of a more fundamental inequality, the *Cauchy-Schwartz inequality,* which can easily be shown to apply to hyperspace. It is succinctly stated in dot-product notation, Eqn.(3.3), and is worth writing down for reference,

$$(\mathbf{x} \bullet \mathbf{y})^2 \leq (\mathbf{x} \bullet \mathbf{x})(\mathbf{y} \bullet \mathbf{y})\,. \tag{3.3}$$

The bold face denotes a vector quantity but need not assume an orthogonal basis. From this inequality, the triangle inequality follows. The Euclidean metric is thus, Eqn.(3.4),

$$\rho = \sqrt{\sum_{(i)} x_i^{\,2}}\,, \tag{3.4}$$

where x_i is the i-th component and i ranges from 1 to n in an n-dimensional space. An n-dimensional Euclidean space is sometimes abbreviated $\mathbf{E}^n$ but little will be lost by overloading that definition onto $\mathbf{R}^n$ as is frequently done. The Euclidean metric coincides with the everyday concept of distance.

In section 3.1 we introduced the concept of isometry before fully discussing the concept of a metric space. An isometry is a 1-1 mapping that preserves the distance between points as determined by some metric. An isometric mapping is a special case of the more general term, *homeomorphism*, which is a big word for more or less a 1-1 mapping. A similar word is *diffeomorphism*, which adds a requirement for sufficient continuity that the mapping be differentiable.

Using the Euclidean metric, we define the filled open sphere of radius r centered at x as $b(x,r)$. In Section 2.5 we found its volume (or more generally its *content*) in an arbitrary number of dimensions; we call it a Euclidean ball. The closed ball, to use one form or notation, $[b(x_0,r)]$ is defined by the relation, $\rho(x_0,x) \leq r$. (Square brackets symbolize the *closure operator*, which transforms an open to a closed set. The notation **clos** is also used and there are others.) It is the set defined by the open ball with all its limit points added. A set which includes all its limit points is a *closed set*. A point is a *contact point* of a set M is a point for which each neighborhood contains at least *one* point of M. Often the open ball itself is used to visualize the concept of neighborhood. Hence $b(x,\varepsilon)$ represents the oft-mentioned *ε-neighborhood* used in proofs.

3.2 Convexity

A set is said to be *convex* if whenever it contains two points, **x**,**y**, it also contains all points of the line segment joining **x** and **y**. A *convex linear subspace* (see Appendix) is similarly defined in Euclidean space as one where if **x** and **y** are two points in the space, any point defined by $\mathbf{p} = \lambda\mathbf{x} + (1-\lambda)\mathbf{y}$ with λ being a positive scalar between zero and one, also belongs to the same subspace. And, a convex set is called a *convex body* if its interior is non-empty. The intersection of any two convex sets is also a convex set but the same cannot be said for unions. Still there is a ring containing convex sets, the *convex ring*, which has useful properties and is important to stereology. The sets of the convex ring are not all convex but can be decomposed into convex sets. This is important because the bulk of shape-independent stereology applies only to convex bodies. In many cases, non-convex bodies belong to the convex ring and it may be possible to extend a formula to a non-convex body by decomposing it into convex bodies. The notion of convexity conforms to our conventional intuition for sets in the plane and in 3-space. A circular solid is convex. A crescent shaped solid is not since a straight line drawn between the two apices of the crescent is not a part of the crescent.

3.3 Borel Sets and σ-algebras

We return to the subject of rings of sets. Axiomatic probability theory utilizes the concept of the B-algebra (an algebra made of Borel sets) extensively to describe sample spaces. cf. [Woodroofe,1975] Stochastic geometry extends the concept of Borel sets on the real line to include the Cartesian product space, $\mathbf{R}^n$. Closure under the taking of unions and intersections provides one motive for modeling sample spaces with B-algebras. We will be concerned with measuring both the contents of sets and the probabilities associated with them. Both the measures of content, e.g volume or area, and of probability are defined on *σ-algebras.*

3.3.1 Sample Spaces and Axiomatic Probability

Sample spaces contain random events associated with their probabilities of occurrence. The meaning of "random" and of "probability" are surprisingly difficult to define well apart from their isolated existence as mathematical entities. One implies the other. The notion of probability can be defined first and that of randomness follows. Here the notion of "random" was introduced in Section 1.1 in terms of how the term is applied. We will follow this intuitive meaning as a basis for defining probability in terms of a space of random events. This can be done through axiomatizing probability. [see Gnedenko p.62] The earliest such system of axiomatic probability was that of Sergei Bernstein in 1917. Later, A.N. Kolmogorov introduced the set-theoretic approach similar to that presented here which is sometimes termed "modern" probability theory.

Kolmogorov defined a sample space and associated with it an algebra of its subsets. He first considered a set U of *elementary events*. He then considered a family F of subsets of U. The elements of the family are called *random events*. Three conditions are customarily imposed on F:

1) F contains U as one of its elements and also $\varnothing$;

2) In terms of subsets of U, F is closed under the taking of unions, pairwise intersections and complements;

3) F is also closed under the taking of infinite unions and intersections.

We can examine these requirements by studying the discrete sample space of tossing a single die. There are six sides to a die and a remote chance that the die will be lost in the tossing. These are the elementary events, which form the set U. Since, if we lose the die we may possibly get another, we can have any sequence of these in die tossing. We can construct a family F of random events out of subsets of U by taking unions. We may consider the joint probability of getting several events together in some way and view this as taking an intersection in the sample space. This joint event belongs to *F*. Similarly we may look at the possibility of not getting some random event. This then is taking the complement of some subset. This sample space contains everything that can happen in tossing a die and is therefore complete and closed under the operations of taking unions, intersections and complements. Since other operations can be defined in terms of these, it is closed with respect to the taking of differences and symmetric differences as well. It is a ring.

In addition to the definition of the sample space, there are three axioms of probability defined as a measure on the subsets of the sample space.

Axiom 1: For each random event in a family of events, there is a non-negative number termed its probability (measure), *P*.

Axiom 2: $P(U) = 1$.

Axiom 3: If events $A_1, A_2, ... A_n$ have $\bigcap_{(i)} A_i = \varnothing$, then the probability of all these events together at once is $P(A_1, A_2, ... A_n) = P(A_1) + P(A_2) + ... + P(A_n)$.

From Axioms 2 and 3, we see that the range of probabilities is [0,1]. Is this always true? Two special cases are worth discussing. The first is that of "thermodynamic probabilities" which are typically large integers. They do represent ratios but these ratios are more like the ratio of the number of members of the family *F* to the number of elementary events than probability as defined here. It is simply a different use of the word, "probability". [Yourgrau et al., 1982.] p.93

The second may arise when we consider expectation. For example, a grain seller seeks the probability of finding a weevil in a bag of grain. When the number of weevils per bag is very small, the probability of finding a weevil in a bag is simply the number of weevils divided by the number of bags. Each bag is a cell that may or may not contain a weevil. We can use this probability and the binomial distribution to assess the probability of getting weevils in two bags in a row. But, as the weevils multiply, the "probability of finding a weevil" increases until it is possible to find a bag full of weevils. Strictly speaking, the probability of finding *a* weevil never exceeds unity even for a bag of them. But the ratio of weevils to bags is no longer a probability. It is the expectation of the number of weevils in a bag. Trouble arises from identifying (treating as alike) these concepts of expectation and probability with each other. When the form of a given problem creates doubts, it is probably safer to use "expectation" rather than "probability" even if some will not understand what is being said.

Going over the consequences of some of these axioms in the context of sample space helps clarify their meaning. Although we are discussing probabilities, axioms 1 and 3 along with the discussion apply to any measure. Looking at Axiom 3, the probability of disjoint events A **or** B taking place is $P(A+B) = P(A \cup B) = P(A) + P(B)$, using bold type for emphasis. At once we see a notational conflict. The probability of A **and** B taking place is $\varnothing$, since the events are disjoint. We identify the symbol "+" with "**or**" rather than "**and**". We identify "**and**" with (Boolean) multiplication. Furthermore, Let A and B be two random events. Then two sets of elementary events in the family F correspond to these two events. If $A \subset B$ then the probability $P(A) \leq P(B)$. This is because B can be divided into two disjoint sets, A and $B\backslash A$. Hence $P(A) + P(B\backslash A) = P(B)$. Since $P(B\backslash A)$ is non-negative by Axiom 1, $P(A)$ can at most equal $P(B)$. We note that, see Appendix, were the events defined on a linear space, this inequality would allow meeting two of the three requirements for probability to be a convex functional.

Here is another example. If A is any event then the probability of its complement A^c equals 1 - $P(A)$. This is because $P(A^c)$ is the $P(U\backslash A) = P(U)$ - $P(A)$ since the event $A(U\backslash A)$, the simultaneous occurrence of A and $(U\backslash A)$ equals that of $A \cap (U\backslash A) = \varnothing$. Since by Axiom 2, $P(U) = 1$, $P(A^c) = 1 - P(A)$. We can extend this example further to include a formula for the probability of two events whose intersection need not be $\varnothing$. Consider events A and B contained in U. For event A there is $P(A)$ and the same for B. $P(A \text{ or } B) = P(A \cup B) = P(A) + P(B\backslash AB) = P(A) + P(B) - P(AB)$. This kind of additivity is called "*c-additivity*". The measures we will encounter in stereology are generally c-additive. In short, we arrange our events into disjoint sets (classes), sum the corresponding probabilities (Axiom 3), then decompose any intersections and subtract their measures. Care is needed in this last step. Subtraction of probabilities is only justified when the sets associated with the minuends (of the probability ratio) are subsets of the common subtrahend.

The procedures and properties just illustrated suggest that the family of sets F behaves much like a B-algebra. Since a B-algebra is a ring, we expect F to be closed under the taking of finite unions and intersections. This is requirement 2 of sample spaces. A B-algebra is also specifically closed under infinite sequences of unions and intersections. This corresponds to requirement 3 for sample spaces. We may examine this a little closer through an example of nested events. Consider a sequence of events in which each is a subset of a previous event. Because of axiom 3, we may imagine this sequence to be infinite. If A_n is the n-th event in the nest, it is the sum of disjoint terms $A_i A^c_{i+1}$ with i starting at n and going to infinity. If we examine the difference between adjacent terms we see that $P(A_n) - P(A_{n+1}) \leq P(A_n) < P(A_{n-1})$. The terms therefore form a Cauchy sequence converging to zero. The event of probability = 0 is $\varnothing$, the impossible event, which by hypothesis belongs to a ring and hence to a B-algebra. Since the sample space contains itself, it contains the required unit, $P(U)$, so the resulting ring is a B-algebra. We recall that every B-algebra is a σ-algebra. Probability theorists have picked up the term σ-algebra and use it in place of B-algebra when referring to the subsets of sample spaces. Similar to its use of the term, σ-algebra, probability theory has other specialized usages. Events A_i are termed *mutually exclusive* if $\bigcap A_i = \varnothing$, that is if they are constituted of disjoint sets. An event A *implies* an event B if $A \subset B$. A set of events $\{A_i\}$ is *exhaustive* if $\bigcup_{(i)} A_i = U$, where U is again the sample space.

3.3.2 Continuous Sample Spaces and Probability Spaces

The definition of a probability space is fundamental to the study of stochastic geometry. To continue our study we must return to the subject of Borel sets. Since Borel sets are initially defined in terms of an algebra of closed sets on the real line, and most of these are continuous; this is a good place to introduce the concepts of continuous sample spaces. In order for there to be a geometry of probability, stochastic geometry, more is required of the sample space than can be accommodated in the concept of Borel sets on the real line. Thus the concept is extended to the Cartesian product space of arbitrary dimensions. When the sample space is continuous, its family of random events F is describable as a family of Borel sets. These, in extension, are subsets of R^n generated from the set of all closed intervals on and a B-algebra. The B-algebra thus augments the set of closed intervals considerably. A restriction is that the family of Borel sets must belong to the minimal B-algebra. The union of the family F is the unit E relative to the system of subsets in question. Since a B-algebra includes infinite unions, the family of Borel sets includes open as well as closed intervals since these can be represented as the limit of a sequence of closed intervals, e.g an infinite limit or as the complement of closed or half open intervals e.g. the complement of $(-\infty,-1]\cup\{1, \infty)$ is $(-1,1)$. It is thus a large family of sets. Nevertheless, it readily ties back to the definitions imposed on a sample space in the last section. We note that the family of Borel sets of a set X is smaller than the system of all subsets of X termed the *power set* $\mathcal{P}(X)$.

A probability space can be defined as an ordered triplet (U,*F*,*P*) where (to save space) the letters U, *F* and *P* can be identified with the definitions already given. The set of elementary events is *U*. The B-algebra is made up of the family of subsets of U designated *F*. Together, *U* and *F* define the sample space. If the sample space is continuous, than *F* is a family of Borel sets. To the sample space, we add a probability measure, *P*. The probability measure must conform to the three axioms given in Section 3.3.3.1. If the sample space is continuous, it must be absolutely continuous and obey Eqn.(A.9), that is, be integrable. Recalling the previous chapter, we see that the probability measure for the real line conforms to the definition of a distribution function.

A key property of a probability space and its probability measure is termed *σ-additivity*. Emphasizing this property is the reason why the term σ-algebra is used rather than B-algebra. From here on we may use the term σ-algebra rather than B-algebra especially when we are not referring to Borel sets. This property applies to the probability measure *P* in (*U*,*F*,*P*) and is stronger than mere additivity in the it specifies the how. It re-states Axiom 3 in a manner more applicable to continuous sample spaces. Sigma-additivity is the property that a set can be divided into a countable (but not necessarily finite) union of disjoint subsets such that the value of a function of the whole set equals the sum of its values on the subsets. A probability measure is defined on sets having this property. The number of stars in the universe also has this property; so, it is not a property of probability alone. However, this property implies systems of sets that are closed under the operations of union, intersection and complementation as said earlier. Systems of Borel sets are closed under infinite unions. This leads naturally to the idea that a σ-additive function on a system of Borel sets would be an integral of some sort.

3.3.3 Lebesgue Measure

The Lebesgue integral is defined differently from Lebesgue measure, but the two notions coincide in their respective contexts. The Lebesgue measure is taken on a d-dimensional Euclidean subspace. Where *B* denotes a family of Borel sets, locally finite measures on $[\mathbf{R}^d, B^d]$ are unambiguously defined by their value on hypercubes. Lebesgue measure is a measure of a hypercube in Euclidean space-which applies just as well to 3-space. We looked at hypercubes a little is Section 2.5. We define a hypercube as a subspace by the Cartesian product $Q = [u_1,v_1]\otimes[u_2,v_2]\otimes\ [u_3,v_3]\otimes\ ...\ \otimes[u_d,v_d]$. The Lebesgue measure of the rectangular hyperparallelopiped (e.g. hypercube) is

$$\nu_d(Q) = (v_1 - u_1)\cdot(v_2 - u_2)\cdot\ldots\cdot(v_d - u_d)\,. \tag{3.5}$$

The measure ν_d, called the Lebesgue content, is the "volume" of the hypercube in hyperspace. This same measure can be applied to other shapes such as balls, toroids, etc. The process of doing this has already been outlined for the hyperball in Section 2.5. We intuitively determined a measure of the volume of a hyperball to be consistent with that of a hypercube which we defined as a continuation of the customary formula for Euclidean 3-space.

The measure ν_d reduces to area measure when $d = 2$ and length measure when $d = 1$. Because length as determined by the Euclidean metric is invariant under translations and rotations, we expect that Lebesgue measure in general is invariant under such transformations. In fact, the Lebesgue measure in *isometry-invariant.* Isometries were defined in Section 3.1.3 as length preserving mappings. Formally,

$$\nu_d(\mathbf{m}A) = \nu_d(A)\ , \tag{3.6}$$

for all isometries, $\mathbf{m}$ and Borel sets A.

Measure theory develops Lebesgue measure rigorously for the ring of *elementary sets* which are a finite number of disjoint rectangular figures. In the process of extending the measure to a σ-ring, the ring, which includes infinite unions, inner and outer measures are introduced in a fashion reminiscent of development of the Riemann integral. Extending the measure to the σ-ring then allows it to be applied to Borel sets. For sets that are not necessarily elementary sets, the *outer* measure is defined as,

$$\mu^*(A) = \inf \sum_{(k)} \nu_d(R_k) : A \subset \bigcup_{(k)} R_k\ , \tag{3.7}$$

where $\{R_k\}$ is a finite set of not necessarily disjoint rectangular figures covering A. For subsets of a unit having unity measure (to avoid the complication of sets having infinite measure) the *inner* measure is defined in terms of the outer measure of its complement, Eqn.(3.8).

$$\mu_*(A) = 1 - \mu^*(E - A) : \nu_d(E) = 1\,, \tag{3.8}$$

where E is the unit. A is said to be Lebesgue measurable if $\mu^*(A) = \mu_*(A) = \nu_d(A)$.

Equation (3.7) is clearly applicable to a system of elementary sets. In order to prove its applicability to some infinite sets, we can show that it is applicable to Borel sets by demonstrating that v_d is countably additive on a system of Borel sets. A set function ϕ is *countably additive* if

$$\phi(\bigcup_{n=1}^{\infty} A_n) = \sum_{n=1}^{\infty} \phi(A_n) : A_j A_{k \neq j} = \varnothing .$$

This is discussed further in the Appendix.

3.3.4 Measuring Boundary Content and a Minkowski Measure

In stochastic geometry and related fields, boundary is designated with the symbol δ which is treated as an operator defining the boundary of its operand. It reads as "the boundary or surface of..." For regular smooth shapes, the dimensionality of a boundary is one less than the dimensionality of the object being bounded. For objects with exceedingly diffuse boundaries, the dimensionality of the boundary cannot exceed that of the space the entity it bounds occupies. The measurement of boundary is often one of the most difficult tasks. One may recall that the origin of the fractal concept lay in the problem of measuring the length of a coastline. The problem does not especially depend on which coastline is measured. We could arbitrarily select Australia.

Australia is certainly a compact set because it is closed and bounded and because its political divisions constitute a finite subcover. We then look at the boundary of Australia, its coastline, and seek some sort of a uniquely converging cover for that. Those who have studied fractals will know that the length of the coastline becomes a function of the length scale of the cover used in such a way as to suggest that the coastline is not one-dimensional. But even if we have not studied fractals, we could anticipate trouble. If we look at Australia on maps, it is two-dimensional while the length of its coastline is only one-dimensional. We want to measure a "real" one-dimensional object in a two dimensional space. Since the two dimensional space is itself an idealization, that should at least be possible. But, since we seem to live in three-space, how many really two-dimensional objects can you handle that are not idealizations in ordinary three-space? As we shall see, sort of like Parkinson's law of work and time, there is a tendency for the dimensionality of objects to expand to fill the *embedding space* allotted.

To begin the measurement of boundary in arbitrary dimensions, we introduce another definition, that of Jordan content. The concept of Jordan content is a simplification of that of Lebesgue measure. Like Lebesgue measure, Jordan content is found as the limit of an inner and an outer measure. In the case of Jordan content, the inner and outer measures are coverings by sets having plane sides. In the plane, these are polygons. To find the measure of a set A in the plane, we draw two polygons defining regions of the plane A^- and A^+ such that $A^- \subset A \subset A^+$. We then define

$\varepsilon = m\ (A^+) - m(A^-)$ where m denotes the measure, which is well-defined for polygons. Given any initial A^+ and A^-, we may reduce the size of ε by some refinement of the coverings such as filling in the space between A and its inner and outer bounding figures with more polygons, adding their measures to the inner measure and subtracting them from the outer measure. By repeating this process, ε may be reduced arbitrarily; and, in the limit, $m(A^-) = m(A) = m(A^+)$, a finite number. If this fails to happen, the set is termed not *measurable*. (For a slightly different definition of measurability see Mattila [1995] p.8. For sets other than in the plane, the n-dimensional generalization of polygons, called *polytops,* is used. We would expect that in the limit the surfaces of the inner and outer bounding sets would also become equal and approach a finite limit, the boundary content of A, δA.

An example of how the process of finding the Jordan content of the boundary can go wrong can give some insight into the problem of boundary measurement. We proceed to find the Jordan content of a hexagon by coverings of triangles. This process works nicely for finding the Jordan content of the hexagon itself but it fails to find the content of the boundary. In the back of our minds we may note that we are trying to use the boundary of a two-dimensional covering to find the measure of a one-dimensional entity. Let us look at the outer measuring cover in Figure 3.5. In the first approximation, it consists of two triangles having sides of length $a/2$ for each face. In the second approximation, there are four triangles per face, each having sides $a/4$ in length. The resulting outer boundary measure is the perimeter of the covering which $12a$ for all approximations.

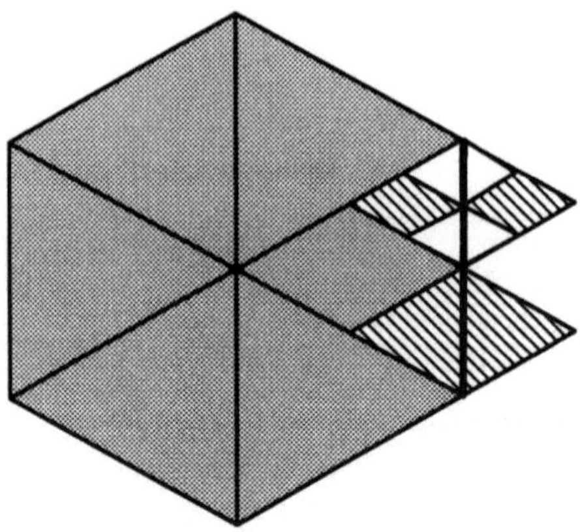

Figure 3.5: Hexagon with sides of length a., boundary of length 6a.

Looking then at the inner cover, we see that it has the same pattern as the outer cover except that due to contact at the corners, it is reduced by the length of one side per hexagonal face. In the first approximation that missing perimeter is $a/2$ per face. Hence the measure of the inner cover is $12a(1-\frac{1}{2}^n)$ where n is the number of the approximation. In the limit, the measure of the inner and outer coverings approach each other, both equal $12a$, which is wrong. We now recognize that the content of the boundary of a limiting cover may not be the same as the content of the boundary.

One way of looking at the failing of the method of Fig. 3.5 is to say that the covering did not follow the contour of the hexagon sufficiently well -- even in the limit. The same problem is seen in that construction which approximates the hypotenuse of a right triangle by a staircase of treads and risers parallel to the other two sides. No matter how fine the division into steps, the approximation of the length of the hypotenuse stays equal to the sum of the lengths of the remaining sides rather than its square being the sum of the squares of the remaining sides. To avoid this contour problem in measuring the surface content of compact convex sets, Minkowski considered the class of all centered sets of the same shape as the set to be measured. This can be described as a continuous class composed of sets each having different Lebesgue measure but all centered with respect to one another. We note that shape problem takes care of itself. The sets may change shape somewhat but not significantly in the vicinity of the set being measured. An additional requirement is imposed as well by way of a construction shown in Fig. 3.6.

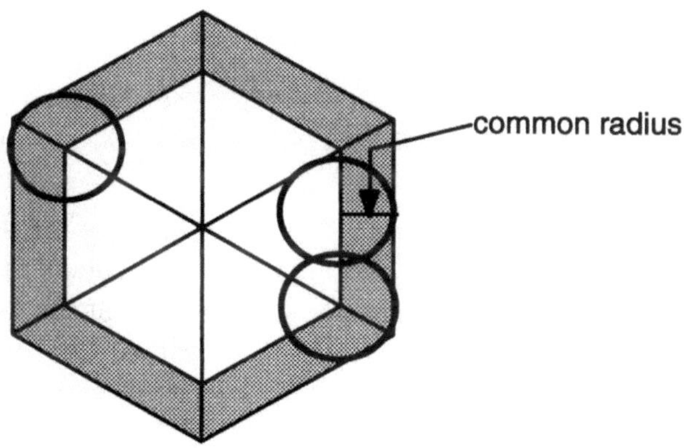

Figure 3.6: Centered hexagon-like shapes having a common radius of separation along their perimeter.

We now require that each member of the class be formable from the set to be measured by a construction in which a ball (of the same number of dimensions as the set being measured) is constructed on each surface point of the inner member. All the balls are given the same common radius. The constructed member (the outer member in Fig. 3.6) is then the union of the inner member and all of the balls. The construction rounds corners but the corner rounding vanishes as the common radius approaches zero. Sets so constructed are called *parallel sets* and may be constructed in any number of dimensions.

We may now use this construction to fabricate an inner and outer set after the fashion of finding the Jordan content. The difference is that we are using a particular derived set rather than a polytop. Having constructed these sets we consider their Lebesgue measure. Students of mystery novels may by now be suspecting that the plot will turn in the direction of that of Section 2.5 where the surface of a hyperball was determined. This would be a very good guess. Leading in that direction we introduce four quantities and take limits.

$$\overline{S}_{+}(A) = \lim_{\rho \to 0^{+}} \sup \left[\frac{v_n(A_\rho) - v_n(A)}{\rho} \right] \tag{3.9a}$$

$$\underline{S}_{+}(A) = \lim_{\rho \to 0^{+}} \inf \left[\frac{v_n(A_\rho) - v_n(A)}{\rho} \right] \tag{3.9b}$$

$$\overline{S}_{-}(A) = \lim_{\rho \to 0^{-}} \sup \left[\frac{v_n(A_\rho) - v_n(A)}{-\rho} \right] \tag{3.9c}$$

$$\underline{S}_{-}(A) = \lim_{\rho \to 0^{-}} \inf \left[\frac{v_n(A_\rho) - v_n(A)}{-\rho} \right] \tag{3.9c}$$

where A is an n-dimensional set as before, v_n denotes it Lebesgue measure, A_ρ is a parallel set to A with radius ρ. A positive radius denotes an increase in the relative measure while a negative radius denotes a decrease in the relative measure. For the boundary to be Minkowski-measurable in this sense, all four quantities in Eqns.(3.9a-c) must be finite and equal. In that case, they are all equal to the boundary content. Equations (3.9) represent derivatives in a sense.

In Section 2.5 we determined the boundary content of a hyperball by taking the derivative of its volume with respect to its radius. We now can see why this worked. Varying the radius of the hyperball generates a family of parallel sets. The role of ρ in Eqns.(3.9) is taken up by dR in Eqn.(2.156). Since a ball is smooth by definition, its parallel sets are also smooth and Eqns. (3.9) are automatically satisfied.

4. MINKOWSKI ALGEBRA

There are a number of image algebras, which have evolved to serve the needs of image processing [Ritter & Wilson. 1996]. The origins of these include the Minkowski operations defined for geometric sets. These operations of Minkowski algebra have an intuitive appeal and have become a part of typical image processing packages.

4.1 Operations on Compact Sets

The first of the Minkowski operations is *Minkowski addition,* Eqn.(3.10). It is how we constructed parallel sets and is defined as

$$A \oplus B = \{a + b : a \in A, b \in B\}. \tag{3.10}$$

Thus, every point of B is added vectorially to every point of A. Compare this with the minus sign discussed in Section 3.1.2. The result of Minkowski addition can be visualized as a convolution when A is a set of discrete points and B is symmetric. For example,

$\int_{-\infty}^{+\infty}\left(\sum_{n=-\infty}^{\infty}\delta(t - n\alpha)\right)\exp\left(\frac{(x-t)^2}{-c}\right)dt$ is the convolution of a Dirac comb with a Gaussian function. The integral of the Dirac comb defines a set of points $\{x : x = n\alpha\}$ on $\mathbf{R}^1$. The convolution with the gaussian, replaces each point in the comb with a gaussian centered on that point forming a series of hills with peaks at $x = n\alpha$. To compare this with Minkowski addition, we make the set A the set of points defined by the Dirac comb and B the gaussian. Repeating this analogy in two dimensions, we could make B a compact set such as the ball., $b_2(o,r)$. Its convolution with the comb would be a string of beads, as would the Minkowski sum.

The construction of parallel sets in Fig. 3.2 can also be interpreted in terms of Minkowski addition. Each point, located at o, of a parent set is vector summed with all the points of a ball, $b_k(o,r)$ where o denotes the origin, and r the radius. The result is the union of the totality of these points. Minkowski addition of a set B to a set A enlarges A and deforms A. If B is not symmetric with respect to its local origin, it may also translate A. Minkowski addition is associative and commutative. Also,

$$A \oplus (B_1 \cup B_2) = (A \oplus B_1) \cup (A \oplus B_2). \tag{3.11}$$

If we define the *translate* of A as $A_x = A \oplus \{\mathbf{x}\}$ where $\{\mathbf{x}\}$ is a single element set , then

$$A \oplus B = \bigcup_{\mathbf{x} \in B} A_x = \bigcup_{\mathbf{x} \in A} B_x \tag{3.12}$$

and

$$A \oplus B = \{\mathbf{x}: A \cap (\check{B})_x \neq \varnothing\}. \tag{3.13}$$

Equation (3.13) states that the Minkowski sum is the set of translation vectors that give a non-null intersection between the set A and the translation of the reflection of B. A sketch provides some clarification. Set A is a ball at the origin.

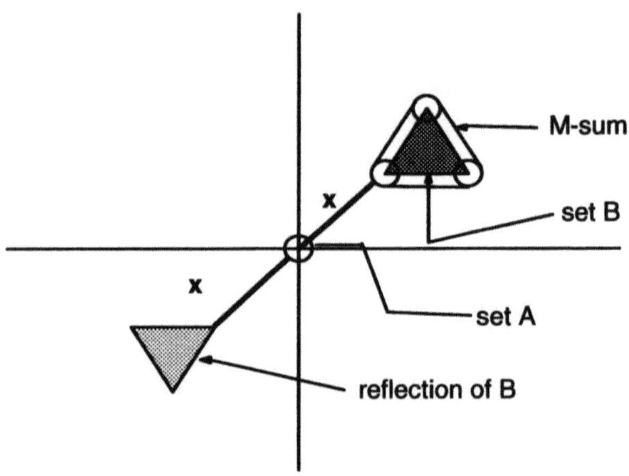

Figure 3.7: The geometry of Eqn.(3.13).

The translation vector x connects the set A with the reflection of B. Translated to the origin it connects the origin with the Minkowski sum. In this case the sum and the set B are parallel sets because set A is a ball.

There is also a *Minkowski difference*. It is defined as

$$A \ominus B = \left(\bigcap_{y \in B} A_y \right), \qquad (3.14)$$

or

$$A \ominus B = \left(A^c \oplus B^c \right)^c, \qquad (3.15)$$

where the complement is taken with respect to the subspace containing A and B.

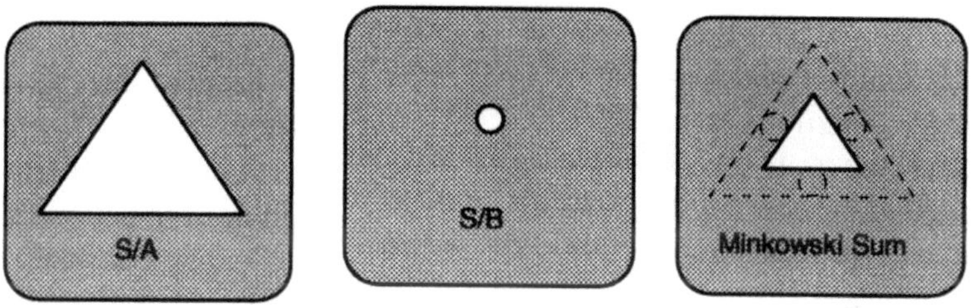

Figure 3.8: Minkowski subtraction as related to the sum of complements.

The rounded rectangle is S. The sharp corners in panel three show that Minkowski addition and subtraction are similar to but not inverse operations.

There is an inequality relationship between Minkowski addition and subtraction,

$$\left(A \ominus \breve{B} \right) \oplus B \subseteq A \subseteq \left(A \oplus \breve{B} \right) \ominus B. \qquad (3.16)$$

Referring to Figure 3.8 we see that if we take the origin to be the center of B then $\breve{B} = B$, and the construction corresponding to the operation in parentheses on the left would then yield the set shown in white on the third panel. Minkowski addition of that set to B would result in the triangle A but now with rounded corners. Since its corners would be "filed away" it would be smaller than A.

4.2 Dilation, Erosion and Parallel Sets

Two important operations are derived from the sum and difference operations just discussed. These are called dilation and erosion. Both operations are defined using $B = b_k(o,r)$. *Dilation* is then defined as

$$A \xrightarrow[dil]{} A \oplus \check{B} . \tag{3.17}$$

Erosion is defined as

$$A \xrightarrow[ero]{} A \ominus \check{B} . \tag{3.18}$$

Since the location, o, is a dummy variable in that it is swept over all members of the set A, B is usually written rb_k which denotes multiplication of the unit k-ball by the scalar, r. Since multiplication by a scalar multiplies all coordinate values, the ball's radius is set to r.

Both dilation and erosion form parallel sets. Because the parallel set formed by dilation is larger and smoother, small irregularities disappear and gaps are filled in. If such features as irregularities and gaps are interpreted as noise, dilation can be viewed as a sort of filter. For instance, some techniques for revealing grain boundaries rely on "decoration" of the boundary with the precipitate of a foreign solute. To a first approximation, decoration can be imagined to be a Poisson process. It is thus natural to suppose gaps and irregularities in a grain boundary revealed in this way. Dilation, by filling gaps and removing previous irregularities, filters out the noise at the expense of making the boundary thick.

Erosion shrinks the set and may even break it into fragments. It is useful in particle counting when particles may be in contact with each other. Particles may cluster to a greater degree due to capillarity than random overlap would predict. In 3-space, totally solid hard convex particles with sharp boundaries would not be expected to favor clustering because they may only contact each other at a single point in any plane containing them. In any case, erosion can be used to estimate clustering properties and, using different radii for **B**, the particle size distribution itself. Although dilation and erosion are customarily performed with balls, other structuring elements can be used as well in procedures to estimate parameters such as covariances. Recall the analogy between Minkowski addition and correlation. (See Appendix if you don't recall.) Correlation and convolution are identical processes when one of the sets is symmetric.

Two operations derived from dilation and erosion are opening and closing. These are the same operations that comprise the left and right portions of Eqn.(3.16). *Opening* is defined by the mapping,

$$A \xrightarrow[op]{} (A \ominus \check{B}) \oplus B . \tag{3.19}$$

When B is symmetric, it amounts to erosion followed by dilation. Since Minkowski addition and subtraction are not inverse operations, the opening of A is not usually the same as A. Since small disconnected fragments disappear under erosion, they are not available for a subsequent dilation to build on. Opening thus cleans up the page removing features smaller than B. The opening of a set A by a set B is sometimes abbreviated A_B.

Closing is defined by the mapping,

$$A \xrightarrow[cl]{} (A \oplus \check{B}) \ominus B \ . \tag{3.20}$$

When B is symmetric it amounts to dilation followed by erosion. The closing of a set A by another set B is sometimes abbreviated A^B. Closing fills in holes and causes proximate but separate subsets to be lumped together. It smoothes boundaries. Neither opening nor closing can be relied upon to reduce noise without loss of information. As is the case with almost all image processing operations, their success depends on some a priori knowledge as to what the "improved" image should look like guiding a choice of operations which alters the "raw" image to meet the expectation. It is erroneous to think that a processed image is more "true to nature" rather it is more true to some idealization. This may seem like abstract philosophy to a student who has not yet tried out these and similar operations. Try them. They are integrated into some scanning electron microscopes. More operations of this variety are treated in the subject of mathematical morphology.

4.3 The Hausdorff Metric

Minkowski algebra is performed upon non-empty compact sets. The Hausdorff metric allows the family of non-empty compact sets to be made into a metric space. The Hausdorff metric measures the distance between two compact sets. It is not the only possible distance between two compact sets, but it one which is usually without ambiguity. In many instances the distance between sets is defined as the infimum of all the distances between all the points.of the two sets. Once the sets overlap, this distance goes to zero. The Hausdorff metric does not go to zero as a result of mere contact. It satisfies the postulates for a good metric of Section 3.3.1. Using **b** to indicate a ball, we define the Hausdorff metric as

$$\rho_H(A,B) = \inf\left(r : A \oplus r\mathbf{b} \supset B, B \oplus r\mathbf{b} \supset A\right) . \tag{3.21}$$

Figure 3.9 illustrates this concept.

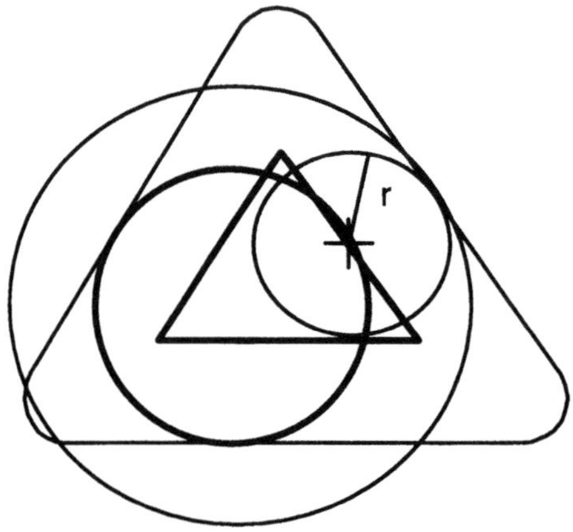

Figure 3.9: The Hausdorff metric.

The initial sets, the circle and triangle, are shown in heavy lines. Two parallel sets are formed. The controlling condition happens to be the containment of the circle by a parallel set of the triangle since a smaller parallel set of the circle would have contained the triangle but not visa versa. The Hausdorff metric is *r*, the radius of the circle which when Minkowski added to the triangle contains the circle.

Of course the Hausdorff metric implies the Euclidean metric in that the radius *r* is measured between the center and the boundary using it. The concepts of sequence and limit may be applied to the Hausdorff metric. Because of its relation to the Euclidean metric, a space defined with the Hausdorff metric inherits the properties that it is a complete (see Appendix) and separable metric space. A compete and separable metric space is sometimes called a *Polish space*. The ordinary spaces of concern to practical stereologists inherit at least separability. A space containing all its limit points is complete. So, ordinary closed spaces are Polish spaces.

5. THE CONNECTIVITY NUMBER

Connectivity is an important geometric property but not an easy parameter for stereologists to measure because the process of sectioning makes it unclear whether or not there actually was a connection between features prior to slicing. This problem becomes worse as distances increase. Connectivity is largely an unmet challenge to stereologists. To overcome this problem in studying the geometry of neural structures, the neurophysiologist, Wendell Krieg, introduced a technique of selective staining to trace nerves in the brain from one slice to another based of the fortuitous way in which myelination takes place. Such techniques are the exception and are only barely "stereology". Nevertheless, in spite of the problem with measuring it, connectivity is important to the theory of stereology.

The connectivity number is also called the *Euler-Poincaré characteristic.* It is defined on a ring of convex compact sets, which we will now define. We have already seen that a metric space can be formed from compact sets under at least the Hausdorff metric. We now consider a subspace comprised of convex bodies and those sets which can be formed by finite unions of convex bodies. Augmenting this subspace with the requirement that it include those sets formed under the taking of intersections of convex bodies, it becomes the *extended convex ring, S.*

We next define an additive functional, h: $S \to \mathbf{R}^1$, as being one that has these properties:

1) $h(\varnothing)=0$.

2) $h(A \cup B)+h(A \cap B)=h(A)+h(B): \quad A, B \in S$.

Property 2 is just an expression of c-additivity. Condition 2 implies the σ-additivity of h were it strictly a measure. We also posit that h is isometry-invariant. Within the family of motion-invariant and additive functionals on S, is the connectivity number χ. For a convex body it takes the value of unity. If we build up a set out of convex bodies through taking a finite number of unions, the result may no longer be convex but should remain compact and therefore still be a body. For such a set, the above conditions lead to what is called the *inclusion-exclusion formula,*

$$\chi(B)=\sum_{(i)}\chi(A_i)-\sum_{(i,j<i)}^{2}\chi(A_i \cap A_j)+\ldots+(-1)^{n-1}\sum_{(i_1,\ldots i_n)}^{n}\chi(A_{i_1} \cap \ldots \cap A_{i_n}), \tag{3.22}$$

where $B=\bigcup_{i=1}^{n} A_i$ and $A_i \in$ the set of convex bodies. Equation (3.22) represents a series of sums: singly, in pairs, in triplets etc. We can see how this works in a simple example.

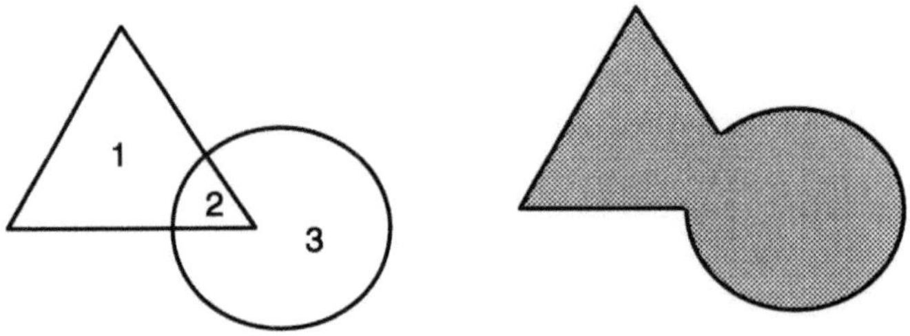

Figure 3.10: Set with connectivity number equal to one.

The set on the right is not convex but is composed of two convex bodies in R^2 as shown on the left. The two bodies are a triangle composed of regions 1 and 2 and a circle composed of regions 2 and three. Region 2 is convex, since it is the intersection of two convex bodies, and therefore has a connectivity number of 1. The connectivity for the set is thus unity, the sum of the connectivity numbers of the triangle and circle less that of region 2. It has been shown that the connectivity number of a set is independent of how it is broken down into convex subsets. When one set is subtracted from another, the connectivity numbers subtract.

Figure 3.11, Sets with connectivity numbers 1,0.-1 respectively.

Figure 3.11 further illustrates Eqn.(3.22).

6. BIBLIOGRAPHY

B.V. Gnedenko, *The Theory of Probability*, trans. B. Seckler (Chelsea Publishing Company, New York, 1967).

A.N. Kolmogorov and S.V. Fomin, *Introductory Real Analysis,* trans. R.A. Silverstein (Prentice-Hall, Englewood Cliffs, NJ, 1970 reprinted Dover Publications Inc. New York, 1975).

G.X. Ritter and J. Wilson, *Handbook of Computer Vision Algorithms in Image Algebra,* (CRC Press Boca Raton, FL USA, 1996).

D. Stoyan and H. Stoyan, *Fractals, Random Shapes and Point Fields* (John Wiley and Sons, NY, 1994).

D. Stoyan, W.S. Kendall and J Mecke, *Stochastic Geometry and Its Applications, 2nd Edition* (John Wiley and Sons, NY, 1995).

P. Mattila, *Geometry of Sets and Measures in Euclidean Spaces* (Cambridge University Press, Cambridge UK, 1995).

M. Woodroofe, *Probability with Applications* (McGraw-Hill, New York, 1975).

W. Yourgrau, A. van der Merwe and G. Raw, *Treatise on Irreversible and Statistical Thermophysics,* (Dover, New York, 1982).

Chapter 4

RANDOM PROBES

In this chapter we will study the elementary stereological probes by themselves. We will do this primarily by deriving the probability density functions for the intersections of probes with the boundaries of hypothetical objects immersed in a sea of random probes. Later, in Chapter 5, we will use what we have derived to find measures for the content and similar properties of specimens based on random probing. *The basic idea behind stereology is that the statistics of interaction between a higher-dimensional unknown object and a well-characterized lower-dimensional "probe" can characterize that object.*

1. RANDOMNESS, INVARIANCE AND OTHER MATTERS

Stereological analyses involve the interaction between a test element and structural features. Following customary usage, we will term the test element a "probe" and will distinguish between zero-, one- and two-dimensional probes which we will refer to as point, linear and planar probes, respectively. It is not usually[1] necessary in general for a linear[2] probe to be straight or for a planar probe to be flat although, as a matter of practical convenience, the latter usually *is* flat at least in metallurgical and geological microscopy. In this chapter we will be studying the foundations of stereology. For pedagogical reasons, we will follow its historical development rather than

[1] The exception is when a feature having a non-integer dimensionality is probed. This is discussed in Chapter 10. Even then a mathematically-flat probe need not always be flat in the common parlance sense.

[2] In this respect the terms "lineal" and "areal" would have been preferable, but "linear" and "planar" are well-established in the literature. "Lineal" and "areal" will be used to describe features.

continue immediately along the more rigorous and admittedly stronger mathematical lines begun in Chapter 3. But, the reader should keep in mind what has been said about sets, manifolds and the like. The material in this chapter serves as examples of these concepts. And when later we extend the subject, we will need these more generalized concepts.

Measurements are most-commonly made on a two-dimensional section through the structure being examined. They may also be made on projected images through slices of finite thickness or on the projections of surfaces. In one sense, therefore, a practical measurement always involves the use of a planar probe. However, we will restrict the use of the term to the case where the measurement procedure absolutely requires an entire two-dimensional view of the image, as would be necessary, for example, if we had to determine the diameters of a collection of circular profiles. On the other hand, if we required the distribution of linear intercept lengths in these profiles, it would only be necessary to observe the image along a line; consequently the measurement can be made with a linear probe. Similarly a point probe implies that only zero-dimensional views of the structure are necessary.

We will be deriving relationships for "random" interactions between a probe and structural features. In this regard, the definition of "randomness" raises problems that are not immediately obvious. An excellent illustration is the paradox posed by Bertrand[3], which concerns the distribution of random chords in a circle. We consider three processes for introducing a chord at "random". (Other processes can be devised, but they usually turn out to be equivalent to one of the three listed.)

Process I: (Fig. 4.1a) Select any circle diameter (by symmetry, its orientation is unimportant) and divide it into a large number of equally spaced points. Then, using a table of random numbers, select one of these points randomly and draw a chord through it perpendicular to this fixed diameter.

Process 2: (Fig. 4.1 b) Divide the circumference into a large number of equally spaced points. Select two points randomly and draw a chord connecting the two points.

Process 3: (Fig. 4.1c) Superimpose a fine uniform grid on the circle. Select the coordinates of a grid point randomly and then draw the chord having this point as its center.

[3] This illustration has also been used by Kendall and Moran [1963]. An interesting discussion of Bertrand's paradox is to be found in M. Gardner, *Scientific American*, **201** (1959) p. 174.

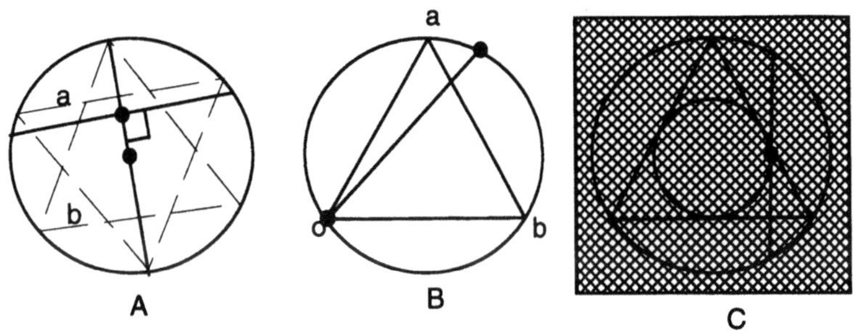

Figure 4.1: Bertrand's Paradox

These processes are all random. Are these three processes equivalent? Or more specifically: do they yield the same distribution of chord lengths? A simple test is to determine the probability that the chord length will exceed the length of the side of an inscribed equilateral triangle. For process 1 this corresponds to the probability that the center of the chord perpendicular to the diameter shown lies between lines *a* and *b*; this is evidently ½, since the sides of an inscribed equilateral triangle bisect perpendicular radii. For process 2 we can assume without any loss of generality that one end of the chord lies at the vertex, o, of the inscribed triangle. For the chord length to exceed the side length, the other end of the chord must lie between points *a* and *b* in the diagram for the second process; the probability is therefore 1/3. For process 3 the center of the chord must lie within the circle inscribed within the triangle and the probability of this is equal to the ratio (1/4) of the area of the inscribed circle to that of the outer circle (again by the perpendicular bisection argument).

Thus the probabilities yielded by the three processes of the chord length exceeding the side length of the inscribed triangle are respectively 1/2, 1/3 and 1/4. We must therefore conclude that the three processes are *not* equivalent. At this point it is tempting to ask[4], which process, then, is the most random of them all? Further reflection shows, however, that the question does not make any sense. All three processes involved the random selection of some parameter and, unless we have some prior knowledge of what the chord distributions are going to be used for, there is no reason for preferring one process to another.

It is now apparent that the chord problem is not so much a paradox as an illustration that the expression "at random" is not meaningful unless the randomization process is specified. That still leaves the question of what process we should select for our purpose. Setting this question aside with the

[4] When I have proposed this question in class, students have often expressed surprisingly strong (but differing) opinions. (J.H.)

intention of coming back to it let us continue our examination of the chords of a circle. From dimensional arguments we know that the average chord length will be proportional to the diameter of the circle, but we would expect that the proportionality constant would differ for the three processes. Now, because of the relationship between the average intercept length and diameter of the circle, it would be possible to estimate the latter from a measurement of the former. (We will defer answering the obvious question of why one would not just measure the diameter of the circle directly.) It is clear from our example that the relationship that one would employ to determine the circle's diameter would obviously depend on the process used to generate the random chords (or intercepts). For example, one procedure would be to drop a very long knitting needle from a great height compared with the diameter of the circle. (Of course, trials in which the needle did not land as a chord on the circle would be ignored.) In this case we might expect that, because the resulting chords would be uniformly distributed with respect to their distance from the center, the distribution of chord lengths would be that given by process 1 and we would accordingly use the average chord length for this distribution in determining the circle diameter.

Next, suppose instead of a single circle we have a collection of circles of constant diameter on a plane, Fig. 4.2, and we make a linear traverse across the plane and record the average length of intersection of the traverse inside the circles. Providing the dispersion of the circles is reasonably random (here we have pushed the question of one randomness onto another), we can expect that "hits" of the traverse with circle diameters perpendicular to the traverse will be uniformly distributed over those diameters; so that again the distribution of intercept lengths would follow process 1. (It is also apparent that if we wished to automate the measurement, the determination of the average intercept length would be easier than a direct measurement of circle diameters. This is an example of the general principle that it is easier to automate a linear probe than it is a planar one.)

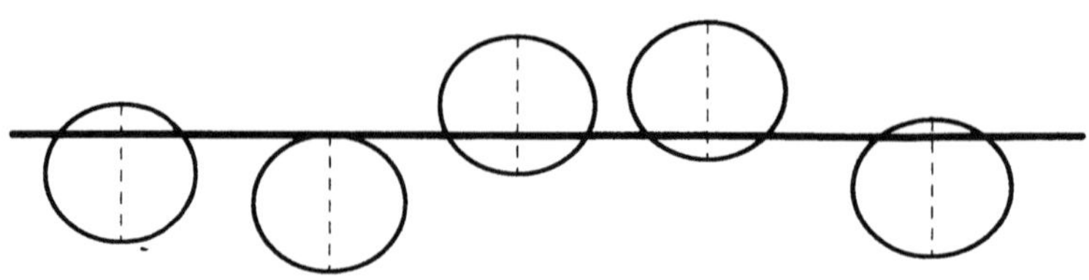

Figure 4.2: Random Interceptions with Circles

Now, of course, it would also be possible to determine the circle diameters from the average length of chords inserted according to process 2 or 3, but these would be more difficult to implement. For example, the selection of two random points on the circumference of each circle is a less convenient operation, whether done manually or automatically, than making a linear traverse. Apart from practical considerations, another important reason for preferring process 1 is that it yields (as we will see later) very powerful general relationships, whereas the other two do not. Those working in geometrical probability theory cannot resort to practical considerations and must therefore find some other basis for selecting a randomizing process. The criterion usually adopted is that the probability density function describing the location and orientation of the probe be invariant under translation and rotation of the coordinate axis. It can be shown that for a given dimensionality of the probe and the space within which it resides, this requirement can be satisfied by one (and only one) function. Fortunately for stereologists, this choice also coincides with, the experimentally convenient one. For example, for a linear probe in two dimensions, it leads to process 1.

We will now consider the distribution of point, linear and planar probes for $N = 1$, 2 and 3 (where N denotes the dimensionality of the space in which the probe lies). As we will see, the first step in each case is to set up a coordinate system which will uniquely define the location and, if necessary, the orientation of the probe. The particular coordinate system that will be selected in each case is the one that by hindsight is known to yield the simplest expression for the probability density function.

2. THE POINT PROBE IN ONE, TWO AND THREE DIMENSIONS

For $N = 1$ the location of a point can be specified by its distance, x, from the origin. For the probability density, $f(x)$, to be finite it is necessary for x to be bounded. We will take the bounding figure to be a line of length, l. A random distribution of the point on this line will be defined as that for which the probability of the point falling in an interval $|dx|$ is $|dx| / l$ and is independent of x. By definition: $f(x)dx = dx / l$,

thus:

$$f(x) = 1/l \quad : 0 < x < l \quad . \tag{4.1}$$

For $N = 2$ we will confine the point to a closed bounding figure of area a which represents the set of points, A, and will define the location of the point by its Cartesian coordinates (x,y). A random distribution is that for which the probability of a point falling within an area $dx\ dy$ is $dxdy/a$. Thus: $f(x,y)\ dx\ dy = dx\ dy/a$, or

$$f(x,y) = 1/a : (x,y) \in A \quad . \tag{4.2}$$

The extension to $N = 3$ is obviously defined by locating a point in a set V belonging to $\mathbf{R}^3$ whose volume is v. In that case,

$$\mathrm{f}(x,y,z) = 1/v : (x,y,z) \in V \ , \tag{4.3}$$

where v is the volume of the bounding figure, V.

It should be evident that v in Eqn.(4.3) is the Lebesgue measure, v_3. In all three Eqns.(4.1-4.3) the probability density function is simply $1/v_d$ where d is the dimensionality of the space being considered. We should by now, having read Chapter 3, be comfortable with the idea that the probability of a point in a set A falling in a subset B of set A is simply $P(B|A) = v_d\,(B)/v_d\,(A)$ where A and B are Borel sets.

3. THE LINEAR PROBE IN TWO DIMENSIONS

Before proceeding it would be helpful to summarize the above through the use of a limited axiom. This axiom applies strictly only to linear probes but is easily generalized to planar probes.

Limited Axiom

A necessary condition for a linear probe to be a uniform random linear probe is:

i) All its orientations are equiprobable. In two-space it is uniformly distributed with respect to angle. In three-space it is uniformly distributed with respect to solid angle.

ii) Its intercepts for any given range of angle with any axis (or flat) are uniformly distributed along that axis (or flat).

Part *i* follows from the requirement that the probability density function for the probe be invariant under rotation. Part *ii* follows from the requirement that the probability density function be invariant under translation. Probes which meet these requirements are termed *independent uniform random* or *IUR* probes whether they are linear or planar. In this section, we use this axiom to define a geometry for a linear probe. Later, under circumstances where this specific geometry becomes cumbersome, we may still refer to this axiom to create another geometry meeting this axiom through a transformation of the distribution about to be derived.

The orientation of the probe will be defined by the angle, that its perpendicular makes with a given reference direction (RD). A random orientation is here defined as that for which all values of θ in the range $0 \leq \theta \leq 2\pi$ are equally probable. Thus:

$$f(\theta)d\theta = d\theta/(2\pi)$$

or

$$f(\theta) = \frac{1}{2\pi}[\![0 \leq \theta \leq 2\pi]\!]. \tag{4.4}$$

If we do not sense the direction of the probe, then:

$$f(\theta) = \frac{1}{\pi}[\![0 \leq \theta \leq \pi]\!]. \tag{4.5}$$

In order to define the location of the probe, it is necessary to introduce a second coordinate, for this we will use ρ, the length of the perpendicular to the probe from the origin, Fig. 4.3.

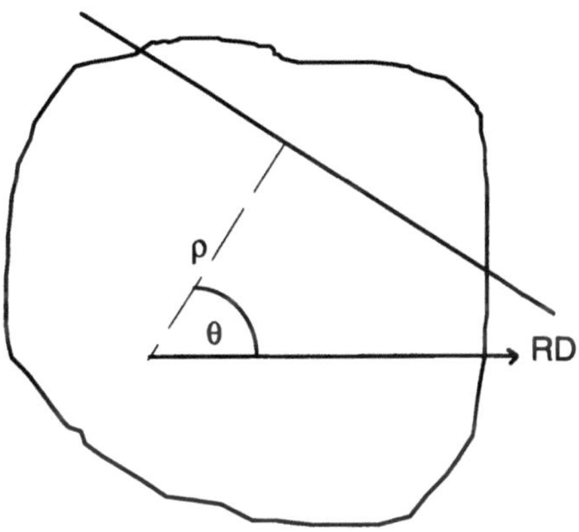

Figure 4.3: Linear Probe and Bounding Figure

But, it should be clear that if we restrict ourselves to $\theta \in [0, \pi]$ that ρ would be required to take negative values. We will define a random location of the probe as being that for which, at a given orientation, all values of ρ are *equally probable.* Since the normalized density function for theta is thus independent of ρ, the joint probability density function is separable and becomes simply a constant times the density function for theta,

$$f(\rho,\theta) = K\, f(\theta), \tag{4.6}$$

where K is that constant. This does not imply that the limits of ρ are independent of θ. For this probability density function to be non-zero it is necessary, as in the case of the point probe, to confine the linear probe to a finite region of the two-dimensional plane. We will therefore assume that the location of the probe is bounded by a closed convex figure, and consider only those probes that intersect the figure. (A convex figure is one for which all chords connecting any two points on its perimeter lie entirely within the figure.)

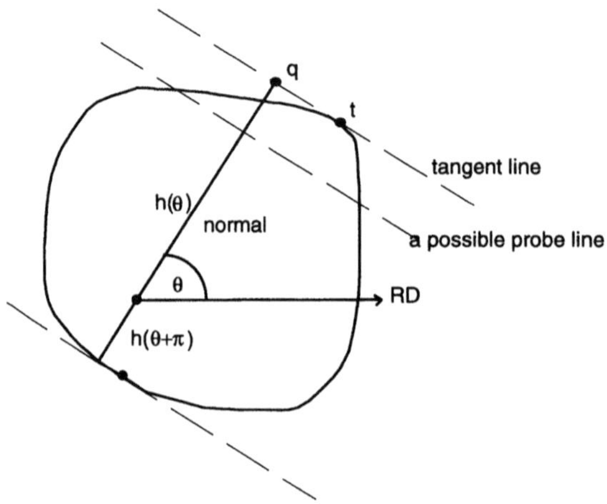

Figure 4.4: Convex Planar Figure

We will now evaluate K for this boundary condition. By the definition of the cumulative distribution function and division by K,
[cf. Eqn.(2.32)]:

$$K^{-1} = \int_{0}^{2\pi} \int_{(\rho)} f(\theta) d\rho d\theta\,. \tag{4.7}$$

Parentheses around ρ indicate that the integral is taken using the appropriate limits for ρ which, in turn, depend on θ. Note the upper limit is now 2π. This is done not because the probe has become directional but rather to limit ρ to positive values.

It is easy to lose track of what we are doing at this point. The form of the integral might suggest that we are doing something like taking the area of the bounding figure. This is absolutely not the case. Let us first consider the integration with respect to ρ the distance from the probe to an arbitrary point contained within the figure. From Fig. 4.4 it is seen that, for a particular value of θ, the maximum value of ρ is $h(\theta)$ the perpendicular distance from the origin to the tangent the line which makes an angle $\theta + \pi/2$ with the reference direction. The choice of origin is random and by this construction arbitrary as well. The only restriction on the probe line is that its normal make the specified angle θ with the RD. Hence to find the maximum value of ρ for any particular θ we may choose any origin along the RD line. But, rather than actually slide the origin along the RD line, which could lead to difficulties, we take an arbitrary origin and use the geometry of Fig. 4.4. Thus:

$$K^{-1} = \int_0^{2\pi} \int_0^{h(\theta)} f(\theta) d\rho d\theta = \int_0^{2\pi} f(\theta) h(\theta) d\theta . \tag{4.8}$$

Now consider a small change $d\theta$ in θ, the point q will move a distance $h(\theta)d\theta$ and an equal shift will occur in the point of tangency t. But this is also equal to an element dc of the perimeter of the figure. Hence, (after substituting $f(\theta) = 1/(2\pi)$) we obtain:

$$K^{-1} = \frac{1}{2\pi} \int_c dc = \frac{c}{2\pi} . \tag{4.9}$$

Substituting in Eqn. (4.6) we obtain:

$$f(\rho, \theta) = \frac{1}{c} : \; 0 \le \theta \le 2\pi , \tag{4.10}$$

or if we restrict the range by exploiting symmetry:

$$f(\rho, \theta) = \frac{2}{c} : \; 0 \le \theta \le \pi . \tag{4.11}$$

Are these equations valid if the bounding figure is not convex? From the derivation just given the answer appears to be no, because the quantity $h(\theta)$ is not uniquely defined for a non-convex figure. In order to see what happens in the non-convex case, it is instructive to consider an alternative derivation

of Eqn.(4.10). (The reader may, in any case, find this derivation more satisfying than the preceding one.)
If we define as in Fig. 4.4 :

$$d(\theta) = h(\theta) + h(\theta + \pi),$$

then $d(\theta)$ is the length of the figure projected onto a line at θ to the RD and we can rewrite Eqn.(4.8) as:

$$K^{-1} = \frac{1}{2}\int_0^{2\pi} d(\theta) f(\theta) d\theta = \frac{d}{2}, \tag{4.12}$$

where d (without any θ) is the mean projected length, or as it is here termed, the *mean-caliper diameter* of the figure. Although it is an average, the bar has been omitted because we will need the symbol $\bar{d}$ for the average of d over a collection of figures.

We can estimate d as follows. Consider an element of the perimeter of length dc at probe angle θ to the RD. The projected length, dc_p, of this element onto a line parallel to the RD is $dc\,|\sin\theta|$ and therefore the average of its projected length is (cf. Fig. 4.4) :

$$\overline{dc_p} = dc \int_0^{2\pi} |\sin\theta| f(\theta) d\theta = \frac{2dc}{\pi},$$

since $f(\theta) = 1/2\pi$. Now, for a convex figure two (and only two) elements of the perimeter will contribute to one element of the projected length; thus, summing over all elements of the perimeter and dividing by 2 we obtain

$$d = \frac{c}{\pi}. \tag{4.13}$$

Substituting into Eqn.(4.12) we find $K = c/2\pi$, which is the same result [Eqn. (4.9)] as that obtained previously.

At this point, it may profit us to examine the properties of this distribution. It is a uniform distribution over a region of two-dimensional space in polar coordinates. The coefficient K has units of reciprocal length giving an appropriately dimensionless result. It may be viewed as a chunk taken out of a larger possibly infinite uniform distribution of probes. The constants K and $f(\theta)$ are normalizing constants adjusting the probability integrated over the figure to unity. Rewriting Eqn.(4.10) as

$dP = \frac{1}{c} d\rho \, d\theta$, we see that it can be restated in terms of an element of area, $dA = \rho \, d\rho \, d\theta$, giving

$$dP = \frac{dA}{\rho c} \, . \tag{4.14}$$

In the case of a non-convex figure more than two elements may contribute to the projected length (Fig. 4.5).

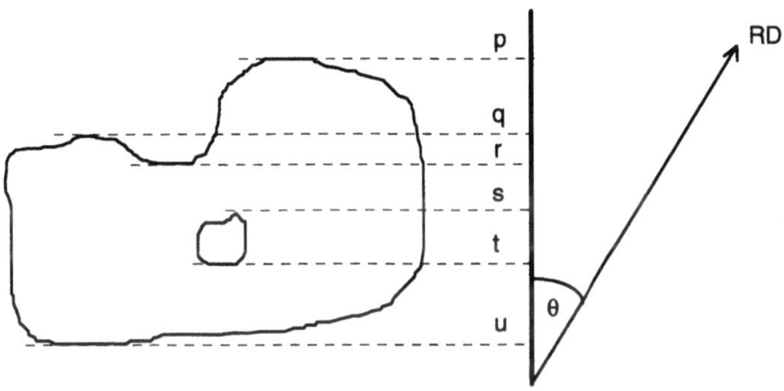

Figure 4.5: Non-convex Planar Figure

For example, between *q* and *r*, and between *s* and *t* the projections of four segments of the perimeter overlap. However, probes of orientation θ that intersect the figure in these regions will be divided into two non-connecting portions. And, in general, if there are 2*n* overlaps in a particular direction, then the corresponding probes will be divided into *n* portions. Counting each portion as a separate probe, it should now be apparent that (as is logical) the distribution defined by Eqn.(4.10) is meaningful for any shape of bounding figure. For a non-convex figure it will correctly predict the probability density of a probe having the coordinates (ρ,θ) since this probability is weighted by the number of separate intersections of the probe with the figure. If the figure contains an internal closed boundary (Fig. 4.5) then *c* is taken as the sum of the interior and exterior perimeters.

It will be seen from Eqn. (4.10) that the probability density function for a linear probe is independent of its orientation, θ. Considering the possibility of a the probe interacting with a long narrow figure, at first sight this may seem puzzling; after all, we have assumed that the probe is uniformly distributed along the projected length of the bounding figure, so why is $f(\rho,\theta)$ not proportional to $1/d(\theta)$? The answer is that we have to allow for the probability that the probe actually "hits" the bounding figure. Now, if one imagines the bounding figure immersed in a uniform flux of parallel probes, it is readily seen that this probability is proportional $d(\theta)/d$ where again d is the mean caliper diameter. The two $d(\theta)$'s in the product thus cancel out. This is very fortunate as otherwise the distribution would depend on the shape of the bounding figure. The same explanation can also be applied to the next two distributions that we will be determining.

4. THE LINEAR PROBE IN THREE DIMENSIONS

In Section 4.3 we introduced a "Limited Axiom" to ease the definition of a random probe. Following that axiom, the linear probe in three dimensions will be defined by its co-latitude, ϕ, and its longitude, θ (Fig. 4.6).

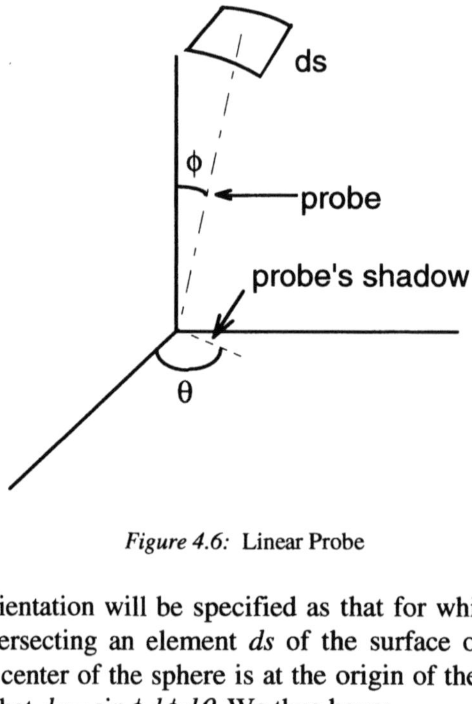

Figure 4.6: Linear Probe

A random orientation will be specified as that for which the probability of the probe intersecting an element ds of the surface of a unit sphere is $ds/4\pi$, when the center of the sphere is at the origin of the probe. From Fig. 4.6 it is evident that $ds = \sin\phi\, d\phi\, d\theta$. We thus have:

$$f(\phi,\theta)=\frac{1}{4\pi}\sin\phi \left| \begin{array}{l} 0\leq\phi\leq\pi \\ 0\leq\theta< 2\pi \end{array} \right. \tag{4.15}$$

or, if we do not distinguish the sense of the probe:

$$f(\phi,\theta)=\frac{1}{2\pi}\sin\phi \left| \begin{array}{l} 0\le\phi\le\frac{\pi}{2} \\ 0\le\theta<2\pi \end{array} \right. . \tag{4.16}$$

No special notation is used here to distinguish between the direction-sensitive and direction-insensitive cases because such a notation would be burdensome. The choice of either Eqn.(4.15) or (4.16) must be determined from the context. Applying Eqn. (2.47) to Eqn. (4.15) we obtain for the marginal distributions:

$$f(\phi)=\frac{\sin\phi}{2}:0\le\phi\le\pi. \tag{4.17}$$

and

$$f(\theta)=\frac{1}{2\pi}:0\le\theta<2\pi \tag{4.18}$$

Since $f(\phi,\theta)=f(\phi)f(\theta)$, the variables ϕ and θ are by definition independent of one another.

We next consider the distribution with respect to location. Again it is necessary to confine the probe and we will therefore only consider probes that intersect a bounding figure. For the time being we will assume the figure is the boundary δ of a convex body.

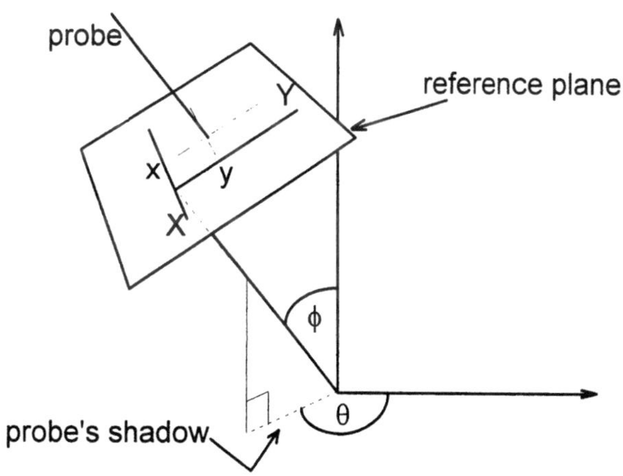

Figure 4.7: Linear Probe and Reference Plane

The location of the probe will be specified by (x,y) the Cartesian coordinates of its intersection with a reference plane. For reasons that will be apparent later, we will assume that this reference plane is free to swivel around a point located outside the bounding figure (Fig. 4.7). Before reading off the coordinates of the probe, the orientation of the reference plane is adjusted so that it is normal to the probe. A random location of the probe will be *defined* (based on the "limited axiom") as that for which all values of x and y within the projected image of the figure on the reference plane are equally probable. We will also assume that (x,y) are independent of (ϕ,θ). A casual glance at Fig. 4.7 could suggest a confusing interpretation, which needs to be dispelled at once. Were the reference plane fixed and the x-y coordinates of the probe dependent on ϕ and θ, a random distribution of ϕ and θ would imply a non-random distribution of x and y. However in the present case all four coordinates are independent. Thus, by virtue of an extension of Eq. (2.48) we can write,

$$f(x,y,\phi,\theta) = K f(\phi,\theta), \tag{4.19}$$

where K is a constant.

We first note that by definition:

$$\int_{(x)}\int_{(y)}\int_0^{2\pi}\int_0^{\pi} f(x,y,\phi,\theta)\,d\phi\, d\theta\, dx\, dy = 1, \tag{4.20}$$

where x and y define the location of the probe while ϕ and θ define the orientation of the probe and where the ranges of integration for x and y are those corresponding to probes intercepting the bounding figure. It follows from these two equations that:

$$K^{-1} = \int_{(x)}\int_{(y)}\int_0^{2\pi}\int_0^{\pi} f(\phi,\theta)\,d\phi\, d\theta\, dy\, dx\ . \tag{4.21}$$

Consider a probe of orientation (ϕ,θ). For this probe to intercept the bounding figure, it must intercept the x,y-coordinate plane within an area which is a projection of the bounding figure in the direction (ϕ,θ). The integration of Eq. (4.21) with respect to x and y therefore yields the projected area $S_p(\phi, \theta)$ for this direction. Therefore:

$$K^{-1} = \int_0^{2\pi}\int_0^{\pi} S_p(\phi,\theta) f(\phi,\theta) d\phi d\theta \tag{4.22}$$

and

$$K^{-1} = \langle S_p \rangle = \overline{S}_p \tag{4.23a}$$

where $\overline{S}_p$ is the average projected area of the bounding figure.

The final step is to relate this to the surface area of the figure. Consider a surface element dS whose normal makes an angle ϕ' to the normal of the plane of projection. The projected area, dS_p of this element is $dS\,|\cos\phi'|$ where ϕ', for a random orientation of the body, has the distribution given by Eqn.(4.17). Averaging over all orientations of the element, we obtain:

$$\langle dS_p \rangle = \frac{dS}{2}\int_0^{\pi} |\cos(\phi')| \sin(\phi') d\phi' = \frac{dS}{2}. \tag{4.23b}$$

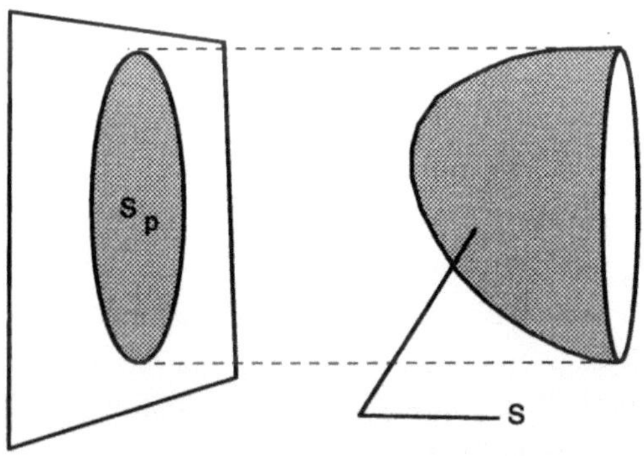

Figure 4.8: Relationship between surface and projected surface

Projection is a mapping of surface elements onto a plane. It follows from Eqn.(4.23b) that the area of the projected surface S_p (Fig. 4.8) averaged over all random orientations of its inverse image S is half the area of that inverse image.

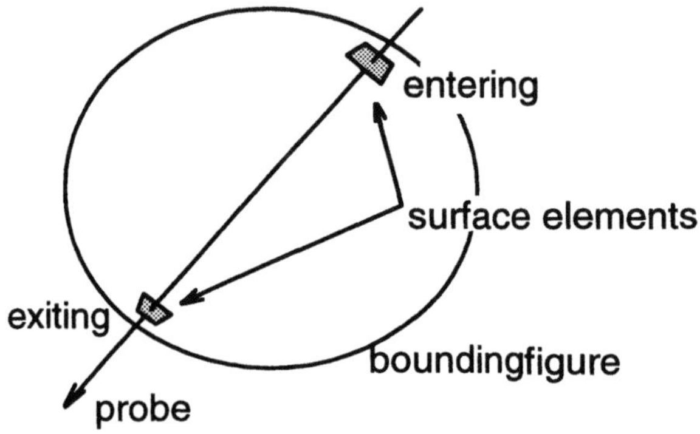

Figure 4.9: Entering and exiting surface elements

Hence, summing over all elements of the surface and remembering that in the projection of the boundary of a *convex* compact set each surface element is overlapped by exactly one other element (Fig. 4.8) we have:

$$\bar{S}_p = \frac{S}{4} \tag{4.24}$$

for convex compact sets where S is the surface area of the boundary. This and Eqns.(4.15), (4.19) and (4.23a,b) yield:

$$f(x, y, \phi, \theta) = \frac{\sin(\phi)}{\pi S}\Bigg|_{0 \le \theta < 2\pi}^{0 \le \phi < \pi} (x, y) \in S_p \tag{4.25}$$

This result also holds for a non-convex bounding figure if, as in the two-dimensional case, multiple intersections are allowed for.

5. THE PLANAR PROBE IN THREE DIMENSIONS

We consider the interaction of a planar probe with a bounding body, which is the boundary of a convex, compact three-dimensional point set, V. We define the uniform random probe in the spirit of the "Limited Axiom" with all orientations equiprobable and a uniform density of interceptions with a flat. The result is an IUR planar probe. When realized as a section through a material, the resulting section is called an *IUR section*. Combining this concept with the A-weighting discussed in Chapter 1 defines the standard *A-weighted IUR section* commonly used by microscopists to which the results of Chapter 5 apply.

In deriving the probability density function for the IUR section, we will be deriving the density function for a random plane. Its use could be for sectioning or for projection. The collection of all such random planes in three dimensions is termed the Grassmann manifold, $G_{2,3}$, which forms a topological space. This space is a metric space because a distance between planes can be defined for the general case $G_{2,n}$ of 2-planes in n dimensions. The selection of a sequence such planes at random connected by sub-sequences of planes, which gradually rotate to form a smooth transition between those randomly selected, is the basis for a type of animation called the *grand tour*[5]. By installing a window in each plane, this animation allows a viewer to see an object from all angles, not only objects in three dimensions but in any number of dimensions. The reader probably has seen three-dimensional examples of this type of animation as part of solid modeling demonstrations. The purpose of random sectioning in stereology is to effect a tour of the specimen from all possible angles and vantage points.

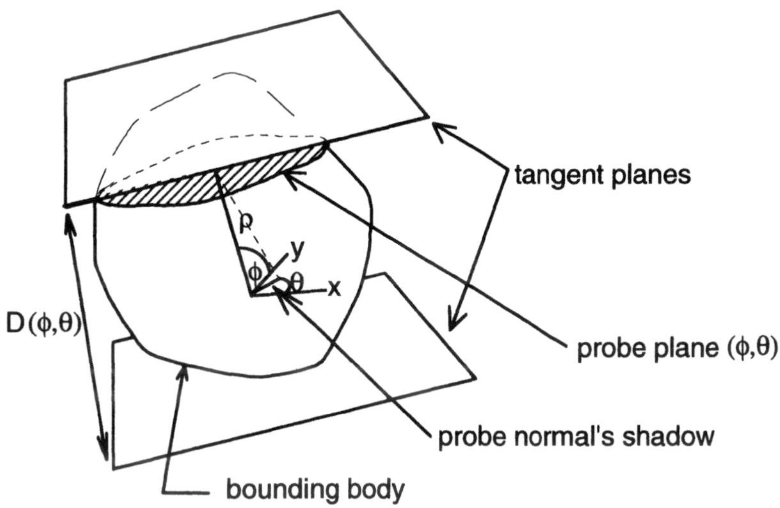

Figure 4.10: Planar Probe

[5] The interested reader should see D. Asimov, "The grand tour: a tool for viewing multidimensional data", *SIAM Journal of Scientific and Statistical Computing*, **6** (1985) 1.

The orientation of a planar probe is defined by angles ϕ and θ of its normal (Fig. 4.10). The distributions of the angles are the same, Eqns. (4.15)-(4.17), as those for a linear probe. The location of the probe within the convex bounding body will be defined by ρ, its perpendicular distance from the origin. Let:

$$f(\rho,\phi,\theta) = K f(\phi,\theta) \tag{4.26}$$

then, by analogy with Eqn. (4.21):

$$K^{-1} = \int_{(\rho)} \int_0^{2\pi} \int_0^{\pi} f(\phi,\theta)\, d\phi\, d\theta\, d\rho \; .$$

Integration with respect to ρ yields:

$$K^{-1} = \int_0^{2\pi} \int_0^{\pi} \rho_{\text{sup}}(\phi,\theta) f(\phi,\theta)\, d\phi\, d\theta \tag{4.27}$$

where $\rho_{\text{sup}}(\phi, \theta)$ is the maximum allowable value of ρ for a probe of orientation (ϕ, θ). This maximum corresponds to the distance from the origin of a tangent plane that is parallel to the (ϕ, θ) plane. Defining:

$$D(\phi,\theta) \equiv \rho_{\text{sup}}(\phi,\theta) + \rho_{\text{sup}}(\pi-\phi,\pi+\theta)$$

in which $D(\phi, \theta)$ is the distance (Fig. 4.10) between a pair of tangent planes of orientation (ϕ, θ) we can re-write Eqn.(4.27) after making the necessary change in limits as:

$$K^{-1} = \int_0^{\pi} \int_0^{\pi} D(\phi,\theta) f(\phi,\theta)\, d\phi\, d\theta = \frac{D}{2} \tag{4.28}$$

in which D is the average distance or "mean-caliper diameter" of the bounding body. (The bar is omitted on the D for the same reason that it was omitted on d in the two-dimensional case.) Finally, substituting from Eqns. (4.15) and (4.28) into Eqn.(4.26) we obtain:

$$f(\rho,\phi,\theta) = \frac{\sin(\phi)}{2\pi D} \left| \begin{matrix} 0 \le \phi \le \pi \\ 0 \le \theta < 2\pi \end{matrix} \right. \rho \in V \; . \tag{4.29}$$

Unlike the two-dimensional case, D cannot be expressed in terms of some simpler property of the bounding figure (we will later find that it involves an integration over the surface). Nor does it appear feasible to extend the result to the case of a non-convex bounding figure.

6. COMBINED SAMPLING OPERATIONS

In practice the application of a linear probe nearly always involves a two-step sampling operation. First a random section is taken through the structure and then a linear or a point probe is applied to this section. Providing the probe is applied only once it can be shown that this process generates for the linear probe the same distribution, Eqn. (4.25), as the random insertion of the probe directly into the three-dimensional structure. However, this is not true if more than one probe is applied to the same section. The implications of this will be discussed in Chapter 7 when we consider sampling procedures. But it is obvious, even at this stage, that the more homogeneous and isotropic the structure is, the less will be the problems posed by nonrandom sampling.

7. SUMMARY AND COMMENTS ON ALTERNATE DERIVATIONS

The probability density functions derived in this chapter are listed in Table 4.1. In subsequent chapters we will be utilizing these functions to derive relationships involving the interception of random probes with structural features. In this regard we will be following the same course as that pursued by Kendall and Moran (1963). An alternative derivation of certain of these relationships used by some authors [for example Weibel (1963) and Underwood (1970)] involves the assumption that the structural feature is located in a cubic box and that there is a uniform flux of probes parallel to an edge or a face of the box. The average density of intersections is then determined for a random orientation of the feature. This scheme has the merit of being easy to visualize but it is not very satisfactory in other respects. For one thing, it is not at all obvious that the same results would be obtained if the probes were not parallel to an edge or a face of the box (in fact, if it were obvious, the assumption would be unnecessary in the first place). Also, the ease of visualization is offset, particularly for a novice, by the ease with which mistakes can be made when using this and similar schemes. (Nor is the danger confined to novices. Several fallacious results in the literature can be traced to this cause.) In contrast, the procedures that we will be following are relatively foolproof and, if anything, are analytically simpler. (J.H.)

8. VERTICAL SECTIONS

Before leaving the topic of random probes some mention should be made of specialized probes used by stereologists. Most of these are composites of the probes we have treated already. The *disector* is a probe consisting of two parallel planes. It is used to estimate number per unit volume and is in effect a volume probe. It will be discussed in Chapter 7. The *nucleator* is a composition of a point probe and a line. It is used to estimate areas and volumes and is associated with the concept of star volume also discussed in Chapter 7. The *orthogonal sketch* mentioned in Chapter 10 is a composite of signed linear probes descended from the spherical harmonic probes treated in Chapter 5. The *vertical section* is a sectioning technique and probe which is the restriction of the $G_{2,3}$ planar probes to those containing a common line. This common line is either in truth or by definition an axis of symmetry. Often specimens are sliced into "french fries" along this axis giving two pairs of orthogonal planes, one of which is free to take any orientation with respect to the specimen itself. They are used to cope with specimen structure problems. Proceeding along the lines of the derivation of Eqn.(4.29) we arrive at Eqn.(4.10), with Eqn.(4.13) substituting d for c, as the probability density function of the vertical section as a probe. In this case d is the mean caliper diameter of the projection of the bounding surface onto the plane perpendicular to the common line. In order to approximate the results of a random linear probe applied to an IUR section, cycloid-curved linear probes are sometimes applied to the surface of the vertical section[6]. Vertical sectioning is an important part of the practical implementation of stereology.

[6] For a discussion of vertical sections see A. Baddeley, H. Gundersen and L. Cruz-Orive, "Estimation of surface area from vertical sections", *Journal of Microscopy*, **142** (1986) 259-276.

Table 4.1:
Probability Density Functions Defining the Random Distribution of Probes

Space Dimensions	Point Probe	Linear Probe	Planar Probe
one	$f(x) = 1/l$ $0 < x < l$	--	--
two	$f(x, y) = 1/a$ $(x, y) \in A$	$f(\rho, \theta) = \frac{1}{c}$ $0 \le \theta \le 2\pi$	--
three	$f(x, y, z) = 1/v$ $(x, y, z) \in V$	$f(x, y, \phi, \theta) = \frac{\sin(\phi)}{\pi S}$ $0 \le \phi \le \pi$ $0 \le \theta < 2\pi$ $(x, y) \in S_p$	$f(\rho, \phi, \theta) = \frac{\sin(\phi)}{2\pi D}$ $0 \le \phi \le \pi$ $0 \le \theta < 2\pi$ $\rho \in V$

Notes to Table:
l = length of one-dimensional bounding body
a = area of two-dimensional bounding body
c = perimeter of two-dimensional bounding body
v = volume of three-dimensional bounding body
S = surface area of three-dimensional bounding body
D = mean caliper diameter of three-dimensional bounding body

9. PROBLEMS

Worked Example: Three of the problems that follow involve the transformation of variables in a multivariate distribution function. As an example of such a transformation of Eqn.(2.157P) of Problem 2.3 for the random distribution of a point in a circle.

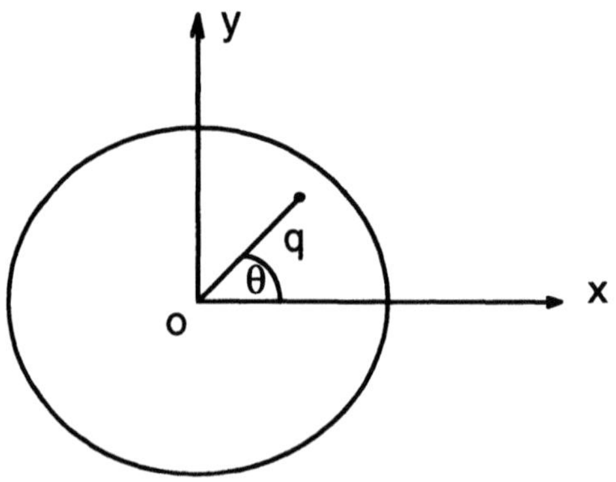

Figure 4.11: Random point in a circle

We define the location of illustrated in Fig. (4.11). of the point:

$$x = q\cos(\theta) \quad \text{and} \quad y = q\sin(\theta).$$

From Eqn. (2.52) it follows that $f(q,\theta)$ is related to $f(x,y)$ by $f(q,\theta) = f(x,y)|J|$, where $J = \begin{vmatrix} \frac{\partial x}{\partial q} & \frac{\partial x}{\partial \theta} \\ \frac{\partial y}{\partial q} & \frac{\partial y}{\partial \theta} \end{vmatrix}$.

Substituting:

$$\frac{\partial x}{\partial q} = \cos\theta; \quad \frac{\partial x}{\partial \theta} = -q\sin\theta;$$
$$\frac{\partial y}{\partial q} = \sin\theta; \quad \frac{\partial y}{\partial \theta} = q\cos\theta$$

and evaluating the determinant we find $J = q$. Also, from Eqn.(4.2) we have $f(x,y) = \dfrac{4}{\pi d^2}$ Therefore: $f(q,\theta) = \dfrac{4q}{\pi d^2}$.

Integrating out θ, we obtain the marginal distribution:

$$f(q)=\frac{8q}{d^2}$$

in agreement with Eqn. (2.157P).

4.1) The distribution of chord lengths generated by process 1, of Bertrand's Paradox discussed in section 4.1, is derived in Section (6.2.1). Using this as a guide, verify that the distributions generated by processes 2 and 3 on a circle of diameter *d* are:

$$f_2(L)=\frac{2}{\pi}\left(d^2-L^2\right)^{-\frac{1}{2}};\quad 0\le L\le d$$

and

$$f_3(L)=\frac{2L}{d^2};\quad 0\le L\le d\,.$$

Also, show that the corresponding average chord lengths are:

$$\overline{L}_2=\frac{2d}{\pi}\quad\text{and}\quad \overline{L}_3=\frac{2d}{3},$$

and that: $1-F_2(L)=\frac{1}{3}:\ L=\frac{\sqrt{3}d}{2}$ and $1-F_3(L)=\frac{1}{4}:\ L=\frac{\sqrt{3}d}{2}$,

in agreement with the values derived from geometrical considerations.

4.2) Show that in terms of the reduced variable, $\lambda = L/d$, the two distributions in the previous problem can be written as:

$$f_2(\lambda)=\frac{2}{\pi}\left(1-\lambda^2\right)^{-\frac{1}{2}};\quad 0\le\lambda\le 1,$$

and

$$f_3(\lambda)=2\lambda$$

4.3) As previously stated, the probability density functions that we are using to define a "random" probe are invariant under rotation and translation of the coordinate axes. Prove that this is so for Eqn.(4.10) defining a linear probe in two dimensions. The demonstration of (isometry) invariance for probes in three dimensions is more difficult. It is to be found in Kendall and Moran [1963] pp. 16-22.

4.4) An alternative choice of coordinates for specifying the orientation and location of a linear probe in two dimensions would be u_1 and u_2; the components of the vector defining the perpendicular from the probe to the origin. Show that the probability density of (u_1, u_2) corresponding to $f(\rho,\theta) = 1/c$ is:

$$f(u_1,u_2)=\frac{1}{c}\left(u_1^2+u_2^2\right)^{-\frac{1}{2}}.$$

(It will be noted that this expression is considerably more cumbersome than that for $f(\rho,\theta)$. In addition, u_1 and u_2 are not independent.)

4.5) Let (x,y) and (x',y') be the coordinates of two random points, P and P', in a convex bounding figure of area a. Since the points are independently distributed, we have from an extension of Eqn. (2.48)

$$f(x,y,x',y')=f(x,y)f(x',y')$$

and hence, from Eqn. (4.2):

$$f(x,y,x',y')=1/a^2 \tag{4.30P}$$

Consider an alternative coordinate system illustrated in Fig. 4.12. A perpendicular, OT is drawn from the origin to the line passing through the two points. Let ρ be the length of the perpendicular, θ its angle to the RD and q and q' the distance of the points from the foot T of the perpendicular. Assuming a suitable convention for the sign of the latter distances, it is easily verified that:

$x=\rho\cos\theta - q\sin\theta$,

$y=\rho\sin\theta + q\cos\theta$ at P, where q is positive;

$x'=\rho\cos\theta - q'\sin\theta$,

and $y'=\rho\sin\theta + q'\cos\theta$ at P', where q' is negative.

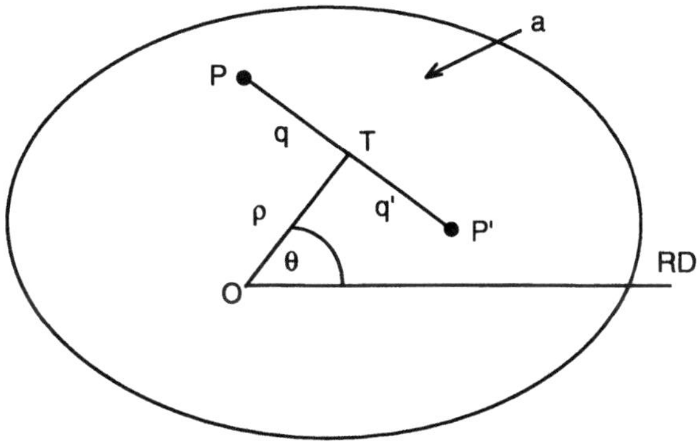

Figure 4.12: Two random points in a convex body

Prove that for the new coordinate system the distribution corresponding to Eqn. (4.30P) is:

$$f(\rho, q, q', \theta) = \frac{|q - q'|}{a^2} \tag{4.31P}$$

$|q\text{-}q'|$ is the distance between P and P'. Although this problem involves evaluating a 4 x 4 determinant, it is not too laborious because four of the elements are zero. We will be using Eqn.(4.31P) in the next chapter.

4.6) You may wish to anticipate the following results that will be derived in the next chapter (In all cases assume that the bounding and enclosed bodies are convex.)

(a) The probability of a point probe intercepting a body of volume V_i inside a body of volume V is Vi/V.

(b) The probability of a linear probe intercepting a figure of perimeter c_i within a figure of perimeter c is c_i/c.

(c) Derive similar relationships for the interception of linear and planar probes with a three-dimensional body.

Chapter 5

GENERAL SHAPE-INDEPENDENT RELATIONSHIPS

In this chapter we will be considering relationships that are general in the sense that they are valid for structures of any form, in contrast to the shape-dependent relationships that will be developed in Chapter 6. The distinction between the two is somewhat arbitrary but not illogical. These general shape-independent relationships involve "total" properties of the structure such as the volume fraction of a constituent, the boundary area per unit volume or the curvature per unit volume. Shape-dependent relationships, Chapter 6, are necessary when more detailed information is required (the size distribution of features, for example). In this chapter, we will be relating these total properties to the statistical expectations of related properties measured on some probe, when that probe takes a "grand tour", in analogy to the term defined in the previous chapter, of the specimen. While the grand tour is officially defined only for projections onto 2-frames, which amount to computer display windows depicting bounded subsets of 2-planes, in this chapter we will be studying intersections with both 2-planes and 1-planes, viz. lines. The grand tour idea capsulizes how we intend to look at our specimen's interior from all angles and possible positions. As we do this, we will be averaging our observations.

Some workers seem to have put a great deal of effort into devising ways of making a single section or a single line particularly random. In cases where only a single section is examined, there cannot be any averaging over anything like a grand tour. So, unless the structure being analyzed is statistically isotropic, the analysis cannot give the same average as would be gotten over a grand tour. If the structure is isotropic, then the choice of section orientation is unimportant.

1. WHAT IS SHAPE-INDEPENDENCE?

In the last chapter we saw that estimating the intersection probabilities of random probes with an object are problems in integration over some manifold. The term "integral geometry" has been applied to the study of general definite integral problems of this kind. Definite integrals map a higher dimensional space into a lower dimensional one. For this reason they are usually not invertible. The choice of integral is often not unique in the sense that the particulars of the integrand may in some way factor out. So, the same integral result can apply to a whole class of integrands (or a whole class of limits of integration or both) allowing an algebraic formula to do the duty of the original integral. For this to be always true, the result should not depend on a specific choice of boundary for integration, so long as it is large enough and not too large. Quantities like the expected surface area per unit volume of some object can be calculated as a function of the length of the line of intersection per unit area of that surface with a planar probe (slicing and measuring the perimeter). Using this functional relationship allows great freedom in the choice of the boundary for the region over which the integral is to be taken.

When an integral over a volume can be written in terms of a definite integral over a line or surface, the resulting formula is said to be shape independent. The reader may be familiar with Gauss's, Stokes's and Green's integrals. These are examples of shape-independent integral relationships, which have nothing to do with stereology. Stereological shape-independent relationships differ in that they are measures of geometric properties.

2. THE INTERACTION MATRIX

The interaction (intersection) of a probe with a structural feature can be specified by the dimensionality of that interaction, I, and this, in turn, is determined by the dimensionalities of the probe, the feature, and the space within which it resides. For example, in three dimensions a linear probe ($P = 1$) intersects a boundary ($F = 2$) at a point ($I = 0$). The various cases for the possible interactions can be represented by an "interaction" matrix. For three-dimensional space Table 5.1 gives the number of dimensions for the various possible intersections.

Table 5.1: Interaction Matrix for Three-dimensional Space

Dimensionality of probe, P	Dimensionality of Structural Element, F: 3	2	1	0
3	3	2	1	0
2	2	1	0	
1	1	0		
0	0			

The elements of the matrix are the dimensionality of the interaction of the probe with the structural element termed the *feature*. Notice that a one-dimensional structural element interacting with a one-dimensional probe results in a blank entry because the result has a dimensionality of -1. This means that there is no interaction unless the space containing the probe and the structural feature is reduced by one. It is impossible (probability zero) for probe lines to intersect feature lines in three-space. Or, if perchance we find that a probe line does intersect a feature line, a constraining surface must be present.

If we define P as the dimensionality of the probe, F that of the structural feature, N that of the space and I that of the resulting intersection, then it is apparent that:

$$I = P + F - N. \tag{5.1}$$

This relationship, which was apparently first pointed out in a stereological context by Weibel [1967], is not new and is essentially the same formula as presented in the well-known text, *Tensor Calculus* [Synge and Schild (1949)]. It correctly predicts all the "impossible" interactions such as a linear probe cannot intersect a point in two dimensions nor a line in three dimensions. Using this formula, the reader can readily develop the corresponding matrices for $N = 1$ and 2.

Equation (5.1) also provides a convenient method of categorizing the general (shape-independent) relationships. For each probe we will derive the interaction formulas for increasing values of N. For a given value of N, these relations will be presented in the order of decreasing F. *In the sub-headings that follow; the subscripts on I, P, F and N denote the values of these variables for the particular case being treated.* Equation (5.1) does not strictly hold for the case where the dimensions are fractional. More will be said about this in Chapter 10.

3. THE POINT PROBE IN ONE, TWO AND THREE DIMENSIONS

$I_0 P_0 F_n N_n : n = \{1,2,3\}$

3.1 Probability of a Point Probe Falling Within a Feature

We wish to determine the probability, p, that a randomly applied point will fall on a lineal feature of length l_i located on a line of length l (Fig. 5.1). Define:

$$u(x) = \begin{cases} 1 \text{ if } x \text{ is inside the feature} \\ 0 \text{ if } x \text{ is not inside the feature} \end{cases}$$

This function is called the *characteristic function* or *indicator function* of a set, if we take the feature as a set in $\mathbf{R}^n$. Other notations for the same function are $\chi(x)$, $\varepsilon_A(x)$ and $\mathbf{1}(x)$. See the Appendix, Section A.3.4.

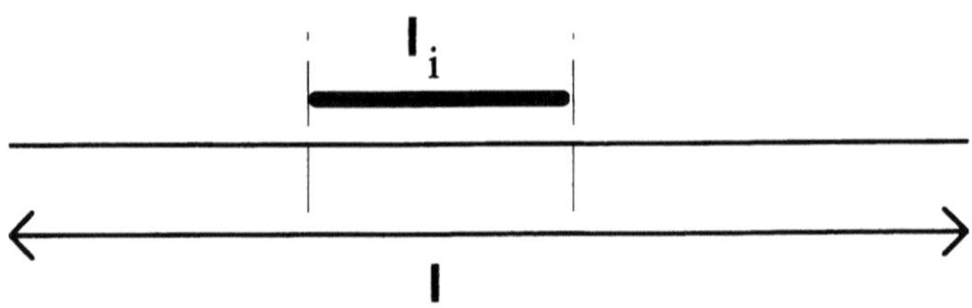

Figure 5.1: A line segment on a probe line.

Then, $p = \int_0^l u(x) f(x) dx$ where $f(x)$ is the probability density function for points on the line. But, for a random distribution of the point, $f(x) = 1/l$, Eqn. (4.1); therefore:

$$p = l_i / l . \tag{5.2}$$

The extension to $N = 2$ and 3 is obviously:

$$p = a_i / a . \tag{5.3}$$

and

$$p = V_i / V \tag{5.4}$$

It is also evident that these expressions are valid for an aggregate of features if l_i, a_i, and V_i are taken to be the total length, area and volume of the features.

3.2 Relationship between Point, Lineal, Areal and Volume Fractions

Consider an analysis in which a number P of point probes are applied at random on a line. The number, P_i falling on i-features will follow a binomial distribution and its expected value, Eqn.(2.58), is therefore:

$$\overline{P_i} = P \cdot p , \tag{5.5}$$

where p is the probability of a single probe falling on an i-feature.

Defining $\overline{P}_p \equiv \overline{P_i} / P$ and $L_L \equiv l_i / l$, Eqns. (5.2) and (5.5) yield:

$$\overline{P}_p = L_L . \tag{5.6}$$

A similar argument for N=2 and 3 leads to:

$$\overline{P}_p = A_A , \tag{5.7}$$

and

$$\overline{P}_p = V_V . \tag{5.8}$$

These expressions are the basis for the principles of Delesse and Rosiwal mentioned in Chapter 1 and derived later in this chapter.

4. THE LINEAR PROBE IN ONE, TWO AND THREE DIMENSIONS

$I_0\, P_1\, F_{n-1}\, N_n : n = \{1,2,3\}$

The intersection of a probe line with a point in $N = 1$ is a point and is scarcely different from the case of a point probe falling on a line.

4.1 The Linear Probe in Two Dimensions, Areal Features

Since we will be primarily concerned with the analysis of three-dimensional structures, the results of $N = 2$ are of less practical interest than those for $N = 3$. However, they do find application in measurements on projected images. In addition, since the geometry is more easily visualized in two dimensions, the derivations for $N = 2$ facilitate the parallel ones for $N = 3$ given in the next section.

4.1.1 Probability of a liner probe intercepting a two-dimensional figure

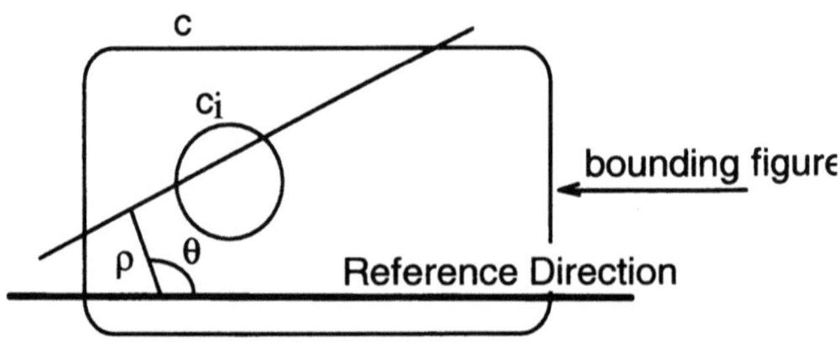

Figure 5.2: Inner convex figure intercepted by probe line.

Let p be the probability that a linear probe, which passes through a bounding figure, will intercept a figure of perimeter c_i located within that bounding figure of perimeter c. We will assume for the time being that both figures are convex (Fig. 5.2). This probability is given by

$$p = \int_{(\rho)} \int_{(\theta)} f(\rho,\theta) d\theta d\rho\,, \tag{5.9}$$

in which the integration is over those values of ρ and θ for which the probe intercepts the inner figure, the i-feature. From Eqn, (4.10), $f(\rho,\theta) = 1/c$, and it is apparent from the derivation of Eqn. (4.10) that

$$\int_{(\rho)} \int_{(\theta)} d\theta \, d\rho = c_i , \tag{5.10}$$

where as before the parentheses indicate that the integration is taken over the domain of the variable.

Combining the last two equations, we have

$$p = c_i / c \quad . \tag{5.11}$$

It follows from Eqn. (4.13) that we can also express p in terms of the mean caliper diameters of the interior and bounding figures

$$p = d_i / d \, . \tag{5.12}$$

The two equations are also valid if one or both figures are not convex. In that case, p is defined as the ratio of the expected number of intercepts to the number of trials. (This is one of those cases where probability merges into expectation. The quantity p is no longer a *proper probability* because for a non-convex inner figure, its value can exceed unity.) The counting convention is illustrated in Fig. 5.3.

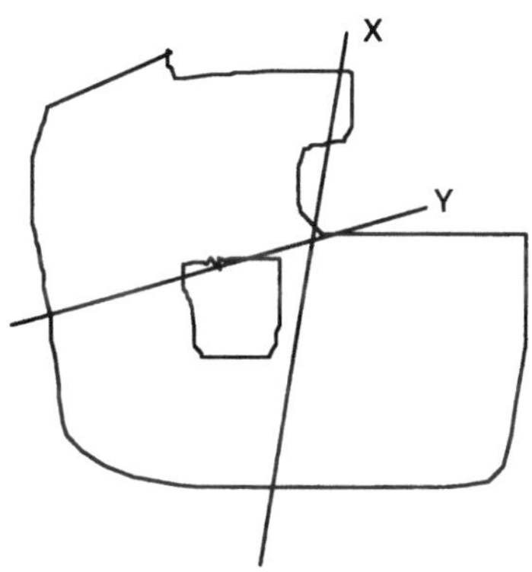

Figure 5.3: Interceptions in non-convex figures.

Probe x would be scored as zero intercepts in two trials and probe y as two intercepts (or 4 point interceptions or 2 lineal interceptions) in one trial.

4.1.2 Average intercept length in a two-dimensional figure

We will next derive an expression for the average intercept length of a linear probe with its own bounding figure of perimeter c and area a (Fig. 5.4) Let $L(\rho,\theta)$ be the intercept length on a probe having the coordinates (ρ,θ). By definition the average intercept length is

$$\bar{L} = \int_0^{2\pi} \int_0^{h(\theta)} L(\rho,\theta) f(\rho,\theta) d\rho d\theta \ . \tag{5.13}$$

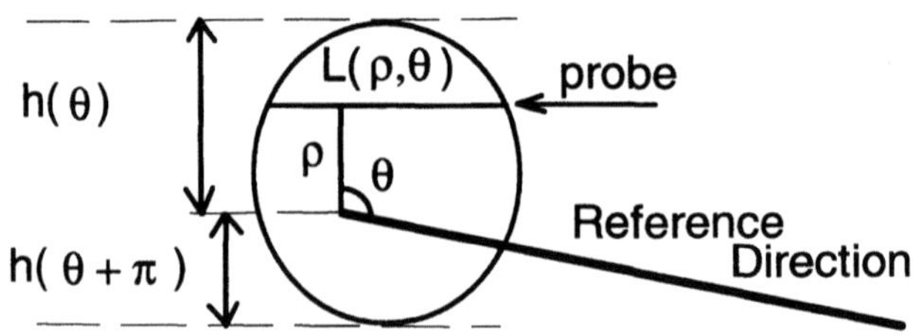

Figure 5.4: Probe in figure.

Changing the limits of integration and setting $f(\rho,\theta) = 1 / c$, we can rewrite this equation as

$$\bar{L} = \frac{1}{c} \int_0^{\pi} \int_{h(\theta+\pi)}^{h(\theta)} L(\rho,\theta) d\rho d\theta \, . \tag{5.14}$$

The first integration with respect to ρ yields the area of the figure. Call this area a. This area is obviously independent of θ. Thus,

$$\bar{L} = \frac{a}{c} \int_0^{\pi} d\theta = \frac{\pi a}{c} \, . \tag{5.15}$$

This result is also valid for a non-convex figure if the previously discussed convention is followed in counting multiple intercepts. This convention is is considered to be "usual" in stereology.

4.1.3 Average Intercept Length in an Aggregate of Figures

Consider a collection of n figures within a bounding figure (Fig. 5.5). We would expect that the average of all intercept lengths in the figures would be equal to some average value of $\pi\, a_i \,/\, c_i$ for the figures. However, there are at least two possible averages that we might take.

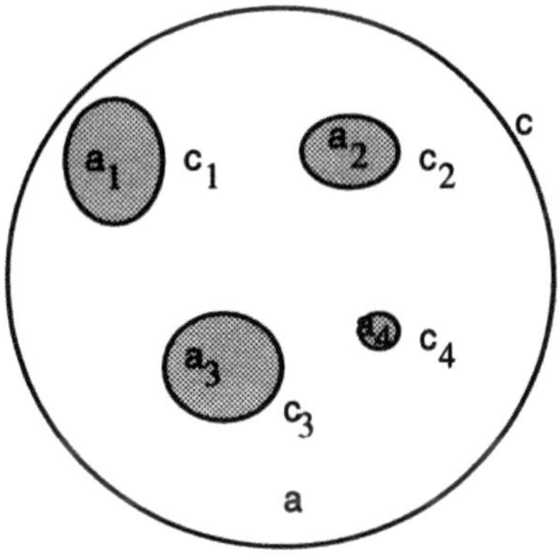

Figure 5.5: Embedded features bounded by perimeters c_i.

The first is the total area of the figures divided by their total perimeter length,

$$\pi \sum a_i \,/ \sum c_i \;;$$

and, the other is the average of the individual area to perimeter ratios of the n figures as given by

$$\frac{\pi}{n} \sum a_i \,/\, c_i$$

These averages are not the same. Which, then, should we use if we wish to determine the average intercept length $\overline{L}$? We first note that in averaging the average intercept lengths $\overline{L}_i$ in the individual figures, we must weight the $\overline{L}_i$'s by a factor that is proportional to the probability that i-th figure will be intercepted. According to Eqn.(5.11) this probability of interception is $c_i \,/\, c$. Thus the ratio of the number of intercepts in figure i to the total number of intercepts in *all* the figures is given by,

$$c_i \,/ (c_1 + c_2 + c_3 + \ldots) = c_i \,/ \sum c_i \;.$$

Weighting the average intercept lengths in the individual figures by this ratio, we obtain,

$$\overline{L} = \frac{1}{\sum c_i}\left(c_1\overline{L}_1 + c_2\overline{L}_2 + c_3\overline{L}_3 + \ldots\right).$$

But, by Eqn.(5.15) $\overline{L}_i = \pi a_i / c_i$.

Thus

$$\overline{L} = \pi \sum a_i / \sum c_i \quad . \tag{5.16}$$

i.e., the average intercept length is proportional to the ratio of the total area to the total perimeter length of the figures. If M is the total number of figures, than $\sum a_i = M\overline{a}$ and $\sum c_i = M\overline{c}$, thus Eqn. (5.16) can be written as

$$\overline{L} = \pi \overline{a} / \overline{c} \, . \tag{5.17}$$

4.1.4 Density of Interceptions of a Linear Probe with an Aggregate of Figures

It follows from Eqn.(5.11) that the expected number of interceptions, $\overline{N}$, of a linear probe (which passes completely through a bounding figure) with an aggregate of features having perimeters c_i located within a bounding figure having a perimeter c is

$$\overline{N} = \sum c_i / c,$$

or if $\overline{c}$ is the average perimeter of the figures and M is their number within the bounding figure,

$$\overline{N} = M\overline{c} / c \ .$$

This is reckoned as the number of interceptions with the aggregate of figures per interception with the bounding figure. From Eqn. (5.15) the average probe length within the bounding figure is $\pi\, a / c$, therefore

$$\overline{N}_L \equiv \overline{N} / \overline{L} = \frac{M\overline{c}}{\pi a} \ .$$

This expression has the advantage that the outer figure, which is arbitrary, has been factored out. This can also be written as

$$\overline{N}_L = \frac{1}{\pi} N_A \overline{c} \tag{5.18}$$

where N_A is the number of figures per unit area[1]. Using Eqn.(4.13) for the mean caliper diameter, we can also express N_L in terms of the mean or average caliper diameter of the features,

$$\overline{N}_L = N_A \overline{d}\ , \tag{5.19}$$

where $\overline{d}$ is the mean caliper diameter of the features.

4.1.5 Third Moment of the Intercept Lengths in a Convex Figure

We will defer discussion of the second moment of the intercept lengths; this has never been solved in a general way. The result that we next derive is (with its three-dimensional equivalent) the most remarkable one in this chapter. By analogy with Eqn.(5.13) the third moment about the origin of the intercept length of a linear probe with its own bounding figure is given by

$$\mu'_3(L) = \int_0^{2\pi} \int_0^{h(\theta)} L^3(\rho,\theta) f(\rho,\theta) d\rho d\theta \tag{5.20}$$

where, as before, $L(\rho,\theta)$ is the intercept length on a probe having coordinates (ρ,θ). At first sight it seems hardly conceivable that there could be any simple expression for $\mu'_3(L)$. The process used to determine $\overline{L}$ is obviously of no avail here since, unlike in the case of the first moment, the integral with respect to ρ is not independent of θ. And, in fact, the starting point is an unlikely one; namely Eqn.(4.31P) of Problem 4.5,

$$f(\rho,q,q',\theta) = |q - q'| / a^2,$$

for the distribution of two random points in a bounding figure of area a, when the coordinates of the points are defined as shown in Figs. 4.12 and $h(\theta)$ is the supremum of ρ at that angle. By definition,

---

[1] John Hilliard used "figure" and "feature" interchangeably when he meant "a feature which is a figure".

$$\frac{1}{a^2}\int_0^{2\pi}\int_0^{h(\theta)}\int_{(q)}\int_{(q')}|q-q'|dq'dqd\rho d\theta = 1 \ . \tag{5.21}$$

Consider first the integral with respect to q and q',

$$I = \int_{(q)}\int_{(q')}|q-q'|dq'dq \ .$$

If s and f are the limits of the q's (Fig 5.6) we can eliminate the absolute value signs in the integrand by splitting the range of integration. We can specify that P lies to the left of P' for the positive condition within the absolute value sign.

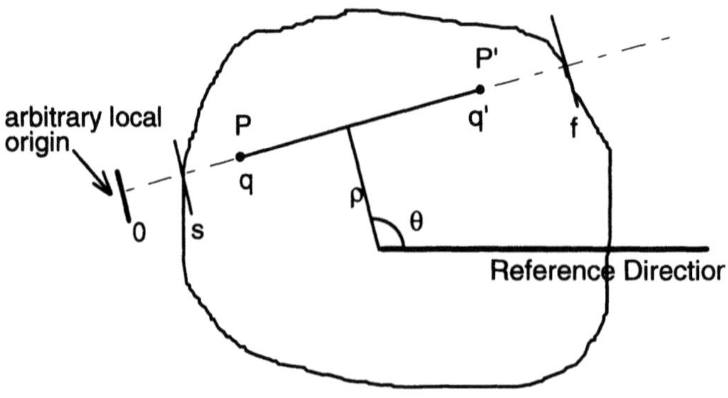

Figure 5.6: A line segment connecting two points within a closed figure.

In Fig. 5.6 we adopt a simpler (for this particular problem) sign convention from that used in problem 4.5. The points P and P' each have respectively coordinates q and q' with respect to some arbitrary local origin at the left of the interconnecting line. Both q and q' occupy independently a line segment between s and f. where $0<s<f$. This makes all values positive. Without any loss of generality, we can draw P' to the right of P. Rho and theta follow the same convention as in Figure 5.4. In anticipation of the final result we notice that the length of the interception of the probe line which passes through P and P' with the figure is $L = f\text{-}s$. We begin by rewriting the integral I,

$$I = \int_s^f \int_s^f |q - q'| dq' dq \,. \tag{5.22}$$

This can be expanded,

$$I = \int_s^f dq \int_s^f |q - q'| dq' = \int_s^f (I_1 + I_2) dq \quad ,$$

where the inner integral's range of integration is split

$$I_1 = \alpha_1 \int_q^f (q - q') dq' \quad I_2 = \alpha_2 \int_s^q (q - q') dq'$$

with respect to some arbitrary value of q. Examining I_1 we see that it is negative unless α_1 is set to -1 to effect the absolute value function. To effect the absolute value for I_2 the factor α_2 is set to unity. Noting that $L^3(\rho,\theta) = (f - s)^3 = f^3 - 3f^2 s + 3fs^2 - s^3$, we obtain

$$I = L^3(\rho,\theta)/3 \,. \tag{5.23}$$

Now, substituting this result into Eqn. (5.21) we obtain

$$\int_0^{2\pi} \int_0^{h(\theta)} L^3(\rho,\theta) d\rho d\theta = 3a^2 \,. \tag{5.24}$$

Since $f(\rho,\theta)$ is a constant equal to $1/c$, we can multiply the integrand by $f(\rho,\theta)$ and the right hand side by $1/c$ to obtain

$$\mu_3'(L) = 3a^2 / c \,. \tag{5.25}$$

This result was first derived by Crofton [1885] and is often referred to as his "second theorem".

Equation (5.24) gives us an invariant with respect to the function $h(\theta)$, in that the integral does not depend on the shape of the figure. But, this invariant does depend on integrating with respect to θ. Equation (5.14) also provided us with another somewhat more trivial invariant

$\int_0^{h(\theta)} L(\rho,\theta)d\rho = a$, which is not dependent on integrating with respect to θ.

These invariants gave us expressions for the first and third moments of the intercept distribution. Pursuing this a little further, dimensional analysis suggests $\mu'_n(L) = K a^{\frac{n+1}{2}} / c$ where K is a function of the entirety of $h(\theta)$. For the first and third moments, K is a constant. For the second moment, the area appears to a fractional power. This in itself hints that K may not be a constant but rather depends in detail on the function h. Searching for useful invariants (e.g. in the fifth moment, area again appears to an integral power; maybe it could be possible to find a shape-independent expression for the fifth moment.) could be done systematically using the methods of differential forms.

4.1.6 Third Moment of the Intercept Lengths for an Aggregate of Figures

For an aggregate of figures we have to weight (as was the case for $\overline{L}$) the individual values of $\mu_3'(L)$ by $c_i / \sum c_i$ to allow for the probability of intercepting the figure. It thus follows from Eqn.(5.25) that

$$\mu_3'(L) = 3\sum a_i^2 / \sum c_i \quad . \tag{5.26}$$

4.1.7 Relationship of Lineal to Areal Fraction

So far we have assumed that a linear probe is randomized with respect to both orientation and translation. In the derivation that follows, the probe will be randomized only with respect to translation only; its orientation will be assumed to be fixed.

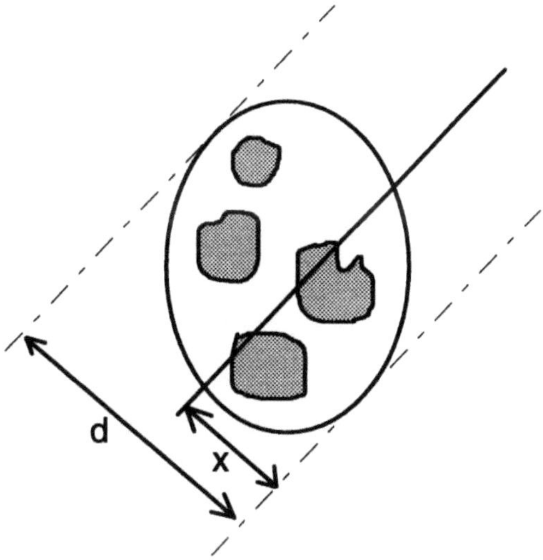

Figure 5.7: Features within a bounding figure.

The position of the probe will be defined by its perpendicular distance, x, from a tangent to the bounding figure that is parallel to the probe (Fig. 5.7). Assuming that all values of x are equally probable,

$$f(x) = 1/d\ , \tag{5.27}$$

where d is the distance between tangent lines. Let $L_i(x)$ be the total intercept length in i-features at a given position x. Then

$$\overline{L_i} = \int_0^d L_i(x) f(x) dx = \sum a_i / d\ ,$$

where $\sum a_i$ is the total area of i-features. Also if $\overline{L}$ is the expected intercept length in the bounding figure of area a, then $\overline{L} = a / d$. Thus the lineal fraction intercepted is

$$\overline{L_i} / \overline{L} = \sum a_i / a\ , \text{or}$$

$$\overline{L_L} = A_A\ , \tag{5.28}$$

where A_A is the areal fraction of i-features. It also follows from this derivation that Eqn.(5.28) is valid even for a non-randomly oriented probe.

It is important to note that $\overline{L}_L$ is the ratio of the expected values and that this is not the same as the expected value of the ratio. This is something worth examining further. If $y = u/v$ while u and v have joint density f, then by making a change of variables and applying the method of Section 2.3.5 we obtain (excluding negative values)

$h(y) = \int_0^{\infty} v f(vy,v)dv$ for the probability density function of y and the expected value of the ratio becomes $E(u/v) = \int_0^{\infty} y \int_0^{\infty} v f(vy,v)dvdy$. Negative values, although not applicable in this immediate case, need not be excluded. If u and v are independently distributed with density functions f and g respectively, then $h(y)$ can be found by first finding $H(y)$, the cumulative distribution function, and then differentiating

$$h(y) = \int_{-\infty}^{\infty} |v| f(vy) g(v) dv . \tag{5.29}$$

However, this expression for the density function of a ratio is not very useful unless the distribution functions f and g are known. The expected value of a product of independent variables was given in Section 2.3.2 and leads to a result in the form $E(y) = \mu_1'(u)\mu_{-1}'(v)$ for a ratio. Finding a negative moment is equivalent to solving Eqn.(5.29). We see that for the expected value of the ratio to equal the ratio of the expected values, then $E(1/v) = 1/E(v)$.

For sampling an extended surface, such as a mixed forest, we are at liberty to choose the shape of the bounding figure and the orientation of our probe. We may choose these so that all the interceptions of the probe with the bounding figure have the same length. Then, $E(1/v) = 1/E(v)$. If we make our bounding figure a narrow strip, then clearly this holds all the time and $L_L = A_A$. Is this a paradox? Not really, because the narrow strip intercepts only a narrow rectangular portion of a feature, not the entire feature. This division into narrow strips is in fact a manifestation of the principle of Cavalieri [Gundersen, 1988], which is not much different from that of a Riemann sum.

4.2 The Linear Probe in Two Dimensions, Lineal Features

Interceptions (intercepts) with lineal features are characterized by:

$I_0\ P_1\ F_1\ N_2$.

4.2.1 Expected Number of Intersections of a Linear Probe with Lineal Features

From Eqn.(5.11) we have $p = c_i / c$ for the probability of a linear probe intercepting a convex figure.

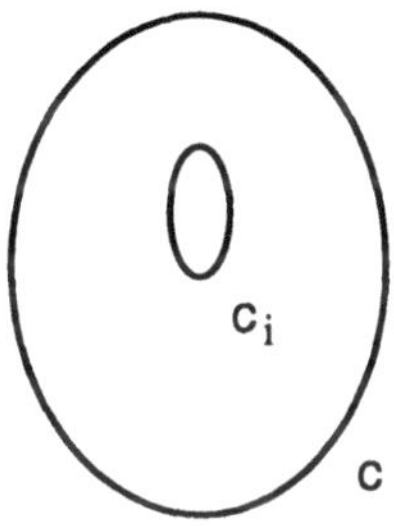

Figure 5.8: A lineal feature within a bounding figure.

For a convex figure each intercept yields two *intersections* with the perimeter of the figure. Thus the expected number of intersections, $\overline{P}$, is:

$$\overline{P} = 2c_i / c \qquad (5.30)$$

Now consider what happens as the inner figure slowly collapses on itself (Fig. 5.8). A stage will eventually be reached when it is not possible to resolve the separation between the two arcs, and the figure will appear as a single arc of perimeter c_i / 2. At the same time, however, the observed number of intersections will also be reduced by a factor of 2. Thus the number of intersections for a given length remains the same. We can therefore conclude that Eqn.(5.30) is valid for a non-closed figure.

4.2.2 Density of Intersections on an Aggregate of Lineal Features

Since one interception of a figure by a linear probe yields two intersections with the perimeter, it follows from Eqn.(5.18) that the number of intersections per unit length of probe on lineal features of length L_A per unit area is:

$$\overline{P}_L = \frac{2}{\pi} L_A \ . \tag{5.31}$$

Since this is a shape-independent relationship, it is valid for a probe of any shape.

4.2.3 Orientation Distribution of a Lineal Array

We will term a set of lineal features such as that shown in Fig. 5.9 a "lineal array".

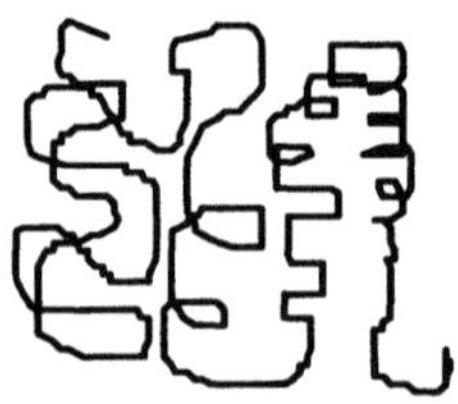

Figure 5.9: A lineal array consisting of only one line.

Apart from dimensionality, no restriction is imposed on the form of the features; they can be straight or curved, continuous or discontinuous. We can define the orientation distribution within the array by the density function $L_A(\theta)$ where $L_A(\theta)\, d\theta$ is the total line length per unit area to which tangents have angles in the range $\theta \pm d\theta/2$ measured anticlockwise with respect to some given reference direction (RD). It will be assumed that the sense of a line element is indeterminate so that $L_A(\theta + n\pi) = L_A(\theta)$ for any integer, n. The total line length per unit area is evidently

$$L_A = \int_0^{\pi} L_A(\theta) d\theta \ . \tag{5.32}$$

In principle, $L_A(\theta)$ could be determined by dividing the array into equal segments, short compared with their curvature, and the measuring the distribution of orientations. Were the array composed of linear segments, the orientation distribution would be characterized by a countable number of discrete angles giving rise to the same number of delta functions in the corresponding continuous distribution. Were the array curvilinear, the density distribution would be continuous. Deciding how to subdivide the array into straight lines would raise the question of "how short a straight segment compared with actual curvature do we use?" There would be no good answer obtainable by simple subdivision since the error would be at best of order δx, the length of the element being summed. If we consider a circle inside a bounding figure, $L = 2\pi r$ and the area is that of the bounding figure. So, $L_A(\theta) = 2r / a$. If the array were not very badly behaved, we could fit curvilinear portions with segments of a circle to an order of δx^2 which for small δx offers a considerable improvement in accuracy. For a single circular segment, it follows that

$$L_A(\theta) \cong [U(\theta - \theta_1) - U(\theta - \theta_2)]\frac{|r|}{a} \ ,$$

where U is the unit step function (Section 2.1.2). For the entire array, $L_A(\theta)$ would be the sum of that for each segment, summing length first and then dividing by total area. One could systematically divide the array with a grid and examine the contents of each square. The shape of the resulting cumulative distribution function for $L_A(\theta)$ would resemble a wobbly staircase. Relatively straight segments would have high risers and narrow treads. Except for certain machine applications, this approach is not very useful[2].

For an alternative there is a Fourier method of analysis; we first have to determine an expression for $P_A(\mu)$, the number of intersections per unit area when a second lineal array (again with no restrictions on its form) with an orientation distribution $L'_A(\omega)$ is superimposed on the first array randomly with respect to translation but with a fixed angle μ between their reference directions. In order to avoid having to worry about edge effects we will suppose that one array is much larger in extent than the other.

By definition, the expected line length in one array with angles in the small, but finite, range $\theta \pm \delta\theta / 2$ is $L_A(\theta)\,\delta\theta$ per unit area. Because the process is linear, without any loss in generality we can assume that this length occurs as a single segment within a unit area. We identify it with fixed line *BB* in Fig. 5.10. We can similarly treat the line length of the other

[2] Of this John Hilliard said in *Stereology, Principles and Practices,* "we will not consider this uninteresting procedure any further". But, the reader should note that direct methods are less prone to false answers than Fourier methods. Reviewing parts of this book, Ivan Saxl stated, "Fourier transforms are not a safe way to obtain the *rose of directions*" which is the expression of $L_A(\theta)$ in polar coordinates. If straight line probes are used and the intersections per length is plotted in polar coordinates, this is the *rose of intersections* [Saltykov, 1958]. Fourier transform methods such as Hilliard's methods are very efficient but must be used with care to avoid aliasing and the exaggeration of noise that occurs when differentiating in the Fourier domain.

array in the range $\omega \pm \delta\omega/2$ as a segment of length $L'_A(\omega)\,\delta\omega$. We identify this length with a relatively movable line *AA*. For this segment to overlap the other, its center or midpoint must fall within the parallelogram formed by the midpoints of line *AA* as it is positioned in the extremes of its contact with line *BB*. In the upper portion of the figure, lines *AA* and *BB* are indicated as solid. Alternative positions of line *AA* at the extreme points of contact with line *BB* are indicated with dashes. The parallelogram is formed by connecting the alternative midpoints with dotted lines. The parallelogram is projected into the lower portion of the figure to indicate the angles and side lengths. The sides have the same lengths as lines *AA* and *BB*. The area of a parallelogram is equal to the product of the lengths of its sides times the sine of the included angle. This area is proportional to the probability of intersection for a random location of line *AA*.

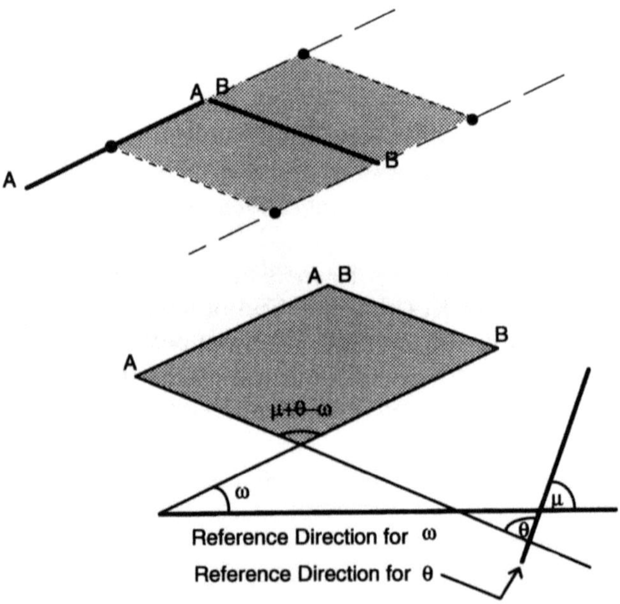

Figure 5.10: Intersection geometry.

Because the area of the bounding figure is taken as unity, the probability of this intersection event for a random location of the center of *AA* relative to *BB* is equal to the area of the parallelogram, namely $L_A(\theta)L'_A(\omega)\left|\sin(\mu+\theta-\omega)\right|\delta\theta\delta\omega$. In order to obtain the total expected number of intersections per unit area it is necessary to sum over all orientations. Proceeding to the limit we thus find

$$\overline{P}_A(\mu) = \int_0^\pi \int_0^\pi L_A(\theta) L_A'(\omega) \left|\sin(\mu + \theta - \omega)\right| d\theta d\omega . \tag{5.33}$$

This is a convolution integral within a correlation integral and a solution can be obtained through taking the Fourier transform. This approach has the advantage of being easily implemented as a computer program. To begin we rewrite Eqn.(5.33) as

$$\overline{P}_A(\mu) = \int_0^\pi L_A(\theta) I(\mu + \theta) d\theta , \tag{5.34}$$

where

$$I(v) = \int_0^\pi L_A'(\omega) \left|\sin(v - \omega)\right| d\omega . \tag{5.35}$$

Although finding the Fourier integral transform of Eqns.(5.34 and 5.35) is simple in principle, the task is made tedious by the facts that neither integral has infinite limits and the limits of integration are not symmetric to the origin. Accordingly, these equations are rewritten,

$$\overline{P}_A(\mu) = \int_{-\infty}^{\infty} W_0^\pi(\theta) L_A(\theta) I(\mu + \theta) d\theta , \tag{5.34'}$$

$$I(v) = \int_{-\infty}^{\infty} W_0^\pi(\omega) L_A'(\omega) \left|\sin(v - \omega)\right| d\omega , \tag{5.35'}$$

where

$$W_0^\pi(x) = \begin{array}{c|l} 1 & x \in [0, \pi] \\ 0 & x \notin [0, \pi] \end{array} . \tag{5.36}$$

The Fourier transform of Eqn (5.36) is a sinc function[3] multiplied by a complex number of absolute value 1 which represents the shift of $\pi/2$ needed to symmetrize the limits. In evaluating the transforms, the explicit use of the sinc function never appears and the product of the two shifting factors goes to unity since L_A and L'_A are periodic with period equal to π. The Fourier integral of $|\sin(x)|$ is

$$G(f)=\frac{2}{\pi}\sum_{n=-\infty}^{\infty}\frac{\delta\left(f-\frac{n}{\pi}\right)}{1-4n^2} \tag{5.37a}$$

where the Fourier integral is here defined, slightly differently from that in Chapter 2, as

$$H(f)=\int_{-\infty}^{\infty}h(x)e^{-i2\pi fx}dx\ , \tag{5.37b}$$

which on account of the periodicity takes the form of Eqn.(5.38). The Fourier integral of Eqn.(5.34') including Eqn.(5.35') turns out to be a product of the Fourier integrals of the several integrands where in the case of $L(\theta)$, the complex conjugate is taken. The presence of the comb of Dirac delta functions in Eqn.(5.37a) restricts evaluation of these other factors to multiples of $1/\pi$. That is, the result intrinsically involves the coefficients of a Fourier series because of the periodicity. Choosing the period to be π, the frequencies become n/π. The coefficients of a complex Fourier series are

$$\alpha_n^{\xi}=\frac{1}{\pi}\int_{-\pi/2}^{\pi/2}f(\xi)e^{-i2n\xi}d\xi\ . \tag{5.38}$$

The expected number of intersections per unit area is then given by

$$\overline{P}_A(\mu)=\sum_{n=-\infty}^{\infty}\frac{2\pi\alpha_n^{\omega}\left[\alpha_n^{\theta}\right]^*}{1-4n^2}e^{i2n\mu}\ . \tag{5.39}$$

Reminder: The star indicates complex conjugation.

In other words, the complex Fourier coefficients of the expected number of intersections per unit area are,

---

[3] Sinc is a term used in signal processing. These functions take the form $\sin(x)/x$.

$$\alpha_n^{\mu} = \frac{2\pi\alpha_n^{\omega}\left[\alpha_n^{\theta}\right]^*}{1-4n^2} \; . \tag{5.40}$$

These are amenable to calculation using the Cooley-Tukey Fast Fourier Transform algorithm, FFT, [Brigham, 1974], which makes automated computation simple. The desirability of using the FFT is the reason for adopting Eqn.(5.37b) as the definition of the Fourier transform.

Philofsky and Hilliard [1969] obtained the result of Eqn.(5.40) for a real Fourier series. While less amenable to machine calculations, this result facilitates constructing test patterns. The real Fourier series is defined[4] as

$$\overline{P}_A(\mu) = a_0 + \sum_{n=1}^{\infty} a_n \cos(2n\mu) + b_n \sin(2n\mu) \tag{5.41}$$

Using the conversions given in Table 5.2, the coefficients can be calculated to be those given by Eqns.(5.42a,b,c).

Table 5.2: Fourier Series Substitutions

Complex → Real	Real → Complex
$a_n = \alpha_n + \alpha_{-n}$	$\alpha_n = (a_n - ib_n)/2$
$b_n = i(\alpha_n - \alpha_{-n})$	$\alpha_{-n} = (a_n + ib_n)/2$

The notation α_{-n} designates the complex coefficient having a negative index with absolute value n. In this table (only), α_n is presumed to have a positive index. So, if a_n^{μ} and b_n^{μ} are the cosine and sine coefficients of the functions indicated by the superscripts, then:

[4] Because the factor $2n$ appears. Philofshy and Hilliard [1969] number their coefficients "$2n$" instead of n.

$$\begin{aligned} a^{\mu}_{n} &= \left[\frac{\pi}{(1-4n^2)}\right](a^{\omega}_{n}a^{\theta}_{n}+b^{\omega}_{n}b^{\theta}_{n})\Big|n>0 \\ a^{\mu}_{0} &= 2\pi a^{\theta}_{0}a^{\omega}_{0} \qquad\qquad (5.42a,b,c) \\ b^{\mu}_{n} &= \left[\frac{\pi}{(1-4n^2)}\right](b^{\omega}_{n}a^{\theta}_{n}-a^{\omega}_{n}b^{\theta}_{n})\Big|n>0 \end{aligned}$$

The procedure for determining the orientation distribution, $L_A(\theta)$, of a lineal array is thus as follows. An array having a known distribution $L_A(\omega)$ is applied at a series of different orientations μ. A count of the number of intersections between the two arrays yields the function $\bar{P}_A(\mu)$ from which can be computed α^{μ}_{n}. Substituting these coefficients into Eqn.(5.40) together with the known α^{ω}_{n} coefficients, one obtains the coefficients of the function $L_A(\theta)$. In order to avoid dividing by zero, the indices must be restricted to the set for which the coefficients of the test array have absolute values significantly greater than the system noise level. In practice, most images to be analyzed are scanned into data arrays; lineal re-scanning using edge detection is probably the most practicable form of orientation testing at this time. The foregoing is based on a treatment by Philofsky and Hilliard [1969]. The special case when one of the lineal arrays consisted of parallel lines was considered by Hilliard [1962][5].

4.2.4 Expected Number of Intersections Generated by Two Lineal Arrays

It is instructive to determine the expected number, $\bar{P}_A$, of intersections yielded by Eqn.(5.33) when the two lineal arrays are superimposed randomly with respect to orientation; i.e., for the case when μ has the distribution $f(\mu)=1/\pi; [0\le\mu<\pi]$. Changing the range of integration so as to eliminate the absolute signs, we can write Eqn (5.33) as:

$$\bar{P}_A=(1/\pi)\int_0^{\pi}\int_0^{\pi}\int_{\omega-\theta}^{\pi+\omega-\theta} L_A(\theta)\; L'_A(\omega)\sin(\mu+\theta-\omega)d\mu d\theta d\omega,$$

which, with the aid of Eqn. (5.32) yields:

[5] The original Hilliard manuscript mentions an additional reference to "Davidson, 1968". Rollo Davidson did do some work on linear probes and anisotropy at that time, but I have been unable to locate a reference to any Fourier treatment by him. The metallurgist David Davidson denied any association with the problem. (L.L.)

$$\bar{P}_A = \frac{2}{\pi} L_A L'_A \tag{5.43}$$

Dividing through by L'_A we obtain for the expected number of intersections per unit length of the probe array:

$$\bar{P}_L = \frac{2}{\pi} L_A ,$$

which is the same as Eqn.(5.31) previously derived. The advantage of the present derivation is that it clearly reinforces that the result is valid for a linear test probe of any form, curved or straight, whereas the previous derivation was confined to a straight probe[6]. Also, Eqn.(5.43) demonstrates the symmetry of the result.

4.3 The Linear Probe in Three Dimensions, Three-dimensional Bodies

In three dimensions, the highest-dimensional object that a probe can intercept is three-dimensional. Interceptions with three-dimensional features are characterized by:

I_1 P_1 F_3 N_3

4.3.1 Probability of a Linear Probe Intercepting a Three-Dimensional Body

Consider a body of surface area S_i located within a bounding body of surface area S. It follows from Eqn.(4.20) that the probability that a linear probe will intercept the interior body is:

[6] Some writers of standards have believed that using a curved test line will underestimate L_A relative to a straight test line. This is not so.

$$p = \int_{(x_i)} \int_{(y_i)} \int_0^{2\pi} \int_0^{\pi} f(x, y, \phi, \theta) d\phi d\theta dy dx, \tag{5.44}$$

in which the range of integration is over values of x and y that yield interceptions with the interior body. Restricting the set over which the integration takes place gives the relative frequency of interceptions. Substituting from Eqn.(4.25) we obtain:

$$p = (1/\pi S) \int_{(x_i)} \int_{(y_i)} \int_0^{2\pi} \int_0^{\pi} \sin(\phi) d\phi d\theta dy dx \ . \tag{5.45}$$

If we integrate Eqn.(5.45) over the boundary of the outer body we obtain,

$$\pi S = \int_{(x_o)} \int_{(y_o)} \int_0^{2\pi} \int_0^{\pi} \sin(\phi) d\phi d\theta dy dx \ .$$

Replacing the boundary S with S_i, we get

$$\pi S_i = \int_{(x_i)} \int_{(y_i)} \int_0^{2\pi} \int_0^{\pi} \sin(\phi) d\phi d\theta dy dx \ , \text{ and therefore,}$$

$$p = \frac{1}{\pi S} \pi S_i = \frac{S_i}{S} . \tag{5.46}$$

The relationship is valid for non-convex bodies if the same convention is adopted as in the two-dimensional case.

4.3.2 Average Intercept Length in a Three-Dimensional Body

If $L(x, y, \phi, \theta)$ is the intercept length on a probe having the coordinates (x, y, ϕ, θ), then the average intercept length is given by:

$$\bar{L} = \int\limits_{(x)} \int\limits_{(y)} \int\limits_{0}^{2\pi} \int\limits_{0}^{\pi} L(x,y,\phi,\theta) f(x,y,\phi,\theta)\, d\varphi d\theta dy dx\,, \tag{5.47}$$

in which the integration is over the bounding figure. Substituting for $f(x,y,\phi,\theta)$ from Eqn.(4.25) and noting that the integration with respect to x and y just yields the volume of the bounding figure, we find:

$$\bar{L} = \left(\frac{V}{\pi S}\right) \int\limits_{0}^{2\pi} \int\limits_{0}^{\pi} \sin(\phi) d\phi d\theta = \left(\frac{V}{\pi S}\right)(4\pi)\,,$$

or

$$\bar{L} = \frac{4V}{S}\,. \tag{5.48}$$

Compare this with $\pi a/c = \bar{L}$ in 2-D.

4.3.3 Average Intercept Length in an Aggregate of Bodies

From Eqn.(5.46) it follows that the probability of a linear probe intercepting a body is proportional to its surface area S_i. The weighting factor in determining $\bar{L}$ for an aggregate of bodies is therefore

$S_i / \sum S_i$, where i is now taken as an index for the several interior bodies. Hence, from Eqn.(5.48) through multiplication by the weighting factor:

$$\bar{L} = 4\sum V_i / \sum S_i\,. \tag{5.49}$$

Compare this with $\pi \sum a_i / \sum c_i = \bar{L}$ in 2-D. Alternatively,

$$\bar{L} = 4\bar{V} / \bar{S}\,. \tag{5.50}$$

This compares with $\pi\, \bar{a} / \bar{c} = \bar{L}$ in 2-D.

4.3.4 Density of Interceptions in an Aggregate of Bodies

Let $\overline{S}_i$ be the average of the surface areas of M bodies in a bounding body of surface area S. It then follows from Eqn.(5.46) that the expected number of interceptions with a linear probe is: $\overline{N} = M\overline{S}_i / S$. This compares with $M\overline{c}_i / c = \overline{N}$ in 2-D. Dividing through by Eqn.(5.48) we obtain for the density of interceptions:

$$\bar{N}_L = \frac{MS_i}{4V} = \frac{N_V \overline{S}}{4}, \tag{5.51}$$

where V is the volume of the bounding body and N_V is the number of bodies per unit volume. Equation (5.51) compares with $N_A\overline{c}/\pi = \overline{N}_L$ in 2-D.

4.3.5 Fourth Moment of the Intercept Lengths in a Convex Body

The derivation of Crofton's second theorem for a three-dimensional body parallels that (Section 5.4.1.5) for the two-dimensional case, We start with the distribution function for two random points with coordinates (x, y, z) and (x', y', z') in a bounding body of volume V:

$$f(\hat{x}, \hat{y}, \hat{z}, \hat{x}', \hat{y}', \hat{z}') = 1/V^2, \tag{5.52}$$

which follows from Eqns.(2.48) and (4.3). We next introduce an alternative coordinate system in which the locations of the points are defined by the coordinates (x, y, ϕ, θ) of the line passing through them and their distances (q, q') from the (x, y) reference plane. This geometry is the combination of Figs. 4.7 and 4.10 with Fig. 5.6. Figure 5.11 shows this combined geometry.

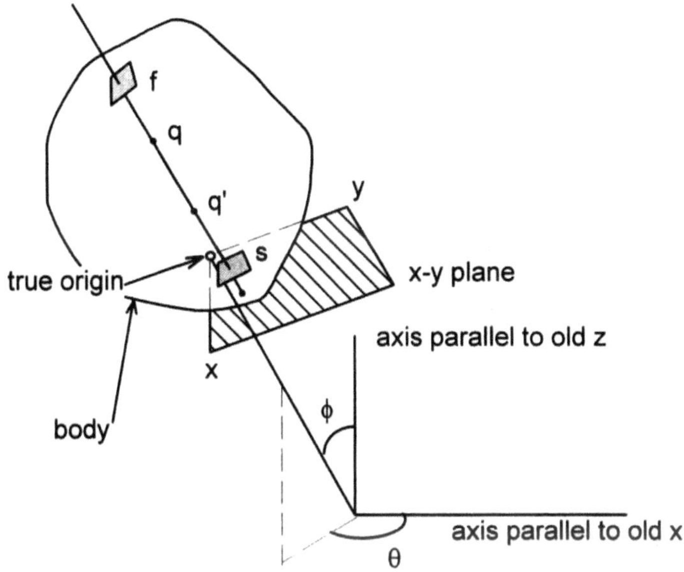

Figure 5.11 Geometry for finding the fourth moment of intercept lengths.

In Fig. 5.11 the true origin, the origin for both the old and the new coordinate systems, is indicated. The new *x-y* reference plane is free to pivot about the true origin. It is pivoted so as to be normal to a line passing through points q and q'. This makes that line parallel to the new z-axis. Hence q and q' are z-coordinates in the new system. The orientation of the new system is indicated by extending the new negative z-axis and drawing in lines parallel to the old x and y axes. This allows giving the orientation of the new z-axis in terms of spherical coordinates. The line passing through q and q' intersects the boundary of the body at s and f. It intersects the *x-y* reference plane at (x,y). The coordinates in the old system $(\hat{x},\hat{y},\hat{z},\hat{x}',\hat{y}',\hat{z}')$ are given in terms of those of the new system (x,y,ϕ,θ,q,q'): The first three old coordinates are given by:

$$\hat{x} = x\cos(\phi)\cos(\theta) - y\sin(\theta) + q\sin(\phi)\cos(\theta)$$
$$\hat{y} = x\cos(\phi)\sin(\theta) + y\cos(\theta) + q\sin(\phi)\sin(\theta)$$
$$\hat{z} = -x\sin(\phi) + q\cos(\phi)$$

The second three old coordinates are found by substituting q' for q. The transformation of variables is somewhat more difficult[7] than in the two-

[7] The transformation follows the rule given in Section 2.3.5. In *Mathematica*™, use Outer[D, old,new] where "old" and "new" are lists. To obtain the symbolic evaluation use FullSimplify[Det[Jacobian]]].

dimensional case, but it can be shown that, in terms of the new coordinate system, the distribution corresponding to Eqn.(5.52) is:

$$f(x,y,\phi,\theta,q,q') = \frac{(q-q')^2 \sin(\phi)}{V^2} \,. \tag{5.53}$$

Therefore, since integrating this over the boundary of the body gives unity,

$$\int\limits_{(x,y,\phi,\theta,q,q')} (q-q')^2 \sin(\phi) dx dy d\phi d\theta dq dq' = V^2 \,.$$

Since integration with respect to q and q' is separable, we can integrate

$$\int\limits_{(q,q')} (q-q')dq dq' = \int_s^f \int_s^q (q-q')^2 dq' dq = (f-s)^4 / 12 \,.$$

Since $(f\text{-}s) = L(x, y, \phi, \theta)$, we have

$$\iiiint L^4(x,y,\phi,\theta) \sin(\phi) dx dy d\phi d\theta = 12V^2 \,,$$

where $L(x,y,\phi,\theta)$ is the length of interception on a probe having coordinates (x,y,ϕ,θ) with the body. Finally, substituting from Eqn.(4.25) we obtain,

$$\iiiint L^4(x,y,\phi,\theta) f(x,y,\phi,\theta) dx dy d\phi d\theta = \frac{12V^2}{\pi S} \,.$$

But, by definition, this integral is just $\mu_4'(L)$. Therefore,

$$\mu_4'(L) = \frac{12V^2}{\pi S} \,. \tag{5.54}$$

4.3.6 Fourth Moment of the Intercept Lengths for an Aggregate of Bodies.

Allowing for the probability [Eqn. (5.46)] of the probe intercepting an individual body we obtain for the average value of $\mu_4'(L)$ in an aggregate of bodies:

$$\mu_4'(L)=\frac{12\sum V_i^2}{\pi\sum S_i}. \tag{5.55}$$

4.3.7 Relationship of Lineal to Volume Fraction

As in the two-dimensional case (Section 5.4.1.7) we will assume that the probe is randomized only with respect to translation. The location of the probe will be defined by the coordinates (x, y) of its intersection with a perpendicular plane, Fig. 5.12.

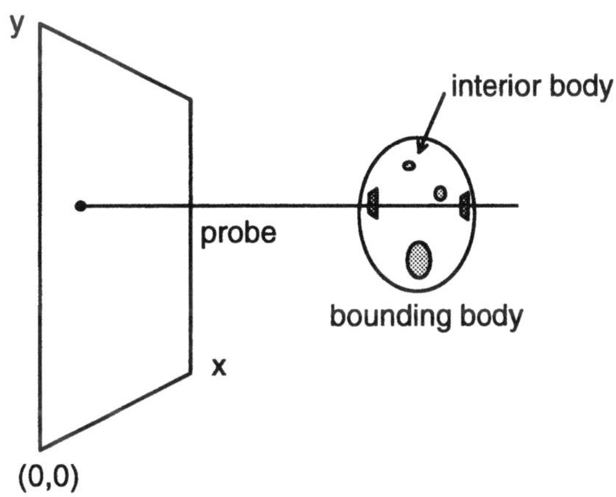

Figure 5.12: Probe, perpendicular reference plane and body.

For a random probe [cf. Eqn.(4.25)] the density of intersection is simply one over the expectation of the projected area since integrating $f(x,y)$ without normalizing gives the projected area. Hence,

$$f(x,y)=4/S, \tag{5.56}$$

where S is the surface area of the bounding body. Let $L_i(x,y)$ be the total length of interception of a probe at (x,y) with interior bodies inside the bounding body. Then:

$$\bar{L}_i = \int_{(x)} \int_{(y)} L_i(x,y) f(x,y) dxdy$$

$$= (4/S) \int_{(x)} \int_{(y)} L_i(x,y) dxdy = \frac{4\sum V_i}{S}.$$

As we have already demonstrated [Eq. (5.48)] the average intercept length of the probe in the bounding body is $4V/S$. Therefore:

$$\bar{L}_i / \bar{L} = \sum V_i / V.$$

or

$$\bar{L}_L = V_V. \tag{5.57}$$

where V_V is the volume fraction of the interior bodies.

4.4 The Linear Probe in Three Dimensions, Areal Features

Interceptions with areal features in three dimensions are characterized by:

$I_0\, P_1\, F_2\, N_3$

4.4.1 Expected Number of Intersections of a Linear Probe with Areal Features

Since each intercept yields two intersections, it follows from Eqn.(5.46) that the expected number of intersections on a body of surface area S_i enclosed within a bounding body with surface area S is:

$$\bar{P} = 2S_i / S. \tag{5.58}$$

Following the same argument as that used in Section 5.4.2.1 we can conclude that this expression is also valid for a non-closed body.

4.4.2 Density of Intersections on an Aggregate of Areal Features

For an aggregate of features where i is now an index, Eqn.(5.58) yields:

$$\bar{P} = 2\sum S_i / S.$$

Dividing through by Eqn.(5.48), where V is the volume of the bounding body, we obtain:

$$\bar{P} / \bar{L} = \sum S_i / 2V$$

or

$$\bar{P}_L = S_V / 2, \tag{5.59}$$

where S_V is the total area of features per unit volume. As in the two-dimensional case we will later demonstrate that this important relationship is valid for probes of any shape.

4.4.3 Orientation Distribution of an Areal Array

Specifying the orientation for an areal array is similar to that for a linear array discussed in Section 5.4.2.3. Again we have an inverse problem and again there are periodicities, which rule out a solution in terms of a continuous transform and instead require a solution in terms of a series of orthogonal functions. In Section 5.4.2.3 we expanded our angular distributions of linear elements in a complex one-dimensional Fourier series. This approach was computationally efficient because it can be implemented with the Fast Fourier Transform algorithm. In principle, the same approach could be used here substituting a two-dimensional Fourier series. However, the historical development of this analysis by Philofsky and Hilliard [1969] was based on spherical harmonics. These are solutions of Laplace's equation in spherical coordinates -- a fact that is not essential to the orientation distribution problem unless perchance the features themselves happen to have been created by some potential field. The reader may ponder this significance. Although the choice of spherical harmonics may seem awkward, they are not as inefficient as one might imagine since recently a fast spherical harmonic transform has been developed along the lines of the FFT [Mohlenkamp, 1999]. And, because the longitudinal variations of functions often tend to fade out at the poles, there is realism to representations using spherical harmonics. By contrast, a two-dimensional Fourier transform might without special restraint would tend to become unrealistically "busy" near the poles. Also, working in spherical harmonics is good exercise.

We can specify the orientation distribution of areal features in three dimensions by the density function $S_V(\psi,\omega)$ defined such that $S_V(\psi,\omega)\sin\psi\, d\psi\, d\omega$ is the total area per unit volume having normals with orientations in the range $\psi \pm d\psi/2$ and $\omega \pm d\omega/2$. The factor of $\sin\psi$ $d\psi$ $d\omega$ is a constant differential of solid angle or, if you prefer, a constant differential area on the unit sphere. $S_V(\psi,\omega)$ so defined will be a constant for a "random" array. We will be considering the interaction of the areal array with a lineal array having the distribution $L_V(\phi,\theta)$ oriented as shown in Fig. 5.13 and defined in a similar manner to $S_V(\psi,\omega)$. The directional sense of a line or areal element will be assumed to be indeterminate, viz. pointing in both directions at once; thus, $0 \le \theta \le 2\pi$ and $0 \le \phi \le \pi/2$. The same relations hold analogously for ψ and ω.

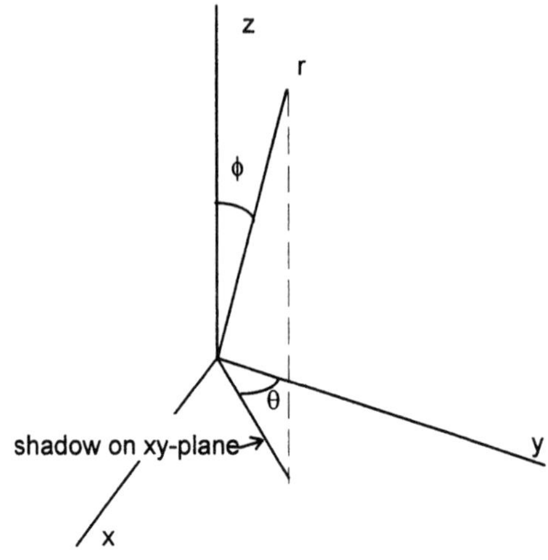

Figure 5.13: Definition of spherical coordinates.

It follows from the definitions that the total area per unit volume is:

$$S_V = \int_0^{2\pi}\int_0^{\pi/2} S_V(\psi,\omega)\sin\psi\, d\psi\, d\omega, \tag{5.60}$$

and the total line length per unit volume is:

$$L_V = \int_0^{2\pi}\int_0^{\pi/2} L_V(\phi,\theta)\sin(\phi)d\phi\, d\theta\ . \tag{5.61}$$

Consider an experiment in which a lineal and areal array are superimposed on one another. They will interact to produce point intersections. The first step is to derive an expression for the expected number, $\bar{P}_V$, of intersections per unit volume generated when the two arrays are superimposed randomly with respect to translation but with their coordinate axes parallel to one another.

Without any loss in generality we can assume that the line length in the range $\phi \pm \left(\frac{d\phi}{2}\right)$ and $\theta \pm \left(\frac{d\theta}{2}\right)$ occurs as a single segment of length $L_V(\phi,\theta)\sin(\phi)d\phi d\theta$ in a unit volume. We can similarly consider an element of area $S_V(\psi,\omega)\sin\psi\, d\psi\, d\omega$ in the areal array. For the line segment to intersect the areal element, its center must fall within the solid figure generated by the line segment as its center is traversed around the circumference of the areal element, Fig. 5.14.

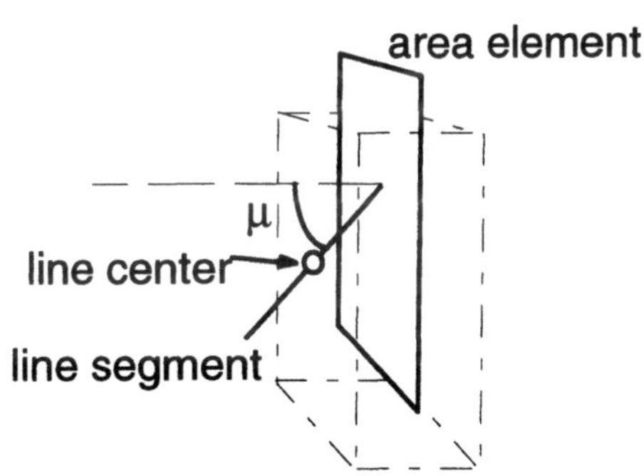

Figure 5.14: Interaction geometry.

Therefore, if the line segment is randomly located within a unit volume, the probability of it intersecting the areal element is just equal to the volume of the solid figure; namely,

$$L_V(\phi,\theta)S_V(\psi,\omega)\sin(\phi)\sin(\psi)|\cos(\mu)|d\phi d\theta d\psi d\omega,$$

in which μ is the angle between the line segment and the normal to the areal element and is given by:

$$\cos\mu = \sin(\phi)\sin(\psi)\cos(\theta-\omega)+\cos(\phi)\cos(\psi). \tag{5.62}$$

The expected total number of intersections per unit volume is found by summing over all orientations in the two arrays. Thus:

$$P_V = \int_0^{\pi/2} d\phi \int_0^{2\pi} d\theta \int_0^{\pi/2} d\psi \int_0^{2\pi} L_V(\phi,\theta)S_V(\psi,\omega)\sin\phi\sin\psi|\cos\mu|d\omega. \tag{5.63}$$

As with the linear array orientation problem, we will expand L_V and S_V in a way that we can solve for S_V algebraically given L_V. As already stated, we will expand these in spherical harmonics. Since spherical harmonics are composed of the product of an associated Legendre function and the exponential of an imaginary argument, the first step in the solution to this equation is to expand $|\cos\mu|$ in a series of Legendre polynomials.

Legendre polynomials are orthogonal with a unit weighting function, viz. when integrating, they do not need to be multiplied by an additional function as do some other orthogonal polynomials, Jacobi's for instance, in order to behave orthogonally. This orthogonality is expressed by

$$\int_{-1}^{1} P_m(x)P_n(x)\,dx = 0 \text{ if } m \neq n \quad \text{and}$$

$$\int_{-1}^{1} [P_n(x)]^2 dx = \frac{2}{2n+1}.$$

While Legendre polynomials are solutions of Legendre's differential equation, they are more conveniently defined by *Rodrigues's formula*,

$$P_n(x) = \frac{1}{2^n n!}\frac{d^n}{dx^n}\left(x^2-1\right)^n.$$

From this formula we can deduce that $P_n(x)$ is even or odd as n is even or odd. Usually x is restricted to the interval [-1,1], and if n is even, the integral over the full interval is simply twice that over the half interval [0,1]. Furthermore $P_n(x)$ is a polynomial of degree n. Examining Rodrigues's formula we see the product, $(2^n n!)$. We will be needing this again. Notice that it equals the product 2n (2n-2) (2n-4) (2n-6) etc. This is like a factorial but terms decrease by increments of two rather than one. This type of function is sometimes called *double factorial.* In double factorial notation, $2^n n! = (2n)!!$; two exclamation points are used. At this time it is also important to note the effect of symmetry on the coefficient of the $P_n()$ term in an expansion. It turns out that the Legendre factor in a spherical harmonic expresses the dependence on the co-latitude, ψ or ϕ as used here. Because the sense of line and area elements are indeterminate, functions of their co-latitudes are even and therefore expressible in terms of only even indices for their Legendre factors. By the same token, the absolute value of the cosine is an even function expressible with only even Legendre polynomials. By using the orthogonality relationship, the absolute value of the cosine can be written as a series of even Legendre polynomials:

$$|\cos\mu| = \sum_{n=0}^{\infty} \alpha_{2n} P_{2n}(\cos\mu). \tag{5.64}$$

To find the coefficients, α_{2n}, we write

$$\alpha_{2n} = \frac{4n+1}{2}\int_{-1}^{1} |v| P_{2n}(v) dv,$$

where $v = \cos(x)$.

Since the integrand is itself even,

$$\alpha_{2n} = (4n+1)\int_0^1 v P_{2n}(v) dv, \tag{5.65}$$

which yields:

$$\alpha_0 = 1/2, \ \alpha_1 = 5/8 \text{ and}$$

$$\alpha_{2n} = (-1)^{n+1}(4n+1)\frac{(2n-3)!!}{(2n+3)!!}. \tag{5.66}$$

Legendre polynomials are discussed in many books on differential equations, astrophysics and quantum mechanics. They have a long and interesting history. Some references given at the end of this chapter are also mentioned below.

We next expand $L_V(\phi,\theta)$ in a series of spherical harmonics. At this point it is probably helpful to look at the definition of a spherical harmonic as it stands today. The definition has undergone changes over the years due to the increased popularity of complex variables especially in quantum mechanics. Hence a good reference is the introductory quantum mechanics text by Liboff. [1993] The definitions for spherical harmonics and associated Legendre functions given in Liboff are consistent with those used in the popular software package, *Mathematica™*. Except for the choice of variable names, they are the same as used here. Here, θ and ϕ, have been reversed in their roles because it is easier to remember and more natural to have the line through the circle represent the axis associated with the variable. Thus ϕ represents the co-latitude since that is taken with respect to the polar axis and θ represents the longitude that associates with the equator.

$$Y_l^m(\phi,\theta)=\sqrt{\frac{(2l+1)}{4\pi}\frac{(l-m)!}{(l+m)!}}P_l^m(\cos(\phi))e^{im\theta} \tag{5.67}$$

The spherical harmonic defined by Eqn. (5.67) is the product of a numerical factor times an associated Legendre function times the exponential of an imaginary function i.e. a Fourier factor. Spherical harmonics are separable with respect to ϕ and θ. The associated Legendre function is defined by a formula similar to Rodrigues's formula,

$$P_l^m(x)=(-1)^m(1-x^2)^{\frac{m}{2}}\frac{d^m}{dx^m}P_l(x)\,. \tag{5.68}$$

It follows that $P_l^0(x)=P_l(x)$. Associated Legendre functions are solutions of a differential equation. But that is not of primary interest here.

One property of associated Legendre functions which is important here is this,

$$\sqrt{\frac{(l-m)!}{(l+m)!}}P_l^m(x)=\sqrt{\frac{(l-|m|)!}{(l+|m|)!}}P_l^{|m|}(x). \tag{5.69}$$

Although the associated Legendre function for negative m does not equal that for positive m, changing the sign of m in the factorial compensates the change in sign in the associated Legendre function. Consequently, when associated Legendre functions appear in spherical harmonics, changing the sign of m has no effect other than that on the factor $e^{im\theta}$. This allows a complex Fourier series to be constructed to represent the longitudinal variation without affecting those that depend on co-latitude. Earlier versions of spherical harmonics did not use this strategy. For example, see MacRobert [1947]. The expression in Eqn. (5.69) leads us to the orthonormality relationship for spherical harmonics:

$$(-1)^m \int_0^{2\pi} \int_{-\frac{\pi}{2}}^{\frac{\pi}{2}} Y_l^m(\phi,\theta) Y_{l'}^{-m'}(\phi,\theta) \sin(\phi) d\phi\, d\theta = \begin{cases} 1 & l = l', m = m' \\ 0 & otherwise \end{cases}. \tag{5.70}$$

In writing Eqn.(5.70), authors often compress the expression taking advantage of the fact that

$$\left[Y_l^m(\phi,\theta)\right]^* = (-1)^m Y_l^{-m}(\phi,\theta),$$

where the star indicates complex conjugation.

Proceeding then,

$$L_V(\phi,\theta) = \sum_{n=0}^{\infty} \sum_{m=-2n}^{2n} \lambda_{2nm} Y_{2n}^m(\phi,\theta), \tag{5.71}$$

in which

$$\lambda_{2nm} = 2(-1)^m \int_0^{\frac{\pi}{2}} \int_0^{2\pi} Y_{2n}^{-m}(\phi,\theta) L_V(\phi,\theta) \sin(\phi) d\theta d\phi. \tag{5.72}$$

Equation (5.72) takes advantage of the fact that spherical harmonics are orthogonal on the half-interval $\phi \in [0,\pi/2]$ when the sum of all the indices is an even number. The function $S_V(\psi,\omega)$ can similarly be expanded in series of spherical harmonics having coefficients $\sigma_{2n\,m}$.

Substituting these series in Eq. (5.63) we obtain:

$$P_V = \int_0^{\pi/2} d\phi \int_0^{2\pi} d\theta \int_0^{\pi/2} d\psi \ldots$$

$$\left\{ \sum_{n_1=0}^{\infty} \sum_{m_1=-2n_1}^{2n_1} \lambda_{2n_1 m_1} Y_{2n_1}^{m_1}(\phi,\theta) \sin\phi \right\} \ldots$$

$$\int_0^{2\pi} \left\{ \sum_{n_2=0}^{\infty} \sum_{m_2=-2n_2}^{2n_2} \sigma_{2n_2 m_2} Y_{2n_2}^{m_2}(\psi,\omega) \sin\psi \right\} \ldots \tag{5.73}$$

$$\sum_{n=0}^{\infty} [\alpha_{2n_3} P_{2n_3}(\cos\mu)] \, d\omega$$

The Legendre polynomial containing μ , $P_{2n}(\cos\mu)$, is complicated on account of μ being complicated; but, it has been studied in detail and is called a Laplacian coefficient. Laplacian coefficients appear in Laplace's theory of spherical potentials, which had its origin in astrophysics. An elementary discussion of them is given in *Mathematical Physics* by Donald Menzel, a former professor of astrophysics at Harvard. [Menzel, 1961] This vade mecum for beginning physics students is still in the Dover catalog for the year 2001.

Because the functions contained in Eqn.(5.73) are all solutions to differential equations, we may presume uniform continuity and change the order of integration at will. Furthermore they are all separable in their variables. So, we can divide Eqn.(5.73) into pieces. The first piece is the integral,

$$I = \int_0^{\frac{\pi}{2}} \int_0^{2\pi} Y_{n_1}^{m_1}(\phi,\theta) \sin(\phi) P_{n_3}(\cos(\mu)) \, d\theta d\phi \, . \tag{5.74}$$

We can expand the Laplacian coefficient in terms of what is sometimes called the "addition theorem for general zonal harmonics",

$$P_l(\mu)=\sum_{m=0}^{l}\frac{2}{a_m}\frac{(l-m)!}{(l+m)!}P_l^m(\cos(\phi))P_l^m(\cos(\psi))\ldots$$
$$\left[\cos(m\theta)\cos(m\omega)+\sin(m\theta)\sin(m\omega)\right] \quad (5.75)$$

In this expression a_m takes the value of 2 when $m = 0$ and 1 for all other values of m. Evaluation of Eqn.(5.74) requires substituting Eqn.(5.67) the definition of a spherical harmonic and Eqn.(5.75) into Eqn.(5.74). Integration with respect to θ sets $m = |m_1|$ and eliminates a_m. Integrating with respect to ϕ sets $n_3 = n_1$.

$$I=\frac{2\pi}{4n_1+1}Y_{2n_1}^{m_1}(\psi,\omega)\ . \quad (5.76)$$

The index m_1 takes the full range of integer values from $-2n_1$ to $2n_1$ including zero. We see that the variables ϕ and θ have been integrated out. Equation (5.73) is now reduced to

$$P_V=\int_0^{\frac{\pi}{2}}\int_0^{2\pi}\frac{2\pi}{4n_1+1}\left[\sum_{n_1=0}^{\infty}\sum_{m_1=2n_1}^{2n_1}\alpha_{2n_1}\lambda_{2n_1m_1}Y_{2n_1}^{m_1}(\psi,\omega)\right]\ldots$$
$$\left[\sum_{n_2=0}^{\infty}\sum_{m_2=-2n_2}^{2n_2}\sigma_{2n_2m_2}Y_{2n_2}^{m_2}(\psi,\omega)\right]\sin(\psi)d\omega d\psi \quad (5.77)$$

The remaining cross-term products vanish on integration of Eqn.(5.77) because of the orthogonality of the spherical harmonics. Hence expanding α_{2n}:

$$P_V=\sum_{n=0}^{\infty}\sum_{m=-2n}^{2n}(-1)^{n+m+1}\left[\frac{\pi(2n-3)!!}{(2n+2)!!}\right]\lambda_{2nm}\sigma_{2n(-m)}\ . \quad (5.78)$$

Notice that the second index of σ is a negative m. This results from the orthonormality relationship Eqn.(5.70). Because of the summation, the choice of whether the negative is applied to the second index of λ or σ is arbitrary.

Equation (5.78) will yield the orientation distribution of an areal array using a lineal array of known orientation distribution. Unlike Eqn.(5.63) it can be inverted giving σ coefficients as functions of the λ coefficients. By superimposing a suitable lineal array on the areal array at a sufficient number of different translations and counting the number of intersections produced, it is theoretically possible to determine by use of Eqn.(5.78) the coefficients of the areal array up to any required degree. It is apparent, however, that a considerable amount of experimental and computational work will be necessary if a high degree of resolution is required in the estimation of the unknown orientation distribution. Even with computed tomography, most three-dimensional distributions are explored through two-dimensional sections since these are easy for humans to view.

In Chapter 1 we discussed some mathematical operations, which can be performed on digitized images. In two-dimensional terms an intersection between a probe line in the plane of the image and a feature plane cutting the plane of the image can be detected by anding the pixels of the probe line with those of the feature plane. The result is a single non-zero pixel or point for each intersection. Counting the points then gives the number of intersections. Similar processes based on intersection counting are used for efficient character recognition [Mori et al., 1999]. Fortunately in practice, certain symmetries are often present which greatly simplify the analysis. We will now consider two such cases.

Processes such as forging and rolling on a sample containing internal boundaries can be expected to produce an orientation distribution with a two-fold axis of symmetry and three perpendicular mirror planes. In crystallography this symmetry is described by the notation (mmm). If the ψ axis is set parallel to the intersection of two of the mirror planes, then:

$$S_V(\psi,\omega) = S_V[\psi,(\pi-\omega)] = S_V[\psi,(\pi+\omega)] = S_V[\psi,(2\pi-\omega)].$$

Under this condition the coefficients σ_{2nm} vanish for odd values of m and $\sigma_{2nm} = \sigma_{2n(-m)}$ for all values of m. Thus the number of coefficients required to describe the orientation distribution is reduced by a factor of four.

Deformation by wire drawing, extrusion etc. results in an areal array having an axis of symmetry. Such symmetry is also often found elsewhere in nature. If the ψ axis is chosen coincident with the axis of symmetry, the orientation distribution will be independent of ω. Thus:

$$S_V(\psi,\omega) = \frac{S_V(\psi)}{2\pi}.$$

Under this condition, only zonal harmonics are present. Figure 5.15 shows the different types of spherical harmonics.

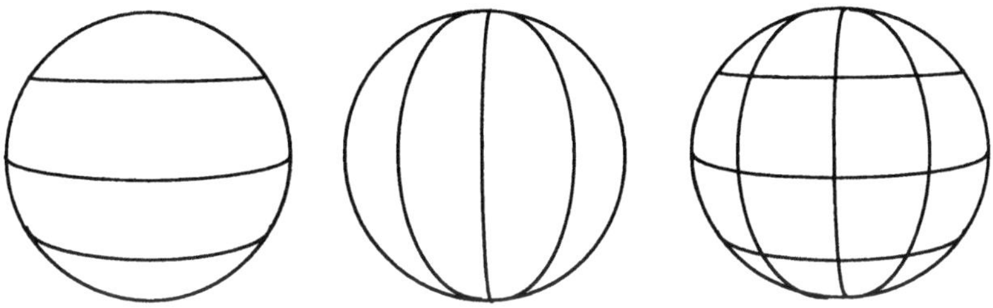

Figure 5.15: Three designations of spherical harmonics showing the lines of their zeros.

Zonal harmonics take the form $Y_l^0(\psi,\omega) = f_n P_l(\psi)$, where f_n is a numerical factor. Sectoral harmonics take the form $Y_l^l(\psi,\omega)$, and tesseral harmonics are of the form $Y_l^m(\psi,\omega)$. Note that $P_l^l(\ldots)$ has no zeros on the open interval (-1,1).

Although Eqn. (5.63) cannot readily be simplified (since the variable μ cannot be easily separated and keeps the problem three dimensional), the associated Legendre functions reduce to Legendre polynomials. For example, $L_V(\phi)$ is represented,

$$L_V(\phi) = \sum_{n=0}^{\infty} \lambda_{2n} P_{2n}(\cos(\phi)). \tag{5.79}$$

Equation (5.78) reduces to:

$$P_V = \sum_{n=0}^{\infty} \frac{(-1)^{n+1}}{4n+1} \frac{(2n-3)!!}{(2n+2)!!} \lambda_{2n} \sigma_{2n}, \tag{5.80}$$

in which:

$$\lambda_{2n} = (4n+1) \int_0^{\frac{\pi}{2}} L_V(\phi) P_{2n}(\cos\phi) \sin(\phi) d\phi,$$

and σ_{2n} is similarly defined with respect to the function $S_V(\phi,\omega)$. The unknown coefficients σ_{2n} can be determined from an application of a two-dimensional lineal array to a single section taken through the areal array parallel to the axis of symmetry.

The problem of evaluating the coefficients for the spherical harmonic expansion in the general three-dimensional case is a knotty one. However, as in the two-dimensional case, it is possible to derive test figures for which the intercept density will be directly proportional to a given Legendre polynomial coefficient of the areal array. We may begin by considering Eqn.(5.79) for the case where there is but one non-zero coefficient λ_{2n}^* , where the star signifies that it's alone. As a first thought we might try to relate λ_{2n}^* to total absolute line length per unit area of the test pattern L_V^{2n} by integrating Eqn.(5.79) for the single term. Except for $n = 0$, the result would be zero, just as would happen with an ordinary Fourier series. From this we can deduce that the test pattern should be composed of signed sections. Since $P_n(x)$ has n zeros on the interval [-1,1] and $P_{2n}(x)$ has n zeros on the interval [0,1], we expect there to be a multiple of n+1 signed segments. Defining λ_{2n}^* positive and taking the integral of the absolute

value, $\lambda_{2n}^* = L_V^{2n} / \int_0^1 |P_{2n}(x)| dx = L_V^{2n} / T_{2n}$, where we hope T_{2n} will not be confused with the old notation for Ferrer's associated Legendre function. (Ferrer's functions have supplanted an earlier definition of the associated Legendre functions and are used here as *the* associated Legendre functions.) If this test figure were applied to a section parallel to the axis of symmetry, then by Eqn. (5.80) the number of intersections for that 2n-test figure would be:

$$P_V^{2n} = \frac{(-1)^{n+1}(2n-3)!!L_V^{2n}}{(4n+1)(2n+3)!!T_{2n}} \sigma_{2n}$$

which can be rewritten:

$$\overline{\sigma}_{2n} = \frac{(-1)^{n+1}(4n+1)(2n+2)!!T_{2n}}{(2n-3)!!} \frac{P_V^{2n}}{L_V^{2n}} , \tag{5.81}$$

in which $\overline{\sigma}_{2n}$ is the expectation of the value of the coefficient and P_V^{2n} is the number of intersections per unit length of the 2n-test figure. It follows from Eqn.(5.79) that the lineal test figure for determining the 2nth coefficient will have an orientation distribution given by:

$$L_V(\phi) = \frac{L_V^{2n}}{T_{2n}} P_{2k}(\cos\phi) .$$

In order to construct the test pattern we consider the following geometry, Fig. 5.16,

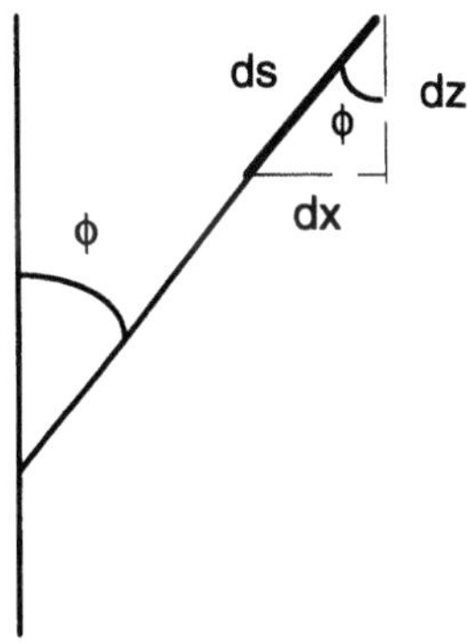

Figure 5.16: Differential geometry of parametric drawing

in which we identify the polar axis with the z-axis and work in the xz-plane. Traversing a differential line segment ds gives rise to the motions dx and dz. Given λ an arbitrary parameter, if we set $ds = L_V(\phi)\sin(\phi)\, d\lambda$, we can draw a curve $s(\lambda)$ having distribution $L_V(\phi)$ so long as λ insures coverage of the domain $[0,\pi/2]$. By setting λ equal to the co-latitude (ϕ is also a co-latitude but here functions as a dummy.), the test figures can be constructed from the following pair of parametric equations:

$$x(\lambda) = \int_0^{\lambda} \frac{L_V^{2n}}{T_{2n}} \left|P_{2k}(\cos\phi)\right| \sin^2(\phi)\, d\phi$$
$$z(\lambda) = \int_0^{\lambda} \frac{L_V^{2n}}{T_{2n}} \left|P_{2k}(\cos\phi)\right| \sin(\phi)\cos(\phi)\, d\phi \quad , \qquad (5.82\text{ a,b})$$

where the sign of the line is given by $\mathrm{sign}\left[P_{2n}(\cos(\lambda))\right]$ and $0 \le \lambda \le \pi/2$.

As an example, the figure for determining λ_6 is given in Fig. 5.17a,b. It should be noted that these figures have properties identical to those of the Fourier figures with respect to signs on the individual arcs, translation of the arcs, and orthogonality. For the reader's convenience some values of T_{2n} is given as Table 5.3.

Table 5.3: Total Length Ratios

2n	T_{2n}
0	1
2	0.3849
4	0.2866
6	0.2385
8	0.2086
10	0.1877
12	0.1721
14	0.1598
16	0.1498
18	0.1415
20	0.1344

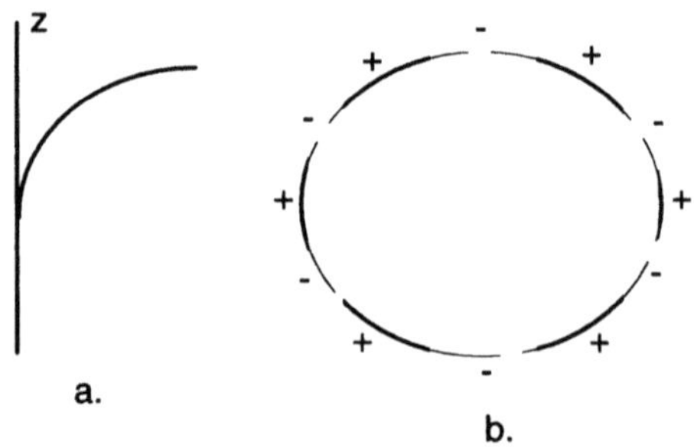

Figure 5.17a,b: Sketch of test figure for λ_6

Fig. 5.17a shows the plot generated by Eqns.(5.82a,b) for $0\leq\lambda\leq\pi/2$. Figure 5.17b shows a closed signed rosette test pattern made from 5.17a by exploiting the axisymmetry. (The true parametric curve is difficult to reproduce and slightly more football-shaped than that here.)

The foregoing treatment is based on an analysis given by Philofsky and Hilliard [1969]. The special cases when the lineal array consists of parallel lines or the areal array parallel planes were considered earlier by Hilliard [1962].

4.4.4 Expected number of Intersections Generated by a Lineal and an Areal Array

Consider an element of area δS and a line segment of length δL inside a bounding body of volume V. Then, following the same argument as used in the derivation of Eqn.(5.63), it is found that the probability of the line segment intersecting the areal element is given by:

$$p = \delta S \, \delta L \left|\cos(\mu)\right| / V \, , \tag{5.83}$$

in which μ is the angle between line segment and the normal to the areal element. The value of p averaged over all values of μ is:

$$\bar{p} = (\delta S \delta L / V) \int_0^{\pi} \left|\cos\mu\right| f(\mu) d\mu \, . \tag{5.84}$$

From Eqn.(4.17) we have for a random distribution, $f(\mu) = (1/2)\sin\mu$ which, on substitution in Eqn.(5.84) and integration yields:

$$\bar{p} = \frac{\left(\frac{1}{2}\right)\delta S \, \delta L}{V} \, .$$

Integrating over all line and areal elements we obtain for the expected number of intersections:

$$\bar{P} = \frac{\left(\frac{1}{2}\right) S L}{V}$$

in which S is the total area and L the total line length. Finally, dividing through by V we obtain for the expected number of intersections per unit volume:

$$\bar{P}_V = \frac{S_V L_V}{2} \, . \tag{5.85}$$

Since $\bar{P}_L = \dfrac{\bar{P}_V}{L_V}$, this expression also yields,

$$\bar{P}_L = \frac{S_V}{2} \, , \tag{5.86}$$

in agreement with the result, Eqn.(5.59), obtained by another method. The advantage of the present derivation is that it holds for linear probes of any shape.

5. THE PLANAR PROBE IN THREE DIMENSIONS

In this section we will for the sake of brevity abbreviate "mean caliper diameter" to "diameter" D or D_i. This announcement is bootstrapping of a sort since "mean caliper diameter" is not yet well defined. Its definition will emerge through the derivation given in Section 5.5.1.1 and be discussed in Section 5.6. Although the results given involving the mean caliper diameter are "shape independent" in a sense, the mean caliper diameter is intimately involved with the shape of the body belonging to it. Hence, much of the discussion of it will be deferred to the chapter on shape-dependent relationships.

5.1 The Planar Probe in Three Dimensions, Three-dimensional Bodies

$I_2\, P_2\, F_3\, N_3$

This configuration is the bread and butter of most practical stereology. Some object such a piece of metal from the *Titanic* or a pickle is carved into "random" sections. The subject of how best to do the carving (showing the pickle) has been reviewed by H. Gundersen et al. [1988] who discuss the "object orientator" a mechanical device for physically replicating the conditions used in the derivation of the probability density function for random slices, Eqn. (4.29) for three-dimensional objects. From a practical standpoint, handling convenience tends to result in correlated rather than truly random slices. But, sometimes a sequence of highly correlated slices is an advantage. These have been used in "fractionator" methods to be discussed later. A set of correlated slices may be regarded to be a new type of probe in a sense, but here will be regarded as an extension of the planar probe.

5.1.1 Probability of a Planar Probe Intercepting a Convex Body

Consider a convex body of diameter D_i inside a bounding body of diameter D. For convenience, place the origin inside (one of the) D_i. The probability, p, that a planar probe will intercept the interior body is:

$$p=\int_0^{2\pi}\int_0^{\pi}\int_{(\rho)}\varepsilon_{D_i}(\rho,\phi,\theta)f(\rho,\phi,\theta)d\rho d\phi d\theta,$$

following the notation conventions developed in the previous section (see Fig. 4.10) in which integration is only over those values of ρ which yield an interception with the bounding body and $\varepsilon_{D_i}(\rho,\phi,\theta)$ is the indicator function introduced in Section 5.3.1 (and also discussed in the Appendix as Eqn.(A.14)). The indicator function takes the value of unity when the plane intersects the set of points comprising the interior body, associated with D_i. Using the probability density function Eqn.(4.29) we may rewrite the above integral as $p=\int_0^{2\pi}\int_0^{\pi}\int_{(\rho)}\varepsilon_{D_i}(\rho,\phi,\theta)\frac{\sin\phi}{2\pi D}d\rho d\phi d\theta$. Since the origin is inside the interior body, we may restrict ρ to values for which an interception with the interior body always occurs. The probability of interception in this case is unity and

$$1=\int_0^{2\pi}\int_0^{\pi}\int_{(\rho)}\varepsilon_{D_i}(\rho,\phi,\theta)\frac{\sin\phi}{2\pi D_i}d\rho d\phi d\theta .$$

Multiplying through by $\frac{D_i}{D}$ recovers the original integral for p. Thus

$$p=\frac{D_i}{D}. \tag{5.87}$$

From the derivation we would expect that Eqn.(5.87) would apply only to convex bodies. So, since an areal element has no 3-D Lebesgue content, even if it is convex, it is not a convex body; and, Eqn.(5.87) ought not to apply. However, Eqn. (5.87) applies more generally. It is possible to define a mean caliper diameter for an areal element. This allows calculation of the probability of a planar probe hitting a particular surface element by tiling a surface with such elements. [Weibel, 1979B]. In this case $D=(\pi/4)d$ where d is the two-dimensional mean caliper diameter. Further clarification of this point will be made in Section 5.6.2.

5.1.2 Average Intercept Area in a Convex Body

Let $A(\rho, \phi, \theta)$ be the area of interception of a planar probe at (ρ, ϕ, θ) with its own bounding body of mean caliper diameter D. The average area of interception is then:

$$\overline{A} = \int_0^{2\pi}\int_0^{\pi}\int_{0(\rho)} A(\rho,\phi,\theta) f(\rho,\phi,\theta) d\rho d\phi d\theta$$

$$= \frac{1}{2\pi D}\int_0^{2\pi}\int_0^{\pi}\int_{0(\rho)} A(\rho,\phi,\theta)\sin\theta d\rho d\phi d\theta \quad .$$

We can change the integral to

$$= \frac{1}{2\pi D}\left[\int_0^{\pi}\sin\theta\int_0^{\pi}\int_{0(\rho)} A(\rho,\phi,\theta) d\rho d\phi d\theta + \int_0^{\pi}\sin\theta\int_0^{\pi}\int_{0(\rho)} A(\rho,\pi-\phi,\pi+\theta) d\rho d\phi d\theta\right]$$

in which the range of θ has been cut in half. Call the first half V_1 and the second V_2. Figure 5.18 clarifies how this is done.

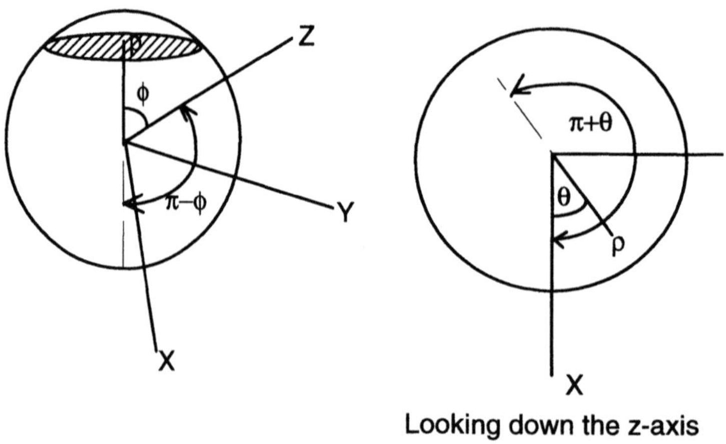

Figure 5.18: Change of angles constitutes negative extension of ρ.

Figure 5.18 is a simplified adaptation of the Fig. 4.10. For any pair of values for ϕ and θ the integration with respect to ρ gives the volumes of the two halves of the body, V_1 and V_2. Thus,

$$\overline{A} = \frac{1}{2\pi D}\left[\int_0^{\pi}\sin\theta\int_0^{\pi} V_1 d\phi d\theta + \int_0^{\pi}\sin\theta\int_0^{\pi} V_2 d\phi d\theta\right] = \frac{V}{2\pi D}\int_0^{\pi}\sin\theta\int_0^{\pi} d\phi d\theta \ .$$

From which it follows that

$$\overline{A} = \frac{V}{2\pi D} 2\pi = \frac{V}{D}. \qquad (5.88)$$

5.1.3 Average Intercept Area in an Aggregate of Convex Bodies

Allowing for the probability of an interception given by Eqn.(5.87) we obtain from Eqn.(5.88) for an aggregate of figures,

$$\overline{A} = \frac{\sum V_i}{\sum D_i}, \qquad (5.89)$$

where V_i now designates the volume of the i-th body in the aggregate and D_i now designates the mean caliper diameter of the i-th body in the aggregate. Dividing both numerator and denominator by the number of bodies gives

$$\overline{A} = \frac{\overline{V}}{\overline{D}}. \qquad (5.90)$$

5.1.4 Density of Interceptions in a Aggregate of Convex Bodies

The density of interceptions is here defined as the expected number of them per unit area of probe plane. Consider again a collection of convex bodies having mean caliper diameters D_i located as before in a bounding body having mean caliper diameter D. If the probability of intercepting a particular body is D_i/D as given by Eqn. (5.87), then the expected number of interceptions on a planar probe is $\overline{N} = \frac{\sum D_i}{D} = \frac{M\overline{D}}{D}$ where M is the number of bodies within the bounding body and $\overline{D}$ is their average mean caliper diameter. Dividing through by the expected area of interception with the bounding body given by Eqn. (5.88) we obtain $\frac{\overline{N}}{\overline{A}} = \frac{M\overline{D}}{V}$. This may be rewritten,

$$N_A = N_V \overline{D}, \qquad (5.91)$$

where N_V is the number of bodies per unit volume.

5.1.5 Relationship of Areal to Volume Fraction

Consider an aggregate of bodies inside a bounding body. (It is *not* assumed that either the bounding or interior bodies are convex.) We again introduce a planar probe of fixed orientation (ϕ,θ) and define its location by its perpendicular distance ρ from a parallel tangent plane (see Fig. 4.10). For a random location of the probe, the distribution of ρ by itself is evidently:

$$f(\rho)=\frac{1}{D(\phi,\theta)}, \tag{5.92}$$

in which $D(\phi,\theta)$ is the distance between tangent planes of orientation (ϕ,θ) to the bounding body. Let $A_i(\rho)$ be the total area of intersection of the interior bodies on the planar probe located at distance ρ. Then the average area of interception is:

$$\bar{A}_i = \int_{(\rho)} A_i(\rho) f(\rho) d\rho = \left[\frac{1}{D(\phi,\theta)}\right] \int_{(\rho)} A_i(\rho) d\rho \ .$$

However, this integral is just equal to the total volume $\sum V_i$ of the interior bodies (cf. Section 5.5.13). In deriving Eqn. (5.88) we found that the average area of intersection $\bar{A}$ with the bounding body of volume V was V/D. It remains to show that $\bar{A}$ is $V/D(\phi,\theta)$. This is actually quite a bit simpler than deriving Eqn (5.88). All we have to do is replace A_i with A in the above integral. The result satisfies our intuitive notion that $\bar{A}\cdot D(\phi,\theta)=V$ and underscores the point that here $\bar{A}=\bar{A}(\phi,\theta)$.

Thus for any given orientation (ϕ,θ):

$$\frac{\bar{A}_i}{\bar{A}}=\frac{\sum V_i}{V}$$

or

$$\bar{A}_A = V_V \,, \tag{5.93}$$

since orientation "divides out". It should be noted that this relationship holds for bodies of any shape; also it is not assumed that the probe is randomized with respect to orientation.

5.2 The Planar Probe in Three Dimensions, Areal Features

$I_1\ P_2\ F_2\ N_3$

Areal features in a three-dimensional solid are its surfaces. They include the inter-grain boundary surfaces of grains in a metal and the free surfaces of mitochondria in an animal cell.

5.2.1 Expected Length of Interception of a Planar Probe with its Bounding Body

The expected length C of interception of a planar probe with its convex bounding body is defined as the average circumference of the two-dimensional figure generated by a random section through the body. Consider a small differential element of area δS on the surface of the bounding body (Fig. 5.19). Without any loss in generality we can assume that this element is normal to the z-axis. Let $\delta C(\rho,\phi,\theta)$ be the length of interception of a probe at (ρ,ϕ,θ) with the element δS .

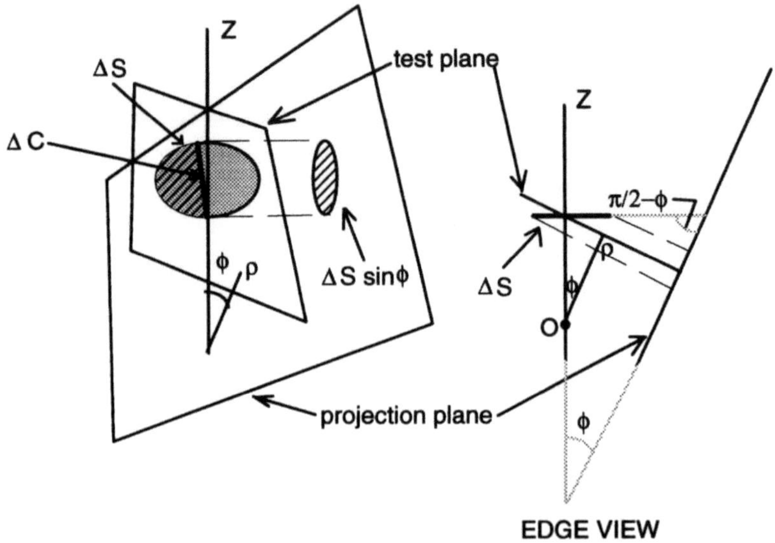

Figure 5.19: Interception Geometry

The average length of interception on δS is:

$$\delta \bar{C} = \left(\frac{1}{2\pi D}\right) \int_0^{2\pi} \int_0^{\pi/2} \int_{(\rho)} \delta C(\rho, \phi, \theta) \sin(\phi)\, d\rho\, d\phi\, d\theta\,, \tag{5.94}$$

in which the upper limit on ϕ has been reduced to $\pi/2$ because probes in the range $\pi > \phi > \pi/2$ will not intercept the surface element. Since this element is sufficiently small that its curvature can be neglected, lines of intersection will be parallel, and integration with respect to ρ over the range in which an intercept occurs yields the area of δS projected onto a plane that is normal to the probe and parallel to the line of interception. (See the edge view in Fig. 5.19.) This projected area is the integral of δC with respect to ρ and is equal to

$$\delta S \cos\left(\frac{\pi}{2} - \phi\right) = \delta S \sin(\phi)\,.$$

Therefore: $\delta \bar{C} = \left(\frac{1}{2\pi D}\right) \int_0^{2\pi} \int_0^{\pi/2} \delta S \sin^2(\phi)\, d\phi\, d\theta\,,$

or:

$$\delta\bar{C}=\left(\frac{\pi}{4}\right)\left(\frac{\delta S}{D}\right). \tag{5.95}$$

Since Eqn.(5.95) is independent of (ρ,ϕ,θ), we may sum over all surface elements and find:

$$\bar{C}=\left(\frac{\pi}{4}\right)\left(\frac{S}{D}\right) \tag{5.96}$$

Example:

The above derivation may not be totally convincing since it is difficult to see how the individual differential length elements sum to a circumference. While a very complicated drawing might make it clear, this drawing might be unreadable on a page and is therefore best drawn in the mind where the bounds of Euclidean space do not apply to labels, arrows and explanatory notes. For further reassurance, here is an example. Consider a sphere with the origin placed at its center. Symmetry reduces the probability density function to $f(\rho,\phi,\theta)$ to simply $f(\rho)=1/D$ where ρ is allowed to take both positive and negative values. Then,

$$\overline{C}=\frac{2\pi}{D}\int_{-D/2}^{D/2}\sqrt{(D/2)^2-\rho^2}\,d\rho=\frac{\pi^2}{4}D.$$

But, by Eqn. (5.96), $\overline{C}=\frac{\pi S}{4D}$. This equals $\frac{\pi}{4}\cdot\frac{4\pi(D^2/4)}{D}=\frac{\pi^2}{4}D$ as expected since D is the mean caliper diameter of a sphere of diameter D.

5.2.2 Average Length of Interceptions on an Aggregate of Convex Bodies

We may denote the act of "hitting" with the arrow $\uparrow$. The *hitting probability* $\Pr(A\uparrow B)$ is the probability $\Pr(A\cap B\neq\varnothing)$, that the intersection of A and B is not the null set. We designate by D or D_i their associated point sets as well as their measures as the context requires. We will use these conventions to work out the associated probability functions for averaging C over an aggregate.

To begin, $\overline{\overline{C}}=\sum\overline{C}_i\cdot p_i$, where i is an index and p_i is the probability that a test plane T hits D_i. The definition of p_i is thus $p_i=\Pr\left(T\uparrow D_i\,|\,T\uparrow \cup D_i\right)$, the probability that the probe T hits D_i given that it hits at least one element of the set of all D_i.

We note that $\Pr(T \uparrow D_i \cap T \uparrow \mathrm{U}D_i) = \Pr(T \uparrow D_i \mid T \uparrow \mathrm{U}D_i) \cdot \Pr(T \uparrow \mathrm{U}D_i)$, while $\Pr(T \uparrow D_i \cap T \uparrow \mathrm{U}D_i) = \Pr(T \uparrow D_i)$.

Now $\Pr(T \uparrow D_i) = D_i / D$ and $\Pr(T \uparrow \mathrm{U}D_i) = \sum D_i / D$ from Eqn. (5.87). Thus by dividing, $p_i = D_i / \sum D_i$.

So, allowing for the probability of interception, we obtain for an aggregate of features of surface area S_i and diameter D_i,

$$\bar{\bar{C}} = \left(\frac{\pi}{4}\right)\frac{\sum S_i}{\sum D_i}, \tag{5.97}$$

or, dropping the extra bar,

$$\bar{C} = \left(\frac{\pi}{4}\right)\frac{\bar{S}}{\bar{D}}. \tag{5.98}$$

5.2.3 Density of Intercept lengths on a planar probe with an Aggregate of Areal Features

Multiplying Eqn.(5.98) through by Eqn.(5.91) we find:

$$N_A \bar{C} = \left(\frac{\pi}{4}\right) N_V \bar{S}.$$

But, $N_A \bar{C} = \bar{L}_A$ the expected length of intersection per unit area of the probe, and $N_V\ S = S_V$ the surface area per unity volume. Hence:

$$\bar{L}_A = \left(\frac{\pi}{4}\right) S_V. \tag{5.99}$$

An alternative derivation of this expression given in the next section demonstrates that it is valid for non-convex surfaces and for non-flat probes.

5.2.4 Expected Length of Interception Generated by Two Areal Arrays

Consider two areal arrays having areas of S_V' and S_V'' per unit volume, which are superimposed randomly with respect to translation and orientation. The two arrays will intercept each other to generate a lineal array. We wish to determine the expected length $\overline{L}_V$ of interception per unit volume.

We start with Eqn.(5.95) and substitute an element of the array S':

$$\delta C = \left(\frac{\pi}{4}\right)\left(\frac{\delta S'}{D}\right) \tag{5.100}$$

for the expected length of interception of a planar probe with an element of area $\delta S'$. Were we to divide this through by the area of interception of the planar probe with its bounding body, we would have the length per unit area of test plane contributed by the interception of areal element $\delta S'$ with that test plane. Now it is apparent that if we make S'' the probe and consider an element $\delta S''$, the expected length $\delta \overline{L}$ of interception of this element with $\delta S'$ is:

$$\delta \overline{L} = \frac{\delta C \, \delta S''}{\overline{A}}, \tag{5.101}$$

where $\overline{A}$ is the average area of the probe. Substituting from Eqns.(5.88) and (5.100) into Eq. (5.101) we find:

$$\delta \overline{L} = \frac{\pi}{4V} \delta S' \delta S'' .$$

Integration over all areal elements and division by V then yields:

$$\overline{L}_V = \left(\frac{\pi}{4}\right) S_V' S_V'' . \tag{5.102}$$

Since $\overline{L}_A = \overline{L}_V / S_V$, this expression also yields Eqn.(5.99).

Somewhat surprisingly the result given in Eqn.(5.102) was apparently not discovered until 1972 when it appeared as a special case of an N-dimensional expression obtained by Miles [1972].

5.2.5 Expected Number of Intersections Generated by Three Areal Arrays

This section is kind of an orphan since if three areal arrays are superimposed they generate point intersections rather than lines. If $S_V^{'}$, $S_V^{''}$ *and* $S_V^{'''}$ are the densities of the three arrays then it follows from a combination of Eqns.(5.85) and (5.102) that the expected number of point intersections per unit volume for a random superimposition is given by:

$$\bar{P}_V = \left(\frac{\pi}{8}\right) S_V^{'} S_V^{''} S_V^{'''} . \qquad (5.103)$$

5.3 The Planar Probe in Three Dimensions, Lineal Features

$I_0\ P_2\ F_1\ N_3$

5.3.1 Expected Number of Intersections of Planar Probe with Lineal Features

Consider a line segment of length δL in a bounding body of diameter D. The probability of a random probe intersecting this segment is:

$$p = \frac{1}{2\pi D} \int_0^{2\pi} \int_0^{\pi} \int_{(\rho)} \varepsilon_{\delta L}(\rho, \phi, \theta) \sin\phi \, d\rho \, d\phi \, d\theta$$

in which by way of the indicator function the integration is over all values of (ρ, ϕ, θ) for probes intersecting the segment. Without any loss in generality we can assume that the segment is parallel to the z-axis and that it is in a positive direction from the origin similar to the geometry of Fig. 5.19. The integration with respect to ρ then yields the projected length, $\delta L \cos\phi$, of the segment on the plane normal to the probe (the "projection plane" in Fig. 5.19). Noting that only probes for which $\phi \le \pi/2$ will intersect the segment, we have,

$$p = \frac{1}{2\pi D} \int_0^{2\pi} \int_0^{\pi/2} \delta L \cos\phi \sin\phi \, d\phi \, d\theta$$

$$=\frac{\delta L}{2D}. \tag{5.104}$$

Integration over all segments then yields for the expected number of intersections

$$\bar{P}=\frac{L}{2D}. \tag{5.105}$$

This is a good point to consider again (previously discussed in Chapter 2) the topic of probability vs. expectation. The symbol p in Eqn.(5.104) is a probability while $\bar{P}$ in Eqn.(5.105) is an expectation. Expectations are found by integrating the quantity to be expected times its probability density function. Equation (5.104) shows that $1/(2D)$ is the probability density function with respect to L. It is a common observation that when an expectation is <<1 it acts a probability. But, probability conventionally means the odds of happening (at least) *once*. The reader may convince herself, if she wishes, by working out the algebra showing that as δL increases beyond a differential, the form of Eqn. (5.104) changes and approaches unity only as an asymptote with increasing L.

5.3.2 Expected Density of Intersections of a Planar Probe with Lineal Features and the Orientation Distribution of a Lineal Array

Dividing Eqn.(5.105) through by the expected area of an interception, Eqn.(5.88), yields an expression for the expected number of intersections per unit area, the density of intersections.

$$\frac{\bar{P}}{\bar{A}}=\frac{L}{2V},$$

or

$$\bar{P}_A=\frac{L_V}{2}, \tag{5.106}$$

where $\bar{P}_A$ is the expected number of intersections per unit area of the probe. We could also have derived this result from Eqn.(5.85) since $\bar{P}_A = \dfrac{\bar{P}_V}{S_V}$.

The results given in Section 5.4.4.3 can be applied without any modification for the estimation of the orientation distribution of an array of lineal features from a measurement of $P_A(\phi,\theta)$, the density of point intersections on a probe of orientation (ϕ,θ). This topic will be explored further in Chapter 10.

In this next section we consider some relationships that do not fit into the classification scheme employed so far.

6. CURVATURE

The integral of the mean over a surface of its principal curvatures is related to the mean caliper diameter of a convex body by way of the fact that all intersections between a planar probe and such a body result in closed convex curves. Curvature itself is of interest because it can be related to the pressure generated by surface tension across an interface and to the number of bodies per unit volume.

6.1 Curvature and Mean Caliper Diameter

We have already used the concept of the mean caliper diameter in a three dimensional context as the average or mean breadth of an object. In fact, some authors use the term mean breadth. For a convex body it is easy to imagine measuring this parameter by taking a micrometer and measuring the varying breadth of the convex body by rolling it with the fingers between the flat anvils of the micrometer. For a plane figure the mean-caliper diameter (or mean-projected height) is defined analogously to what we have already seen in three dimensions by:

$$d = \int_0^{2\pi} d(\theta) f(\theta) d\theta, \qquad (5.107)$$

in which $d(\theta)$ is the distance between parallel tangents of orientation θ as shown in Fig. 4.4. As has already been demonstrated [Eqn. (4.13)]:

$$d = \frac{c}{\pi}, \tag{5.108}$$

where c is the perimeter length.

For a three-dimensional convex body:

$$D = \int_0^{2\pi}\int_0^{\pi} D(\phi,\theta) f(\phi,\theta) d\phi d\theta = \left(\frac{1}{4\pi}\right)\int_0^{2\pi}\int_0^{\pi} D(\phi,\theta)\sin(\phi) d\phi d\theta, \tag{5.109}$$

in which $D(\phi,\theta)$ is defined as shown in Fig. 4.10. The calculation of D from this expression is difficult unless the body is of a simple shape. It is interesting and insightful to compare this expression with the notation of an equivalent one [Stoyan, Kendall & Mecke, 1995],

$$\bar{b}(K) = \frac{1}{2\pi}\int_0^{\pi/2}\int_0^{2\pi} l\left(p_{\phi,\theta}K\right)\sin\phi d\theta d\phi .$$

In this expression K is a convex set, $\bar{b}$ designates ""mean breadth" and $l\left(p_{\phi,\theta}K\right)$ means "the length of the projection of K onto a line passing through the origin having direction coordinates (ϕ,θ). A different expression was developed by Minkowski in 1903:

$$D = \left(\frac{1}{4\pi}\right)\int_S \left(K_1(x) + K_2(x)\right) dS = \left(\frac{1}{2\pi}\right)\int_S m(x) dS, \tag{5.110}$$

in which K_1 and K_2 are the *principal curvatures* of the surface and the integration is over the surface of the body. The quantity $m(x) = \left(K_1(x) + K_2(x)\right)/2$ is called the *mean curvature* at the point x.

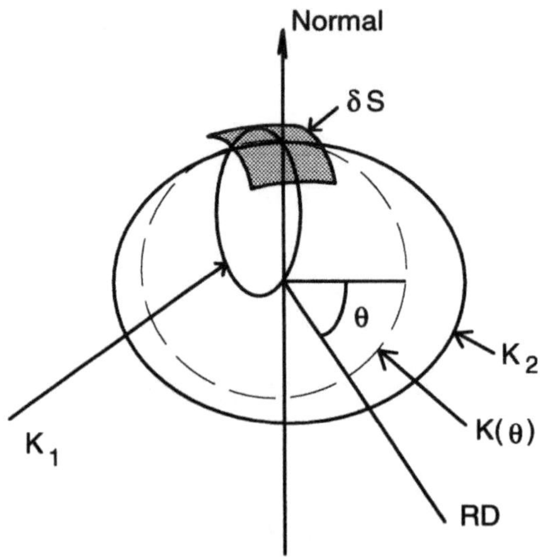

Figure 5.20: Meusnier's theorem.

The principal curvatures are the maximum and minimum curvatures of the curves formed by the interception of planes with the surface, when the planes contain the normal to the surface. It can be shown that the principal curvatures are mutually perpendicular. In fact, by Meusnier's Theorem the curvature at any angle can be expressed as $K(\theta) = K_1 \cos^2 \theta + K_2 \sin^2 \theta$, in a suitable system of coordinates as shown in Fig. 5.20. In that figure we see a differential patch δS of the surface of a convex body. The vertical axis in the figure is the surface normal and is not truly vertical but is tilted out of the page toward the reader for the sake of visibility. The reference direction RD is perpendicular to the axis of the surface normal. The circle labeled K_1 has the larger principal curvature $K_1 = 1/r_1$ where r_1 is its radius. This circle fits the curvature of the patch in the direction shown. The large circle belongs to the small principal curvature, K_2. Its disk is perpendicular to that of K_1. The dashed circle fits the curvature of the surface element at angle θ. In general all surface elements (in three dimensions) have two principal curvatures. A curvature is defined as positive if the center of its radius is located inside the body and negative if located outside it. It follows that the surfaces of convex bodies have only positive curvatures. A general surface can have one principal curvature positive and another negative. This defines a saddle point. Weibel reminds the reader [1979B] that at a *monkey saddle* a surface can have three principal curvatures all of which are zero. A conventional saddle point has two-fold symmetry. A monkey saddle is one that could possess three-fold symmetry: the ordinary saddle has two down-sloping sides for the rider's legs; the monkey saddle has three down-sloping sides -- one to accommodate the monkey's tail. Continuing this line of reasoning one could devise an n-fold saddle. The disk of a ravioli cutter might be an example. But, examining the ravioli cutter with the mind's eye, we see that locally at any point it still has only two principal curvatures rather than n of them. Two principal curvatures of zero are not distinguishable from three or even n zero curvatures.

Although Eqn.(5.110) was derived in 1903, it did not become useful in stereology until much later when it was re-discovered. The derivation of Eqn.(5.110) that follows is based on treatments by Cahn [1967] and DeHoff [1967]. These have been placed on a more formal basis by Miles [1981]. This treatment is considerably simpler than the original one by Minkowski or that reproduced by Kendall and Moran [1963]. The papers by Cahn and DeHoff, which treat the estimation of average mean curvature, were received by the same journal within five days of one another. Although their derivation of the basic relationships is essentially the same, there are appreciable differences in the applications of the results. A comparison of the two papers provides an interesting study of different approaches to the same problem.

The starting point for deriving Minkowski's expression is an expression for the total curvature of an open curve on a plane. Consider a segment, dc, of the curve Fig. 5.21 that is short enough to be assumed circular.

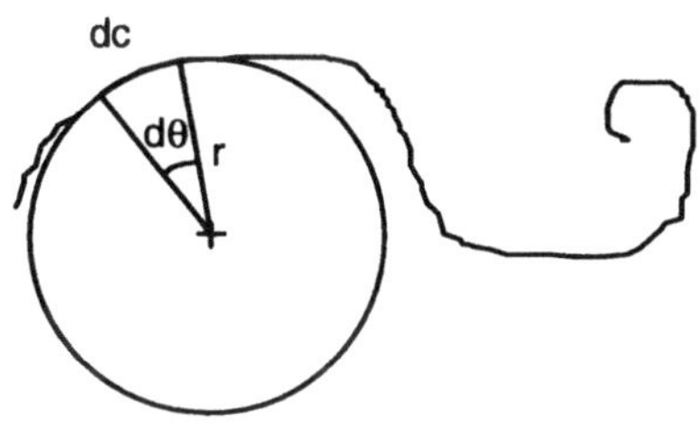

Figure 5.21: Curvature at a point is a limit $d\theta/dC$.

If r is the radius of this segment, then: $dc = r d\theta$,
or, in terms of the local curvature, $K' = 1/r$, of the segment:

$$K' = \frac{d\theta}{dc}. \tag{5.111}$$

The total curvature of the figure is given by:

$$K = \int_C K' dc = \int_\theta d\theta = \Delta\theta, \tag{5.112}$$

where $\Delta\theta$ is the difference between the angle of the tangents to the ends of the curve. For a closed figure (not necessarily convex) Eqn. (5.112) yields:

$$K_{loop}=2\pi\ . \tag{5.113}$$

We can also define a mean curvature of a lineal figure:

$$\overline{K}=\frac{K}{c} \tag{5.114}$$

where c is the length of the figure (open or closed).

The next consideration is to relate K to the average curvature of the surface of a three-dimensional compact body. Consider a small element of area δS as in Fig. 5.20 on the surface of a bounding body of diameter D, Fig. 5.22. The test plane makes a random angle with respect to the

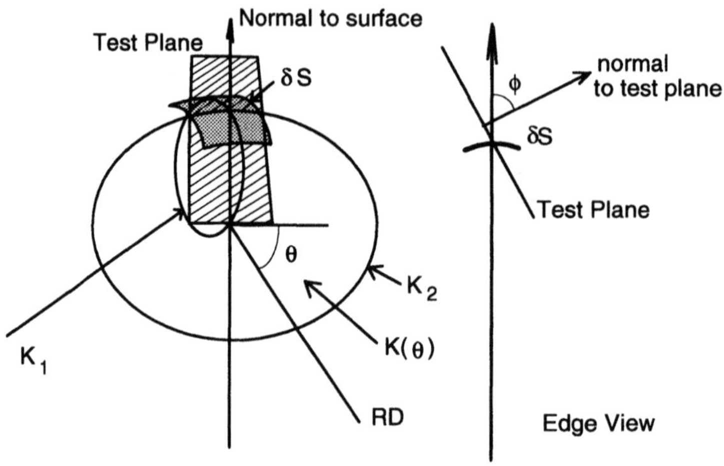

Figure 5.22: Intersection of test plane with area element

normal to the surface element. The angle between the two normals is ϕ. If the normal to the surface were within the test plane then the curvature as a function of θ would be given by Meusenier's Theorem above. Given that the two principal curvatures are 90 degrees apart, proving the formula is easy and has a lot in common with the well-known Mohr's Circle construction for principal stresses -- the formulas come out the same. However, the test plane makes an angle with the normal to the surface element. So, we would like to find the curvature of the trace of the intersection of the tilted test plane with the curved element. While this calculation is most rigorously done in using the formalism of differential geometry, intuition is developed by doing the calculation using ordinary calculus. Not wanting to spend too much time on this dalliance, I will sketch the approach.

Meusnier's Theorem gives us a curvature at θ. Since we are taking a limit, we may construct the partial upper half of a cylinder having the specified curvature and intersecting the test plane as shown on the left-hand side of Fig. 5.22. If the circular outline of this cylinder follows the equation

$x^2 + y^2 = r^2$ in a coordinate system in which y is now identified with the surface normal and x with the direction θ, it will follow the equation of an ellipse $x'^2 + \sin^2(\phi) y'^2 = r^2$ in the test plane where x' is identical to x but y' is the projection of y onto the test plane. The space x',y' is where we want to find the curvature. Differentiating twice and setting x' equal to zero gives $d^2y'/dx'^2 = -1/(r\sin\phi)$ where r is the radius of curvature given by the Meusenier formula. Since the magnitude of the curvature of any plane curve is $\left|(d^2y'/dx'^2)/(1+(dy'/dx'))^{\frac{3}{2}}\right|$ and here dy'/dx' equals zero, we find that

the curvature of the trace of the intersection is equal to that given by the Meusenier formula divided by $\sin\phi$.

For the purposes of integration over the surface, it is convenient to re-lable the axes. Without any loss in generality, we make the surface element's normal parallel the z (or ϕ) axis. We also set the x (or θ) axis, labeled "RD", parallel to the plane containing the maximum curvature, K_1 and the normal to the element. Consider a planar probe, the test plane, whose coordinates (ρ,ϕ,θ) are such that it intercepts the element. We will now calculate the expected total curvature of the intersection of the test plane and the surface of the compact body. By the above discussion the curvature of the intercepted curve on the probe is:

$$K(\phi,\theta) = \frac{\left(K_1\cos^2\theta + K_2\sin^2\theta\right)}{\sin(\phi)}. \tag{5.115}$$

Let δC be the length of the intercepted curve on the element δS; then the expected contribution that this element $\delta K = \left\langle \delta C \overline{K}(\phi,\theta)\right\rangle$ makes (angle brackets here denote expectation) to the total (integral of) curvature of the interception (averaged over all probes intercepting the element) is

$$\delta K = \int_0^{2\pi} \int_0^{\frac{\pi}{2}} \int_{(\rho)} K(\phi,\theta)\delta C\, f(\rho,\phi,\theta) d\rho\, d\phi\, d\theta\,, \tag{5.116}$$

in which the upper limit on the integration with respect to ϕ is $\frac{\pi}{2}$ because interceptions cannot occur for $\phi > \frac{\pi}{2}$. Substituting from Eqns.(4.29) and (5.115) we obtain:

$$\delta K = \frac{1}{2\pi D} \int_0^{2\pi} \int_0^{\frac{\pi}{2}} \int_{\rho} \left(K_1 \cos^2(\theta) + K_2 \sin^2(\theta)\right) \delta C\, d\rho\, d\phi\, d\theta\,. \tag{5.117}$$

The integration with respect to ρ yields as before, following the argument given with Fig. 5.19, the projected area $\delta S \sin(\phi)$ of the element on a plane that is perpendicular to the probe and parallel to the curve of interception. Integration over ϕ and θ then yields

$$\delta K = \left(\frac{1}{2D}\right)(K_1 + K_2)\delta S\,.$$

Summing over all surface elements we obtain for the expected total curvature of the closed planar figure formed by the interception of a planar probe with a body of mean caliper diameter D:

$$\langle K \rangle = \left(\frac{1}{2D}\right) \int_S (K_1 + K_2) dS\,, \tag{5.118}$$

where again angle brackets are used to denote expectation to avoid confusion. However, as has already been shown, Eqn. (5.113), $K = 2\pi$ for a closed figure of any form. Substitution for K in Eqn. (5.118) thus yields Minkowski's result, Eqn. (5.110).

The latter equation is also valid for a body containing singularities (or edges) in its surface because even though the curvature is infinite at such a singularity, the integral remains finite. This can be demonstrated as follows.

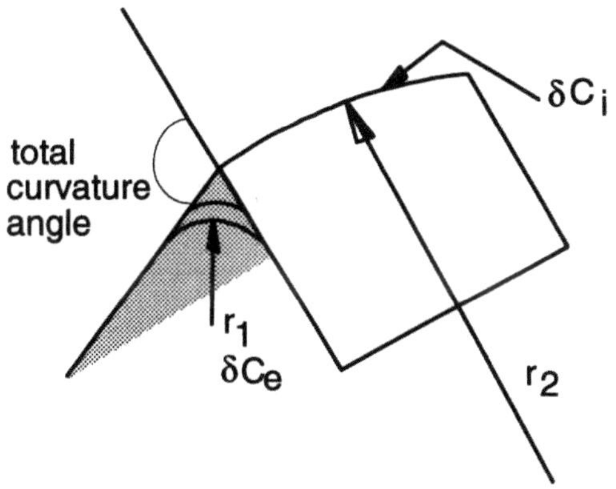

Figure 5.23: Intersection of two planes of a bent polyhedron.

Consider (Fig. 5.23) an element designated by its length, δC_i, of an edge whose radius of curvature in the plane of the edge is r_2. We may "round off" the edge by differing amounts as indicated by the arcs above the radius arrow r_1 so that it has a finite radius of curvature r_1 in a plane normal to the edge. We note that the length of the arc that results from the rounding is $\delta C_e = r_1\omega$ where ω is the "total curvature angle" shown in the figure, which is also equal to that between the normals to the two planes. If α_i is the dihedral angle between the surfaces meeting at the edge, then the total curvature angle is $(\pi-\alpha_i)$. The area of the "bead" formed by rounding is thus $(\pi-\alpha_i)r_1\,\delta C_i$. Using Eqn. (5.110), the contribution δD_i of this surface to D is therefore,

$$\delta D_i=\left(\frac{1}{4\pi}\right)(\pi-\alpha_i)r_1\,\delta C_i\left(r_1^{-1}+r_2^{-1}\right)=\left(\frac{1}{4\pi}\right)(\pi-\alpha_i)\delta C_i(1+r_1/r_2).$$

Thus, as $r_1 \to 0$,

$$\delta D_i=\left(\frac{1}{4\pi}\right)(\pi-\alpha_i)\delta C_i\ . \tag{5.119}$$

Notice the disappearance of r_2 and with it K_2. Summing over all edges and noting that no contribution arises from the flat portions we obtain for a polyhedron,

$$D=\left(\frac{1}{4\pi}\right)\sum_{(i)}(\pi-\alpha_i)C_i\ . \tag{5.120}$$

where the C_i are the lengths of the edges.

We can generalize by grafting our polyhedron onto a general smooth body. Then, allowing for the contribution from the remaining surfaces again by way of Eqn. (5.110) we obtain:

$$D=\left(\frac{1}{4\pi}\right)\left\{\int_S (K_1+K_2)dS+\sum_{(i)}(\pi-\alpha_i)C_i\right\}, \tag{5.121}$$

in which the integration is over the non-singular portions of the surface. This result was first derived by Santalo (1936).

6.2 Densities of Curvature

A length of line in two dimensions has an average curvature $\overline{K}$ (Eqn.(5.114)) which is total curvature K per unit length. It was found by dividing the total curvature by the length of the curve. We could also take the total curvature and divide it by the area of a bounding curve in the plane. We call that density of curvature K_A. We would like to know what its expected value would be in case a surface was intersected at random by a probe plane. If we intersected the surface of a body, the curve of the intersection would be closed and K_A would equal $2\pi / A$, where A is again the area defined by some boundary enclosing the surface. But, what if the surface were not that of a body so that the resulting two-dimensional curve was not closed?

In deriving Eqn.(5.96) we were forced to use a convex bounding figure in order to define the probability density function for the probe. Yet, our integration in Eqn.(5.94) was only over a surface element and summed only line segments. Afterward we had to convince ourselves that the intersection of a random planar probe with this bounding body was indeed a closed curve. Later in deriving Eqn.(5.97) the diameter of the closed bounding surface divides out and there is no longer a need that the surfaces of the elements of the aggregate be closed. In the case that there is only one element inside the closed bounding surface, an aggregate of one, Eqn.(5.97) still applies but becomes:

$$\overline{C}=\frac{\pi S}{4D} \quad . \tag{5.122}$$

Although Eqn.(5.122) looks like Eqn.(5.94) it has a different meaning since the bounding body is now out of the picture and S need not be the area of the surface of a convex body. The intercept length $\overline{C}$ need not be that of a closed curve.

Example:

We may examine the meaning of Eqn.(5.122) through a construction, Fig. 5.24.

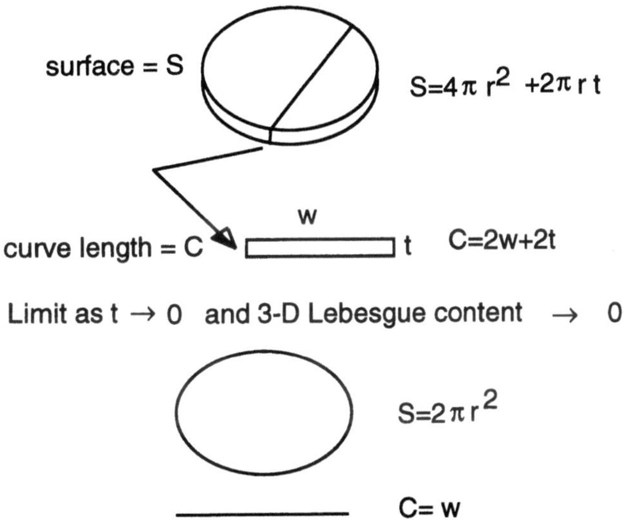

Figure 5.24: A surface element as the limit of a thin disk.

Consider a thin disk as show above. A random cut is made through the disk by a plane producing a plane surface. The plane surface has an average boundary length C, which is the average intercept length. The disk has some mean caliper diameter D. As thickness t becomes very small, D approaches a finite limit. At that limit, as always, $D = (\pi/4)(S/C)$. We may imagine a case where t is infinitesimal. Then the volume, which is the Lebesgue content in three dimensions, is also infinitesimal. The ratio of surface area to intercept length is $(4\pi r^2)/(2w)$. However, we could further imagine a case where the volume goes exactly to zero. In that case there is nothing to bound, so there is no boundary. Instead of the disk having two surfaces, one for each side, it only has one. But, the ratio of surface to intercept length remains the same, $\left(2\pi r^2\right)/w$. This same construction could be applied to any surface tile.

Returning to Eqn. (5.118) we define $\mathsf{M} = \int_S m dS$. Then $\langle K\rangle = \mathsf{M}/D$. We can eliminate D using Eqn.(5.122) giving $\langle K\rangle S\frac{\pi}{4} = \overline{C}\mathsf{M}$. Since S is the total surface within the bounding body and M is the integral of mean curvature over that entire surface, division by the volume of the bounding body gives, $\frac{\langle K\rangle}{\overline{C}}\frac{\pi}{4}S_V = \mathsf{M}_V$. From Eqn. (5.99), $\overline{L}_A = \frac{\pi}{4}S_V$.

So, $\frac{\langle K\rangle}{\overline{C}}\overline{L}_A = \mathsf{M}_V$. We see from Eqn.(5.114) that $\overline{K} = K/C$. To the extent that the ratio of the means estimates the true proportion,

$$\left\langle \overline{K}\right\rangle \overline{L}_A = \left\langle K_A\right\rangle = \mathsf{M}_V. \tag{5.123}$$

The quantity M_V is a property of the specimen under study while $\langle K_A\rangle$ is an expectation of the total curvature per intercept area of that specimen when sampled.

Equation (5.123) has been used to estimate M_V by a tangent count method in the papers of Cahn and DeHoff mentioned earlier. In this method carefully oriented random sections are taken, and lines are swept across the specimen. The number of tangent points is counted subtracting the count of tangents to negative curvature from that to positive curvature. The observed K_A is π times the net tangent count.

6.3 Gaussian Mean Curvature

The mean curvature $m = (K_1 + K_2)/2$ is not the only average curvature. The Gaussian curvature $G = K_1 \cdot K_2$ is also a form of average curvature and is sometimes (confusingly) referred to as the total curvature. Analogously with $K' = d\theta/dc$,

$$G = \frac{d\Omega}{dS}, \tag{5.124}$$

where $d\Omega$ is an element of solid angle. Integration of Eqn.(5.124) over a closed surface gives 4π.

7. SUMMARY OF RELATIONSHIPS

7.1 General Relationship

$$I=P+F-N. \tag{5.1}$$

I is dimension of intercept; P is dimension of probe; F is dimension of feature; N is dimension of embedding space.

7.2 Properties of Two-Dimensional Figures

c is perimeter length; a is area.

(i) Mean projected length (applicable to non-convex figures if multiple overlaps are allowed for) or mean-caliper diameter of a convex figure.

$$d=\frac{c}{\pi} \qquad (4.13)(5.108)$$

(ii) Expected intercept length in an aggregate of figures.

$$\bar{L}=\frac{\pi\bar{a}}{\bar{c}} \tag{5.17}$$

(iii) Expected number of figures per unit length.

$$\overline{N}_L = N_A\bar{d} \tag{5.19}$$

(iv) Third moment of intercept length in an aggregate of convex figures.

$$\mu_3'(L)=\frac{3\sum a_i^2}{\sum c_i} \tag{5.26}$$

7.3 Properties of Three-Dimensional Bodies

A is surface area; *V* is Volume; *S* is surface area. *K* is the integral of curvature (5.112).

(i) Mean-caliper diameter of a convex body having local principal curvatures of K_1 and K_2 and with edges of length C_i and dihedral angles α_i.

$$D=\left(\frac{1}{4\pi}\right)\left\{\int_S (K_1+K_2)dS+\sum_{(i)}(\pi-\alpha_i)C_i\right\} \tag{5.121}$$

(ii) Average projected area.

$$\bar{S}_p=\frac{S}{4} \tag{4.24}$$

(iii) Average intercept length in an aggregate of bodies.

$$\bar{L}=\frac{4\bar{V}}{\bar{S}} \tag{5.50}$$

(iv) Fourth moment of intercept length in an aggregate of bodies.

$$\mu_4'(L)=\frac{12\sum V_i^2}{\pi\sum S_i} \tag{5.55}$$

(v) Average intercepted area in an aggregate of bodies.

$$\overline{A} = \frac{\overline{V}}{\overline{D}} \qquad (5.90)$$

(vi) Average profile circumference in an aggregate of bodies.

$$\overline{C} = \left(\frac{\pi}{4}\right)\frac{\overline{S}}{\overline{D}} \qquad (5.98)\ \text{q.v.}(5.122)$$

7.4 Density of Interceptions in One- and Two-dimensional Spaces

(i) Point fraction.

$$\overline{P}_p = L_L \qquad (5.6)$$

$$\overline{P}_p = A_A \qquad (5.7)$$

(ii) Lineal fraction.

$$\overline{L}_L = A_A \qquad (5.28)$$

(iii) Intersections of a linear probe.

$$\overline{P}_L = \frac{2}{\pi} L_A \qquad \text{(Buffon's principle) } (5.31)$$

(iv) Intersections generated by two lineal arrays.

$$\overline{P}_A = \frac{2}{\pi} L_A L_{A'} \qquad (5.43)$$

7.5 Density of Interceptions in Three-Dimensional Space

(i) Point fraction.

$$\overline{P}_p = V_V \tag{5.8}$$

(ii) Lineal fraction.

$$\overline{L}_L = V_V \qquad \text{(Rosiwal's principle) (5.57)}$$

(iii) Areal fraction.

$$\overline{A}_A = V_V \qquad \text{(Delesse's principle) (5.93)}$$

(iv) Intersections on a linear probe.

$$\overline{P}_L = \frac{S_V}{2} \qquad (5.59)(5.86)$$

(iv') $$N_L = S_V / 4$$

(v) Intersections generated by a lineal and an areal array.

$$\overline{P}_V = \frac{S_V L_V}{2} \tag{5.85}$$

(vi) Interceptions on a planar probe.

$$\overline{N}_A = N_V \overline{D} \tag{5.91}$$

(vii) Intersections on a planar probe.

$$\bar{P}_A = \frac{L_V}{2} \tag{5.106}$$

(viii) Intercept length on a planar probe.

$$\bar{L}_A = \left(\frac{\pi}{4}\right) S_V \tag{5.99}$$

(ix) Intercept length generated by two areal arrays.

$$\bar{L}_V = \left(\frac{\pi}{4}\right) S_V^{'} S_V^{''} \tag{5.102}$$

(x) Intersections generated by three areal arrays.

$$\bar{P}_V = \left(\frac{\pi}{8}\right) S_V^{'} S_V^{''} S_V^{'''} \tag{5.103}$$

(xi) Expected intercept curvature on a planar probe.

$$\langle K_A \rangle = \mathsf{M}_V \tag{5.123}$$

M is the integral of mean curvature (5.123).

8. BIBLIOGRAPHY FOR CHAPTERS 4 AND 5

Citations

W. Bitterlich, “Die Winkelzählprobe”, *Allgemeine Forst und Holzwirtschaftliche Zeitung*, **59** (1/2) (1948) 4-5.

E.O. Brigham, *The Fast Fourier Transform*, Prentice-Hall, Englewood Cliffs, NJ (1974).

G.L. Buffon, "Essai d'arithmetique morale", *Supplement a l'Histoire Naturelle (Paris),* **4** (1777).

J. W. Cahn, "The significance of average mean curvature and its determination by quantitative metallography", *Trans. AIME (Met. Soc.),* **239** (1967) 610 -616.

H.W. Chalkley, J. Cornfield and H. Park, "A method for estimating volume-surface ratios", *Science*, **110**, (1949) 295.

M. W. Crofton, "Probability", *Encyclopedia Britannica*, 9th Edition Vol. 19 (1885)[8].

R.T. De Hoff, "The quantitative estimation of mean surface curvature", *Trans. AIME (Met. Soc.)*, **239** (1967) 617-621.

M.A. Delesse, "Procédé méchanique pour déterminer la composition des roches", *Comptes Rendus Acad. Sci. (Paris)*, **25** (1847) 544-545.

H. Elias, A. Hennig and P. Elias, "Contributions to the geometry of sectioning V, some methods for the study of kidney structure", *Zhurnal fur die Wissenschaft des Microskopie,* **65** (1961) 70.

H.F. Fischmeister, "Shape factors in quantitative microscopy”, *Zeitschrift für Metalkunde,* **65** (1974) 558-562.

M. Gardner "Mathematical Games", *Scientific American*, **201** (1959) 174.

H.J. Gundersen et al. (11 authors), "Some new, simple and efficient stereological methods and their use in pathological research and diagnosis", *Acta Pathologica Microbiologica et Immuniologica Scandinavica*, **96** (1988) 379-881.

J. Hilliard, "Specification and measurement of microstructural anisotropy", *Trans. AIME (Met. Soc.)*, **224** (1962) 1201-1211.

M.G. Kendall and P.A Moran, *Geometrical Probability*, Charles Griffin, London (1963).

R.L. Liboff, *Introductory Quantum Mechanics*, 2nd edition, Addison-Wesley, Reading, MA (1992).

[8] The date 1885 is often quoted. Portions of Crofton’s work appear in the 11th Edition (1911). In the 11th Edition the date of the 9th Edition is given as 1889. The integral called “Crofton’s Second Theorem” appears as a corollary in the encyclopedia articles to another integral given in M.W. Crofton, *Comptes Rendus* (1869) p.1469.

T. M. MacRobert, *Spherical Harmonics*, Dover, New York (1948).

D. H. Menzel, *Mathematical Physics*, Dover, New York (1961).

R.E. Miles, "Multidimensional perspectives on stereology", *Journal of Microscopy*, **95** (1972) 181-195.

R.E. Miles, "Stereological formulae based upon planar curve sections of surfaces in space", *Journal of Microscopy*, **121** (1981) 211-237.

M. J. Mohlenkamp, "A fast transform for spherical harmonics", *Journal of Fourier Analysis and Applications*, **5** (1999) 159-184.

S. Mori, H. Nishida and H. Yamada, *Optical Character Recognition*, John Wiley & Sons, New York (1999).

E. Philofsky, *Specification an measurement of line and boundary anisotropy* M.S Dissertation, Northwestern University, Evanston, IL (1966).

E. Philofsky and J. Hilliard, "On the measurement of the orientation distribution of lineal and areal arrays", *Quarterly of Applied Mathematics*, **27** (1969) 79-86.

A. Rosiwal, "Ueber geometrische Gesteinanalysen, *Verhandlungen K. K. des Geologische Rechsanstalt, Wien*, **13** (1898) 143-175.

S.A. Saltykov, *Stereometric Metallurgy* (Moscow, 1958; translated as document AD267700, Armed Services Technical Information Agency, Arlington, VA, 1961) better title translation: *Stereological Metallography.*

D. Stoyan, W.S. Kendall and J. Mecke, *Stochastic Geometry and its Applications*, 2nd Edition, John Wiley & Sons, Chichester (1995).

J. L. Synge and A. Schild, *Tensor Calculus*, University of Toronto Press (1949); Dover, New York (1978).

E.E. Underwood, *Quantitative Stereology*, Addison-Wesley, Reading, MA, (1970).

E. R. Weibel, "Structure in space and its appearance on sections", *Stereology*, Proceedings of the 2nd International Congress for Stereology, H. Elias, ed., Springer Verlag, New York, (1967) p. 15.

E. R. Weibel, *Stereological Methods*, Volume 1, Academic Press, London (1979A).

E. R. Weibel, op cit., Volume 2 (1979B).

Additional General Reference

S.A. Saltykov, *Stereometric Metallurgy* (title translated), Moscow (1958) US Gov't translation AD267700 (1961).

9. PROBLEMS

1) Verify any of the following expressions that are not obvious to you:

(a) $S_V = N_V \overline{S}$

(b) $V_V = N_V \overline{V}$

(c) $L_L = N_L \overline{L}$

(d) $P_L = 2N_L$ (for non-contiguous particles);

(e) $P_L = N_L$ (for contiguous particles, e.g. grains in a metal)

2) An analysis was made on a specimen containing spherical nuclei that were known to be of *constant* size. The largest profile seen on a section had a diameter 0.01 mm and the observed number of profiles was 1630 per mm^2. Determine the number of nuclei per unit volume.

3) Determine the expected area of interception of a planar probe with a sphere of diameter *D*.

4) A specimen was traversed under the cross-hairs of a microscope. The following were the distances traversed in each of the three constituents:

$\alpha - 4.23mm$; $\beta - 2.61mm$; $\gamma - 3.09mm$.

Determine the volume fraction of α.

5) Irradiation of lithium fluoride with neutrons causes the formation of cubic voids. Measurements were made on a cleaved section through an irradiated crystal. The density of void outlines was 53 per mm^2 and the average intercept length on a linear probe was 1.8×10^{-2} mm. Assuming that the voids were randomly oriented (which actually was not the case) determine: (a) Their edge length and (b) The number per unit volume.

6) In an estimation of the total length of glomerular blood channels in the kidney of a 7 year old boy, Elias and Hennig [1967] counted the number of intersections of the glomeruli falling within a square of 35 mm edge length applied to an image at a magnification of 700x through a very thin section.

A total of 362 intersections were counted in 42 applications of the square. In a separate experiment the total volume of the kidney occupied by the glomeruli was found to be 1.5×10^3 mm^3. Determine the total length of glomeruli.

7) Continuing with the same kidney, Elias and Hennig applied test lines to images at a magnification of 700x and counted the number of intersections with the traces of the membranes adjacent to the blood channels. In a total length of 1.24×10^4 mm a total of 1390 point intersections were counted. Determine the total membrane area in the kidney.

8) A dislocation is a line defect that is found in crystalline materials and in its simplest form consists of an incomplete plane of atoms -- the plane ends along the line and the adjacent planes close together around it. In an electron transmission image, dislocations appear as dark lines. A test circle (inscribed on a plastic sheet) 100 mm in diameter was applied to a series of micrographs of an aluminum foil and the number of intersections of the circumference of the circle with images of the dislocations were counted. The total number of intersections in 30 applications of the test circle was 83. Given that the magnification was 30,000x, the foil thickness 100 nm and assuming all dislocations were parallel to the plane of the foil, determine the total dislocation length per unit volume (1nm = 10^{-9} mm).

9) Two twigs 100 mm in length are blown at random into a backyard 10 x 20 m. What is the probability that the two twigs will overlap?

10) Show that for space filling contiguous grains:

$$\bar{L} = \frac{2}{S_V}. \tag{5.125P}$$

Note the factor of 2 difference between this relationship and Eqn.(5.48) which is for non-contiguous particles.

11) Consider the application of short linear probes of length l applied to a section through an aggregate of non-contiguous particles (of any shape). Show that:

$$\frac{\bar{S}}{\bar{V}} = \frac{4P}{pl} \tag{5.126P}$$

where $\overline{S}$ and $\overline{V}$ are the average surface area and volume of the particles, P the number of intersections of the probes with the perimeter of the particle profiles and p the number of end points of the probes that fall within the particle profiles.

(It is important to note that $\frac{\overline{S}}{\overline{V}}$ is not the same as S_V ; the latter is the surface area of the particles per unit volume of the *specimen*.)

[Chalkley, Cornfield and Park (1949)].

12) Consider an aggregate of particles, of the same shape but differing in size, embedded in a matrix. Let T be some linear measure of the size (for example, the maximum chord length) and α, β, and γ the mean caliper diameter, surface area and volume of a particle of unit size. For example, if the particles are spherical and we select the radius as the measure of their size, then $\alpha = 2, \beta = 4\pi$ and $\gamma = \left(\frac{4\pi}{3}\right)$.

Show that :

$$\frac{\mu_3'(T)}{\mu_1'(T)} = \frac{\alpha}{\gamma} \frac{P_P}{N_A} \tag{5.127P}$$

and

$$\frac{\mu_2'(T)}{\mu_1'(T)} = \frac{\alpha}{\beta} \frac{2P_L}{N_A}. \tag{5.128P}$$

13) In chapter 6 it is shown that the distribution of intercept lengths, L, in a sphere of diameter D is:

$$f_D(L) = \frac{2L}{D^2}; \; [0 \le L \le D]$$

Determine the first and fourth moments of this distribution and verify that they satisfy the general relationships: $\overline{L}=\frac{4V}{S}$ and $\mu_4^{'}(L)=\frac{12V^2}{\pi S}$.

14) Show from first principles that the distribution of the projected height h of a randomly oriented square of edge length a is

$$f(h)=\frac{4}{\pi}\left(2a^2-h^2\right)^{-\frac{1}{2}},[a\leq h\leq a\sqrt{2}]\ . \tag{5.129P}$$

Determine the average height $\overline{h}$ (the same as d in our previous notation) and verify that it is in agreement with the general relationship $d=c/\pi$.

15) For an aggregate of convex particles we can define an average surface curvature by

$$\overline{K}_m=\frac{1}{2S}\int_S\left(K_1+K_2\right)dS\ , \tag{5.130P}$$

where S is the total surface area of the particles and the integration is over the surface of all particles.

Show that:

$$\overline{K}_m=\frac{\pi N_A}{P_L}\ , \tag{5.131P}$$

where N_A is the number of particle profiles per unit area of a planar probe and P_L is the number of intersections of a subsequent linear probe with profile perimeters per unit length of linear probe.

16) One of the earliest results in geometric probability was obtained by Buffon in 1777. A needle of length l is dropped onto a set of parallel lines of spacing d.

(a) Show from first principles that if $l<d$, the probability p of the needle intersecting a line is

$$p=2l/(\pi d). \tag{5.132P}$$

(b) Derive the same result by use of Eqn.(5.31).

(It follows from the second derivation that Eqn.(5.132P) is valid for $l > d$ if p is replaced by $\overline{P}$ the expected number of intersections.)

17) Consider a modification of the Buffon needle problem in which a ring of radius r is dropped at random onto a square grid of lattice spacing d.

(a) Determine from first principles the probability, for $2r < d$, of getting 0, 2 or 4 intersections and show that the expected number of intersections is:

$$\overline{P} = 8r / d . \tag{5.133P}$$

(b) Derive the same result using Eqn. (5.31).
(The second derivation shows that the result is also valid for $2r > d$ and also if the ring is distorted into any other plane shape with the same perimeter.)

(c) Take a sheet of millimeter graph paper and divide it into squares 30 mm in edge length. Flick a penny 20 times (or more if you have the patience) across the paper and record the number of 0, 2 and 4 intersections with the sides of the 30 mm squares.

(d) Estimate the diameter of the penny from the average number of intersections.

18) Consider the special case of the analysis of the orientation distribution of a lineal array (Section 5.4.2.3) in which the test array is a straight line or a set of parallel lines.

(a) Let it be given that a_n^λ and b_n^λ are the Fourier series cosine and sine coefficients of $\overline{P}_L(\lambda)$, where $\overline{P}_L(\lambda)$ is the expected number of intersections of test line applied at angle λ to the RD of the unknown array. (We may choose the reference direction of the line to yield the simplest result.) If a_n^θ and b_n^θ are the Fourier coefficients of the $L_A(\theta)$ of the unknown array, show that:

$$a_n^\theta = \frac{1-4n^2}{2} a_n^\lambda , \tag{5.134a}$$

$$b_n^\lambda = -\frac{1-4n^2}{2} b_n^\lambda . \tag{5.134b}$$

(b) Check your work. A circle in a patch has a uniform density function $L_A(\theta)$. Use a circle and Eqns.(5.134) to derive Buffon's principle. Hint:

Make sure the density function for the circle matches your limits of integration in deriving Eqns.(5.134).

19) A hole is drilled with a wobbly bit. Its cross-section takes the form of a triangle with bulged sides. How can a bit make a triangular hole without changing diameter? At first thought it might appear that a circle would be the only plane figure for which the caliper diameter $d(\theta)$ is independent of θ. However there are other members of the class for which this is true. These are generated from regular polygons by drawing an arc from each vertex.

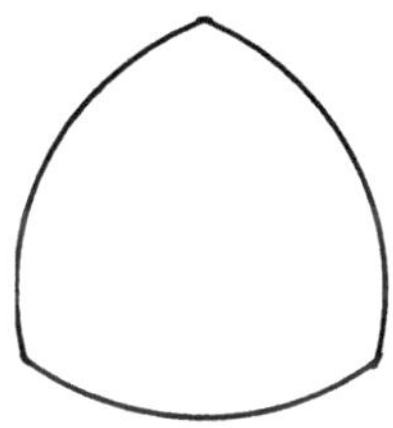

Figure 5.25P: Rotor of a Wankel Engine

An example for the case of a triangle is shown in Fig. 5.25P and is the shape of the rotor of a Wankel engine. Verify that this figure satisfies the general relationship $d = c / \pi$.

20) Why are manhole covers usually circular? What other shape(s) would satisfy the same practical requirement of not falling down the manhole and being easy to place?

21) In some types of solid-state transformations (such as the formation of pearlite in steel) a constituent nucleates on an existing grain boundary and grows as a hemispherical colony into the grain. When a section through such a specimen is viewed under the microscope, most of the patches of the transformation product are seen to be attached to a grain boundary. However there are occasional patches that are unattached to a boundary. This has led some observers to claim that some nucleation must have occurred within the grains themselves.

Show that this claim is unjustified unless the fraction of unattached patches exceeds approximately 12 per cent. In performing the calculation, assume that the colonies do not impinge upon each other or onto other grain boundaries than their origin. Hint: This problem can be solved in two or three lines.

22) Consider a non-convex plane figure composed of two circles with equal diameters that are just touching. Demonstrate that Eqn.(5.15) for the average intercept length is valid for this figure but that Eqn.(5.25) for the third moment is not valid. (This establishes that the latter holds only for convex figures.)

23) Show: (a) from first principles and (b) by use of Eqn.(5.121) that the mean caliper diameter of a thin disk of radius r is $\pi r / 2$. Also, (c) perform the same calculations to demonstrate that the mean caliper diameter of a rod of length l and radius r approaches $l/2$ as $r/l \to 0$.

24) As was mentioned in Chapter 4, the application of a linear probe in three dimensions can be considered a as a two stage sampling process: (1) The structure is sectioned with a planar probe. (2) The linear probe is applied to the two-dimensional section. By considering such a two-step process, derive $\overline{L} = 4V/S$, Eqn.(5.48) from Eqns.(5.15), (5.88) and (5.96).

25) One function of measured properties that has been used to characterize the average shape of of particles embedded in an opaque matrix is:

$$F_1 = \frac{3\pi}{2} V_V N_A / P_L^2, \tag{5.135P}$$

where V_V is the volume fraction of the particles, N_A is the number of particle profiles per unit area and P_L is the number of intersections per unit length of a linear probe with profile boundaries.

(a) Verify that F_1 is dimensionless. (This is an important requirement if a shape parameter is to be independent of the average particle size.)

(b) Show that for an aggregate of particles of the same size and having a surface area S and volume V:

$$F_1 = 6\pi V D / S^2. \tag{5.136P}$$

(c) Show that for an aggregate of particles of different sizes:

$$F_1 = 6\pi \frac{\sum V_i \sum D_i}{\left(\sum S_i\right)^2} . \tag{5.137P}$$

(d) Determine F_1 for (i) a sphere, (ii) a tetrahedron and (iii) an infinitely thin disk [Fischmeister, 1974].

26) A second shape parameter is defined in terms of the moments of the intercept length distribution in the particle profiles as follows:

$$F_2 = \sqrt{\frac{27\overline{L}^4}{16\mu_4'(L)}} . \tag{5.138P}$$

(a) Show that for an aggregate of particles having the same size and (of course) shape:

$$F_2 = 6\pi^{\frac{1}{2}} V / S^{\frac{3}{2}} , \tag{5.139P}$$

and for an aggregate of of non-uniform sized particles:

$$F_2 = 6\sqrt{\pi \frac{\left(\sum V_i\right)^4}{\left(\sum S_i\right)^3 \sum V_i^2}} . \tag{5.140P}$$

(b) By using the isoparametric inequality $V \le S^{\frac{3}{2}} / (6\pi^{\frac{1}{2}})$, prove if possible that F_2 is a maximum for a sphere.

(c) Determine F_2 for (i) a sphere, (ii) a tetrahedron and (iii) an infinitesimally thin disk.

27) Consider yet a third shape parameter defined in terms of the other two by:

$$F_3 = \sqrt{F_1 F_2} . \tag{5.141P}$$

See if you can show that in terms of the experimental quantities,

$$F_3 = \sqrt{\frac{9\pi(3)^{\frac{1}{2}}\,\overline{L}^3 N_A}{8N_L\left(\mu_4'(L)\right)^{\frac{1}{2}}}}\ .$$

28) Consider the projection of a line of length l onto a plane. Show that for a random orientation of the line in three dimensions, the expected projected length is:

$$\bar{l} = \frac{\pi}{4} l\ . \tag{5.142P}$$

Compare this result with the corresponding result in two dimensions. Which gives the longer expected projection and why?

29) Figure 5.26P shows a means of counting the density of circles using a combined sampling operation. Random points are placed on the plane. At each point, those circles are counted which subtend an angle greater than a certain angular threshold, θ. Show that the expected areal fraction A_A of circles equals the number of circles counted times $\theta^2/4$. This result does not depend on the distribution of circle diameters, but a good choice of θ may require that there be no significant overlap. This is the basis for a method of measuring the areal density of trees and bushes [Bitterlich, 1948]. Hint: Consider circles of constant diameter first; then show that diameter is immaterial.

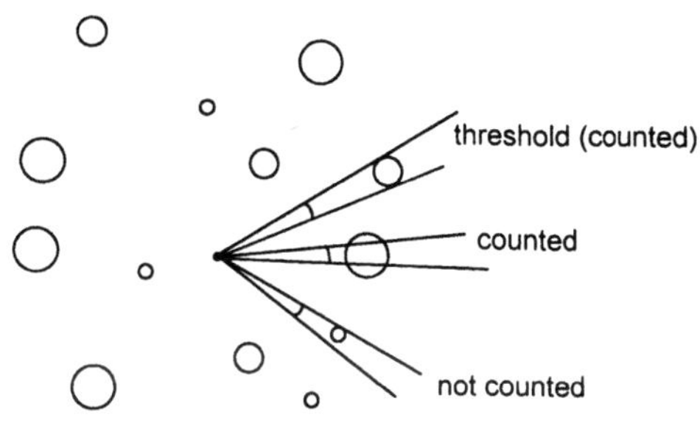

Figure 5.26P: Counting circles from a random point.

Chapter 6

SHAPE-DEPENDENT INTERCEPT DISTRIBUTIONS

Intercept distributions depend on the shape of what is being intercepted. In Chapter 5 we examined shape-independent relationships. These are the most general and perhaps the most powerful relationships in the entire subject of stereology. They are what people think of when the word "stereology" is said. In this chapter we examine relationships that are less extensive and more intensive.

1. INTRODUCTION

In Chapter 5 we derived a series of relationships that can be used for estimating such properties as volume fractions and the density of lineal and areal features. These quantities were referred to in Chapter 1 as indices since almost all of them belong to $\mathbf{R}^1$ and often represent ratios. They summarize general properties in a way not unlike econometric indices, but they are limited in the amount of information they convey. Unless restrictive assumptions are made, these relationships do not even provide any means for determining the number per volume of second phase bodies. Nor do they provide a means of determining the size distribution of an aggregate of bodies. In this chapter we will be concerned with setting up expressions for estimating the latter properties. Since these relationships are specific to the shape of the body, we will classify them by shape. We will be restricted to basic shapes. In many cases these can be used a building blocks to construct more generalized ones. For each shape we will first determine the distribution of intercept sizes in a single feature and then the distribution in an aggregate of features. This (in most cases) yields an integral equation, which will be solved for the size distribution of the features in the aggregate. The resulting expressions form the basis for estimating particle size distributions. In most cases we shall also derive a relationship between the moments of the feature's size distribution and those of the measured function. These relationships can be used for parameterizing particle size distributions.

2. INTERCEPTIONS OF A LINEAR PROBE WITH CIRCLES

We will begin with two-dimensional distributions of which the line-probe with the circle is possibly the most elementary example.

2.1 Linear Intercept Distribution for a Single Circle

In this section we are going to derive the probability density function $f_d(L)$, where $f_d(L)\,dL$ is the probability that a random linear probe will produce an intercept of length $L \pm dL/2$ in a circle of diameter *d*. We could start with the distribution of a linear probe in two dimensions defined by Eqn.(4.11), $f(\rho,\theta)=2/c$. However, for this highly isotropic case it is easier to note that the intercept length L is a function only of the perpendicular distance ρ of the probe from the center of the circle, Fig. 6.1.

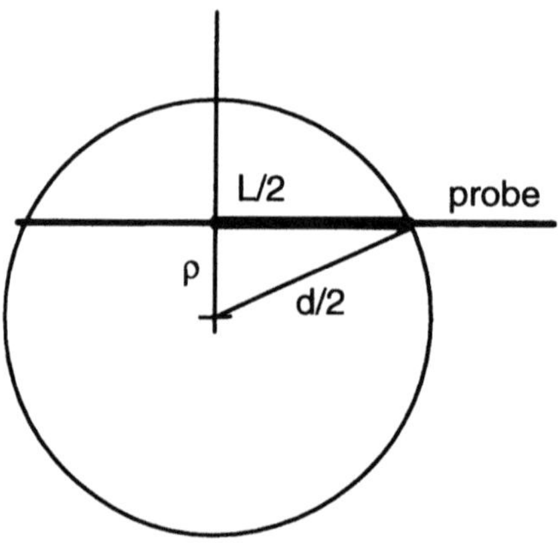

Figure 6.1: Linear probe intercepting a circle.

For a uniform distribution of ρ,

$$f(\rho)=2/d \quad :0\le\rho\le d/2. \tag{6.1}$$

This gives us a distribution for ρ when what we need is a distribution for L, the intercept length. Intercept length and ρ are related:

$$\rho^2 = \left(\frac{d}{2}\right)^2 - \left(\frac{L}{2}\right)^2 . \tag{6.2}$$

From this relation we can transform Eqn.(6.1).
Thus,

$$\frac{d\rho}{dL} = -\frac{L}{4\rho} . \tag{6.3}$$

Using Eqn. (2.51) for change of variable we obtain for a specific diameter d,

$$f_d(L) = f(\rho(L)) \cdot \left|\frac{d\rho}{dL}\right| . \tag{6.4}$$

Since $f(\rho(L)) = 2/d$,

$$f_d(L) = \frac{L}{d}\left(d^2 - L^2\right)^{-\frac{1}{2}} . \tag{6.5}$$

Needless to say, $0 \le L \le d$. The first moment about the origin, Eqn.(2.35) is

$$\overline{L} = \int_0^d L f_d(L) dL = \frac{1}{d}\int_0^d \frac{L^2}{\sqrt{d^2 - L^2}} dL = \frac{\pi d}{4} . \tag{6.6}$$

The third moment about the origin of the intercept distribution is related to the area and circumference of the circular feature by Eqn.(5.25). We can calculate this moment directly (by substituting $L = d \sin\theta$) into the integral; we obtain,

$$\mu_3'(L) = \int_0^d L^3 f_d(L) dL = \frac{3\pi}{16} d^3 . \tag{6.7}$$

Equation (5.15) gives $\overline{L} = \pi a / c$. Substituting πr^2 for the area and $2\pi r$ for the circumference gives $\overline{L} = \pi r / 2$ in agreement with Eqn.(6.6). Equation (5.25) gives $\mu_3'(L) = 3a^2 / c$. Making the same substitutions results in $\mu_3'(L) = 3\pi r^3 / 2$ in agreement with Eqn. (6.7).

For the cumulative distribution, Eqn. (6.5), integrated, with the suggested substitution, yields,

$$F_d(L) = 1 - \frac{1}{d}\sqrt{d^2 - L^2} \quad : 0 \le L \le d, \tag{6.8}$$

which satisfies the requirements that $F_d(0) = 0$ and $F_d(d) = 1$. The astute reader may remember seeing what amounts to this expression before in Problem (2.4).

These distributions can also be written in terms of the reduced variable $\lambda = L / d$. Noting that $f(\lambda) = f(L(\lambda)) \cdot dL / d\lambda$ we obtain from Eqn.(6.5)

$$f(\lambda) = \frac{\lambda}{\sqrt{1 - \lambda^2}} \quad : 0 \le \lambda \le 1. \tag{6.9}$$

Since the change of variable formula was derived by setting two cumulative distributions equal to each other and then differentiating, we may note that for the cumulative distribution only direct substitution is required to effect the transformation.

$$F(\lambda) = 1 - \sqrt{1 - \lambda^2} \quad : 0 \le \lambda \le 1. \tag{6.10}$$

2.2 Linear Intercept Distribution in an Aggregate of Circles

Consider a collection of circles in a plane. It will be assumed that the arrangement is such that if a linear probe were applied to the plane the length of intercepts on the circles are independent. That is, the circles do not touch or overlap; a close-packed arrangement would be a definite "no-no". We wish to determine the relationship between the density of distribution $N_A(d)$ of circle diameters per unit area of the plane and the resulting density of intercept lengths $N_L(L)$ per unit length of the probe.

Since diameter *d* is a variable, confusion between *d* as a variable and *d* as indicating a derivative or differential may arise in this section. To minimize this confusion, we will follow the convention of Mathematica™ and use curly-∂ for differentiation in this section since the distinction between partial and total differentiation is not important here. We define $N_L(L,d)\partial L\partial d$ as the number of intercepts of length $L \pm \partial L/2$ per unit length of probe on circles of diameter $d \pm \partial d/2$. This function is the product of two factors: (1) The expected number of interceptions on circles of diameter $d \pm \partial d/2$ per unit length of the probe; and, (2) the probability that, if a circle of the specified diameter is intercepted, the length of that interception would be $L \pm \partial L/2$. It follows from Eqn.(5.19) that the first factor is equal to $dN_A(d)\partial d$. By definition, the second factor is $f_d(L)\partial L$. Thus, $N_L(L,d)\partial L\partial D = dN_A(d)f_d(L)\partial L\partial D$, which with Eqn.(6.5) yields,

$$N_L(L,d) = \frac{LN_A(d)}{\sqrt{d^2 - L^2}} \,. \tag{6.11}$$

Now, the use of a linear probe implies a one-dimensional view of the image, since linear probes are one-dimensional. We can therefore determine only an intercept length distribution $N_L(L)$ and not and not the diameter of the circle(s) on which it occurred. In order to make Eqn.(6.11) more useful, we therefore integrate it with respect to d to obtain the marginal distribution as defined by Eqn.(2.47).

$$N_L(L) = L \int_L^{\sup d} \frac{N_A(d)}{\sqrt{d^2 - L^2}} \partial d \,, \tag{6.12}$$

where the lower limit of integration is L since a circle cannot yield an intercept larger than its diameter.

Now $N_L(L)$ is what we can measure, and $N_A(d)$ is what we would most likely want to know. This sort of inverse measurement problem is what distinguishes stereology apart from stochastic geometry in general. Equation (6.12) is related to one of the first inverse integral problems to be solved. That problem was studied by Abel (who died of illness from starvation and poverty[1]) and its general form was named *Abel's (integral) Equation* (of the first kind) after him. In Abel's Equation the variables under the radical appear to the first power. Equation (6.12) is worth studying apart from this

---

[1] It has been reported that when Abel sent Gauss some of his work, Gauss denounced it without actually reading it. (The unopened envelope was found among Gauss's effects.) Abel eventually did get offered a university appointment, but the news reached him only after he had died.

application since it illustrates a classic inverse problem in which the solution is obtained by a method of radicals.

Inverting Eqn.(6.12) is a long process. We begin by concocting a new integral obtained by multiplying $N_L(L)$ by $\left(L^2-d^2\right)^{-\frac{1}{2}}$ and integrating with respect to L from d to sup d. From the left had side we obtain,

$$I(d)=\int_{d}^{\sup d}\frac{N_L(L)}{\left(L^2-d^2\right)^{\frac{1}{2}}}\partial L. \qquad (6.13)$$

Applying this same manipulation to the right hand side of Eqn.(6.12) and introducing the dummy variable ξ to avoid confusion,

$$I(d)=\int_{d}^{\sup d}\frac{L}{\left(L^2-d^2\right)^{\frac{1}{2}}}\int_{L}^{\sup d}\frac{N_A(\xi)}{\left(\xi^2-L^2\right)^{\frac{1}{2}}}\partial\xi\partial L. \qquad (6.14)$$

The next step is to reverse the order of integration. This requires looking carefully at the region over which the integral is taken.

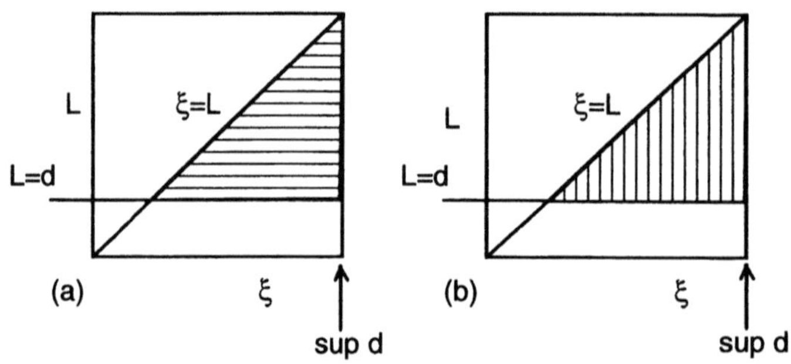

Figure 6.2: Regions of integration.

Equation (6.14) integrates over region as described in Fig. 6.2 part (a). The integral is first taken along the ξ-direction from $\xi = L$ to sup d, the maximum value of d. Then the second integration is taken along the L-direction from $L = d$ to sup d. When the order of integration is reversed, the first integral is taken in the L-direction from d to ξ. The second integration is now taken along the ξ-direction from d to sup d. Although the same region is covered, the limits change in this way when the order of integration is reversed.

$$I(d) = \int_{d}^{\sup d} N_A(\xi) \int_{d}^{\xi} \frac{L}{\left(L^2 - d^2\right)^{\frac{1}{2}} \left(\xi^2 - L^2\right)^{\frac{1}{2}}} \partial L \partial \xi \tag{6.15}$$

The inner integral when evaluated yields a somewhat surprising result. One path to evaluating that integral is first to make the substitution $u = \sqrt{L^2 - d^2}$. This transforms the inner integral to

$$\int_{0}^{\sqrt{\xi^2 - d^2}} \frac{\partial u}{\sqrt{(\xi^2 - d^2) - u^2}} .$$

Then making the substitution $u = \sqrt{\xi^2 - d^2} \sin\theta$ gives

$$\int_{d}^{\xi} \frac{L}{\left(L^2 - d^2\right)^{\frac{1}{2}} \left(\xi^2 - L^2\right)^{\frac{1}{2}}} \partial L = \frac{\pi}{2} . \tag{6.16}$$

And, $I(d) = \int_{d}^{\sup d} N_A(\xi) \frac{\pi}{2} \partial \xi$, which effects a considerable simplification,

$$\int_{d}^{\sup d} N_A(\xi) \partial \xi = \frac{2}{\pi} \int_{d}^{\sup d} \frac{N_L(L)}{\left(L^2 - d^2\right)^{\frac{1}{2}}} \partial L . \tag{6.17}$$

We could reasonably argue that the integrands are related, but substitution of d for L results in an undefined expression. Multiplying both integrals by (-1) to reverse the limits lets us see that argument of importance is d, which appears in the limits but unfortunately is also a parameter. Differentiating both sides would, in principle at least, give us what we seek, N_A as a function of N_L. Differentiation of an integral of this type depends on seeing it as a function of three variables: the two limits and the internal parameter. Following the discussion on page 220 of Courant [Courant, 1959], where F denotes the integral of f with respect to y and F' its derivative (The subscript x denotes partial differentiation.)

$$F'(x) = \int_{\psi_1(x)}^{\psi_2(x)} f_x(x,y)dy - \psi_1'(x) f(x,\psi_1(x)) + \psi_2'(x) f(x,\psi_2(x)) \,. \tag{6.18}$$

Applying this formula directly to Eqn.(6.17) again results in an undefined expression due to the singularity at *d*. This singularity can be removed by writing the Eqn.(6.17) as

$$\int_d^{\sup d} N_A(\xi)\partial\xi = \frac{2}{\pi} \int_d^{\sup d} \frac{N_L(L)}{L} \frac{L}{\left(L^2 - d^2\right)^{\frac{1}{2}}} \partial L \,. \tag{6.19}$$

Integration of the right hand side by parts gives

$$\left[\frac{N_L(L)}{L} \left(L^2 - d^2\right)^{\frac{1}{2}} \right]_d^{\sup d} - \int_d^{\sup d} \left(L^2 - d^2\right)^{\frac{1}{2}} \frac{\partial}{\partial L}\left(\frac{N_L(L)}{L}\right) \partial L \,.$$

The left hand term equals zero since it is obviously zero at the lower limit, and $N_L(\sup d)$ must also go to zero at least as the limit taken from above. The remaining integral now has an integrand which goes to zero at $L = d$ rather than becoming undefined. This integral can be differentiated using Eqn.(6.18). Only the first term in Eqn.(6.18) is non-zero when it is applied. The final result of differentiating both sides of Eqn.(6.19) is

$$N_A(d) = -\frac{2d}{\pi} \int_d^{\sup d} \left(L^2 - d^2\right)^{-\frac{1}{2}} \frac{\partial}{\partial L}\left(\frac{N_L(L)}{L}\right) \partial L \,. \tag{6.20}$$

We see that the success of Eqn.(6.20) depends on the differentiability of N_L.

2.3 Relationship Between Circle Diameter and Intercept Length Moments

As we shall now proceed to show, there is a one-to-one correspondence between the moments of the linear intercept distribution and those of the circle diameter distribution. Multiplying Eqn.(6.12) by L^{n-1} and integrating with respect to L we obtain

$$\int_0^{\sup L=\sup d} L^{n-1} N_L(L) \partial L = \int_0^{\sup d} L^n \int_L^{\sup d} \frac{N_A(\xi)}{\left(\xi^2 - L^2\right)^{\frac{1}{2}}} \partial\xi \partial L . \tag{6.21}$$

As in Eqn.(6.14) we have introduced the dummy variable ξ and continue to use curly-∂ to indicate a differential since it avoids worse confusion with the variable *d*. Again we change the order of integration and the limits.

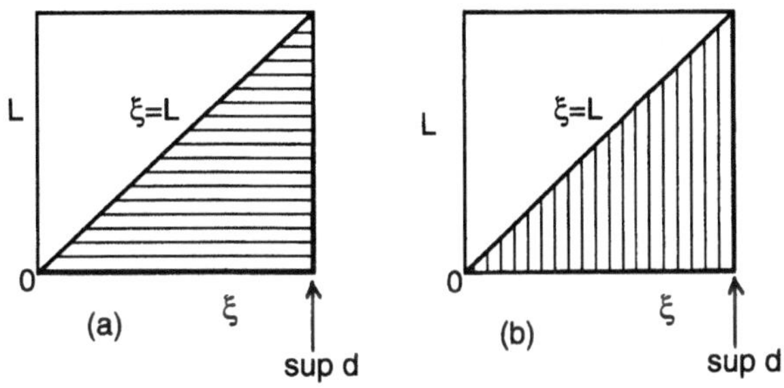

Figure 6.3: Regions of integration.

Originally the first integration proceeded along the ξ-direction followed by integration along the *L*-direction. These are now reversed and the consequences are the same as discussed above. The result is written

$$\int_0^{\sup L=\sup d} L^{n-1} N_L(L) \partial L = \int_0^{\sup d} N_A(\xi) \int_0^{\xi} \frac{L^n}{\left(\xi^2 - L^2\right)^{\frac{1}{2}}} \partial L \partial\xi . \tag{6.22}$$

Substituting $L = \xi \sin\theta$ gives on the right hand side

$$\int_0^{\sup d} N_A(\xi) \int_0^{\frac{\pi}{2}} \xi^n \sin^n \theta \partial\theta \partial\xi \text{, which equals } \int_0^{\sup d} N_L(\xi) \xi^n \frac{\sqrt{\pi}}{2} \frac{\Gamma\left(\frac{n+1}{2}\right)}{\Gamma\left(\frac{n}{2}+1\right)} \partial\xi .$$

As a consequence, Eqn (6.21) can be rewritten as

$$\int_0^{\sup d} d^n N_A(d) \partial d = \frac{2}{\sqrt{\pi}} \frac{\Gamma\left(\frac{n}{2}+1\right)}{\Gamma\left(\frac{n+1}{2}\right)} \int_0^{\sup d} L^{n-1} N_L(L) \partial L . \tag{6.23}$$

Noting that $N_L = \int_0^\infty N_L(L)\partial L$ and $N_A = \int_0^\infty N_A(d)\partial d$ we obtain

$$N_A \frac{\int_0^{\sup d} d^n N_A(d)\partial d}{\int_0^{\sup d} N_A(d)\partial d} = \frac{2}{\sqrt{\pi}} \frac{\Gamma\left(\frac{n}{2}+1\right)}{\Gamma\left(\frac{n+1}{2}\right)} N_L \frac{\int_0^{\sup d} L^{n-1} N_L(L)\partial L}{\int_0^{\sup d} N_L(L)\partial L},$$

which reduces to

$$N_A \mu_n'(d) = \frac{2}{\sqrt{\pi}} \frac{\Gamma\left(\frac{n}{2}+1\right)}{\Gamma\left(\frac{n+1}{2}\right)} N_L \mu_{n-1}'(L), \tag{6.24}$$

as a consequence of the definition of moments about the origin, Eqn.(2.35).

With $n = 1$, the formula $N_L = N_A \bar{d}$ results, which is Eqn.(5.19). Setting $n = 0$, we obtain

$$N_A = \frac{2}{\pi} N_L \mu_{-1}'(L), \tag{6.25}$$

where μ_{-1}' is the reciprocal of the harmonic mean of the intercept lengths. We can substitute Eqn.(5.19) into Eqn.(6.24) in order to eliminate N_A and N_L.

$$\mu_n'(d) = \frac{2}{\sqrt{\pi}} \frac{\Gamma\left(\frac{n}{2}+1\right)}{\Gamma\left(\frac{n+1}{2}\right)} \bar{d}\, \mu_{n-1}'(L) \tag{6.26}$$

3. INTERCEPTIONS OF A LINEAR PROBE WITH STRIPES

In this section we will be considering linear interceptions in a two-dimensional stripe. We will restrict ourselves to stripes which are long enough that compared with their width that for it to be justifiable to ignore interceptions that pass through the end of the stripe. The results will also be applicable, at least as approximations to "non-linear" curved stripes provided the radius of curvature is large compared with the width of the stripe. Stripes would seem to be a simple concept. But, using the form of the probe developed in section 4.3 turns out to be very inconvenient. Some of this inconvenience can be illustrated by a straw-man derivation which goes as follows. Consider a stripe of length l and width t located in a plane as shown in Fig. 6.4.

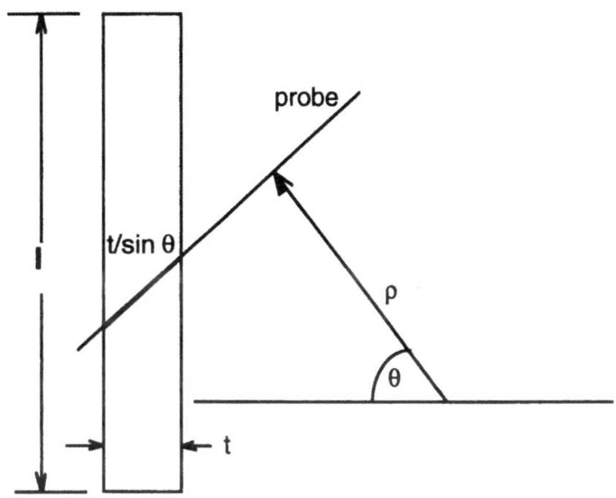

Figure 6.4: A linear probe intersection a stripe.

It is assumed that l >> t. We wish to determine the distribution $f_t(L)$ of intercept lengths on a random linear probe. We define the orientation of the probe by the angle θ between the perpendicular to the probe and that to the stripe. By the geometry of Fig. 6.4 $L = t/\sin\theta$. Ignoring end effects, translation of the probe does not change the intercept length. We need therefore only consider randomization with respect to the orientation θ, which has the distribution $f(\theta) = 2/\pi : 0 < \theta < \pi/2$. Since $f_t(L) = f(\theta)|d\theta/dL|$, we have $f_t(L) = \dfrac{2t}{\pi L\sqrt{L^2 - t^2}} \quad : L > t$. The lower limit is t because this is the shortest intercept a stripe of width t can produce. We may now proceed to find the average intercept length. We have: $\overline{L} = \int_t^{\infty} L \dfrac{2t}{\pi L\sqrt{L^2 - t^2}} dL$, which becomes infinite. On the other hand we have from Eqn.(5.15),

$$\bar{L} = \frac{\pi a}{c} = \frac{\pi l t}{2(l + t)} \underset{l \to \infty}{\to} \frac{\pi t}{2}. \tag{6.27}$$

This should alert us to a common error in stereology. As to the source of the error, it is said: "The error in the derivation was due to the failure to allow for the fact that the probability of the probe intercepting the stripe is a function of orientation".-J.H. It would seem true that were end effects to be neglected a vertical probe could not intercept the stripe. On the other hand, we are only interested in probes that intercept the stripe. Another point that should have alerted us to potential trouble is that a two-variable expression is needed for the probability density, and we used only a one-variable distribution saying that the other variable was meaningless. At this point we must begin anew. We could begin with the geometry of Fig 6.4. It turns out to be very awkward. It might be convenient to place the origin of the probe coordinate system outside the feature being probed. But, we might notice that when using (ρ, θ) as a probe coordinate system in the past, the origin was always inside the feature being tested, even in the case of aggregates. This is not a necessary assumption. In problem 6.2 the verification that the origin may be either inside or outside the figure is addressed.

3.1 Linear Intercept Distribution for a Single Stripe

Figure 6.4 was used to derive a false distribution of probe intercept lengths for the infinite stripe. It is possible to use the geometry of Fig. 6.4 to derive a correct distribution of intercept lengths, but it is not easy. For the case of the finite stripe, it is very tedious since there are many special cases giving rise to many separate integrals with complicated limits of integration. Although many of these cases can be discounted and broad generalizations made, the simplified derivation that results is not convincing. For the case of the infinite stripe, the derivation using the Fig. 6.4 geometry, which is associated with the Section 4.3 distribution, seems as much like hand waving, as did the false derivation just given. So, instead of proceeding directly using the probe in the form derived in Section 4.3, we will map its distribution onto a simpler geometry and use the transformation method of Section 2.3.5 to define a new probability density function for it. This form of the distribution may not appear to explicitly meet the requirements of the "limited axiom" of that section. However, since it is a mapping of all the probe lines of the distribution of Section 4.3, which does meet the axiomatic requirement, it functions as a good distribution.

The probe of Section 4.3 intersects a positioning line at right angles. The positioning line itself is placed at an angle θ, and the point at which the probe line intersects the positioning line is measured as ρ units from the origin. What would be more convenient for a stripe would be to reverse the order. The positioning line would remain fixed and the probe line would intersect it at angle θ with respect to the normal, at a distance ρ from the origin. This geometry introduces an obvious distortion: the positioning line must be infinite to capture all the probe lines passing through even the smallest finite stripe; some form of angular weighting is likely to appear. However the convenience of having the positioning line identified with one side of the stripe more than compensates for the angular weighting.

Figure 6.5 illustrates the mapping the two distributions onto each other. Each system has its own origin. The coordinates (ρ_1,θ_1) represent those of the probe in the system of Section 4.3. The coordinates (ρ_2,θ_2) represent the probe in the new system. The ranges of the coordinates are $-\pi$ to π for θ_1 since ρ_1 takes only positive values; θ_2 takes values in the range $-\pi/2$ to $\pi/2$ while ρ_2 ranges from $-\infty$ to ∞. Note that in Fig. 6.5, θ_2 takes a negative value.

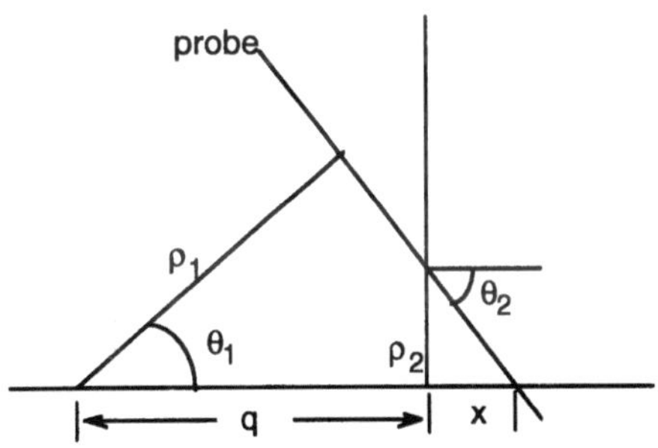

Figure 6.5: Two defining geometries for the same probe line.

We would like to define the transformation between the two coordinate systems. To begin, we will examine the relation between θ_1 and θ_2. By similar triangles we see that

$$\theta_2 = \theta_1 - \frac{\pi}{2}. \tag{6.28}$$

We also see from the large triangle on the left that $\cos\theta_1 = \dfrac{\rho_1}{q+x}$ while from the small triangle on the right, $\tan\theta_1 = \dfrac{x}{\rho_2}$. It follows that,

$$\rho_2 = \frac{\rho_1}{\sin\theta_1} - q\cot\theta_1 , \tag{6.29}$$

where q is the distance between the origins of the two coordinate systems. The Jacobean of the transformation (Eqn. 2.52) is thus

$$\begin{vmatrix} \frac{\partial \rho_1}{\partial \rho_2} & \frac{\partial \rho_1}{\partial \theta_2} \\ \frac{\partial \theta_1}{\partial \rho_2} & \frac{\partial \theta_1}{\partial \theta_2} \end{vmatrix} = \begin{vmatrix} \cos\theta_2 & -\rho_2 \sin\theta_2 - q\cos\theta_2 \\ 0 & 1 \end{vmatrix} . \tag{6.30}$$

It follows that the probability density function stated in terms of the variables corresponding to the new geometry is,

$$f(\rho_2, \theta_2) = f(\rho_1, \theta_1)|\cos\theta_2| = \frac{|\cos\theta_2|}{c} . \tag{6.31}$$

We can test the goodness of this result by finding the average intercept length for the infinite stripe. In order to do this we need to lay out the appropriate geometry. Figure 6.6 shows a single stripe.

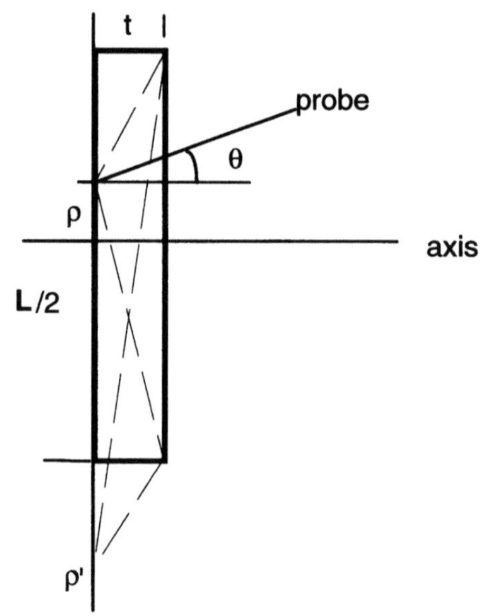

Figure 6.6: A single stripe.

The stripe is divided in the middle by the axis. Each half has length L /2 and width t. There are three basic cases for intercepts. The first is where the probe enters and leaves through the side of the stripe. This is the case as shown with the probe shown at ρ when its angle lies between the two dashed lines. The second case is that where the probe makes contact with a side and an end. This is illustrated for the probe at position ρ' having an angle between the two dashed lines. The third case is that where the probe enters and leaves through the end of the stripe. This is illustrated by that for the probe at ρ' having an angle between the vertical axis and the dashed line.

Considering the first case, the dashed lines emanating from the intercept point ρ define the angular limits of that case. The upper limit is given by $\theta_{\max} = \arctan\frac{\frac{L}{2}-\rho}{t}$ while that for the lower limit is given by $\theta_{\min} = \arctan-\left(\frac{\frac{L}{2}+\rho}{t}\right)$. Integrating intercept length times the density function between these two limits accounts for all possible ways a probe line can pass through both sides. Considering the second case, ρ must be either greater than $L\ /2$ or less than the negative of it. (The same is true for the third case.) When ρ' is negative, as shown in Fig. 6.6 for the second case, $L = \frac{t}{|\cos\theta|} - \frac{|\rho'| - \frac{L}{2}}{|\sin\theta|}$. Returning to the first case, $L = \frac{t}{|\sin\theta|}$ for all possible orientations. From these arguments, it is possible to calculate the intercept lengths for all possible cases. For the case in point, the infinite stripe, we see that as the stripe becomes very long $\arctan\frac{\frac{L}{2}-\rho}{t}$ approaches $\pi/2$ for all values of ρ squeezing out the other cases. That is, the probability in the sense of relative frequency of the probe passing through the ends falls to zero when the ends of the stripe contribute insignificantly to its circumference. In terms of the density function, we see that $\cos\pm\frac{\pi}{2} = 0$. The studious reader may like to see on her own whether the possible contribution of an infinite length times a zero probability would affect some higher moment of the length distribution. Here we are only considering the first for which we have an easy check.

For an infinite stripe, the first case discussed above seems to be the only relevant one. Integrating the product of the intercept length times the probability density function we obtain,

$$\bar{L} = \int_{-L/2}^{L/2} \int_{\arctan-\left(\frac{\frac{L}{2}-\rho}{t}\right)}^{\arctan\frac{\frac{L}{2}-\rho}{t}} \frac{t}{|\cos\theta|}\frac{|\cos\theta|}{c} d\theta d\rho = \frac{\pi t}{2} \tag{6.32}$$

as expected from Eqn.(6.27).

Finding the average length does not give us the distribution of lengths, which is really what we set out to find. In order to find the distribution of lengths we must establish a relationship between the single variable, intercept length, and the probability of its occurrence. The difficulty is that both length and its probability are functions of two variables. Let us first examine the general case where the points (ρ, θ) form a continuous curve for a fixed value of intercept length L. For each point (ρ, θ) the intercept length is given by $L(\rho,\theta)$ and the probability of a probe falling into that point's neighborhood is $f(\rho,\theta)d\rho d\theta$. If all the values of ($\rho$, θ) which belong to L form a continuous curve, we can parameterize it in terms of s and call it $S_L(s)$. At any point, since L is constant, dL is normal to the tangent to S. Hence the curves $S_L(s)$ and those of L form a coordinate net. We can therefore transform $f(\rho,\theta) \rightarrow f'(s,L)$ and integrate out s to find $f(L)$, the distribution function for L that we seek.

Implementing the above strategy,

$$f(L) = \int_{(s)} f(\rho(s,L),\theta(s,L)) \cdot |J| ds . \tag{6.33}$$

We now need to look at the case of the stripe in particular. Consider Fig. 6.6, the range of θ over which probe lines enter and leave through the sides of the stripe. Since $L = \frac{t}{|\cos\theta|}$, curves of constant L are lines of constant θ. We can diagram the region of integration. Qualitatively, it resembles Fig. 6.7 below.

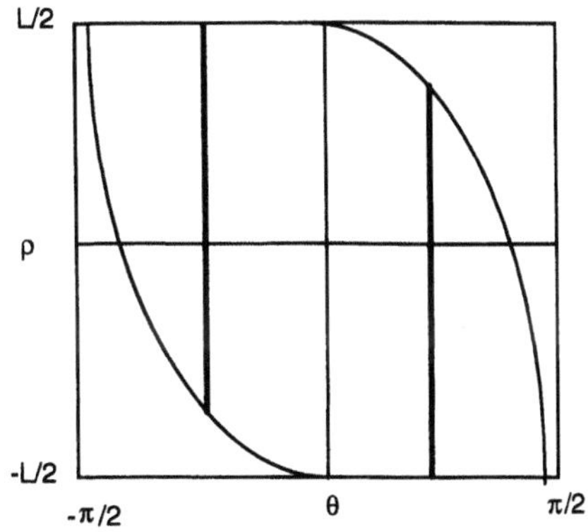

Figure 6.7: Region of integration showing two curves for the same L.

The significance of the absolute value sign in the expression for L is seen through the fact that there are two lines of constant intercept length (heavy lines) in Fig. 6.7 for each value of L. Since the cosine is even, the absolute value would seem gratuitous. Fig. 6.7 also indicates the general appearance of the limits of integration (light curves) for ρ. On account of the symmetry, we need only integrate over a region s_u in which θ is positive.

It follows that, $f(L)=2\int_{s_u} f(\rho(s,L),\theta(s,L))\left\|\begin{matrix}\frac{\partial\rho}{\partial s} & \frac{\partial\rho}{\partial L}\\ \frac{\partial\theta}{\partial s} & \frac{\partial\theta}{\partial L}\end{matrix}\right\| ds$.

Since $\theta=\arccos\frac{t}{L}$,

$$f(\rho(s,L),\theta(s,L))=\frac{t}{Lc}. \tag{6.34}$$

The Jacobean evaluates to $\left|\begin{matrix}1 & 0\\ 0 & \frac{t}{L\sqrt{L^2-t^2}}\end{matrix}\right|=\frac{t}{L\sqrt{L^2-t^2}}$ since ρ and s are parallel, and therefore ρ and L are perpendicular in the map.

The next step is to evaluate what is left of the integral, Eqn.(6.33), simply the length of s which equals the length of the stripe less the projection of the intercept onto the side of the stripe, $\mathrm{L} - t\tan\theta = \mathrm{L} - \sqrt{L^2 - t^2}$. In the limit of L approaching infinity, the perimeter c approaches 2L. As a result,

$$f(L) \xrightarrow[\mathrm{L}\to\infty]{} f_t(L) = \frac{t^2}{L^2\sqrt{L^2 - t^2}}\ , \tag{6.35}$$

where $t \le L < \infty$. The reader can verify that the average for this distribution satisfies Eqn.(6.32).

Apart from the matter of not explicitly taking account of the two-variable nature of the problem, the first derivation had a more specific error. While this error was eliminated by being systematic, examining it gives one a qualitative feeling for a quantitative aspect of the geometry. After all, we did not take much account of the same two variables when we examined circles; yet our derivation turned out all right. And, it must be possible to do the stripe derivation without recourse to the Jacobean since the Jacobean matrix is diagonal. The error was again "due to the failure to allow for the fact that the probability of the probe intercepting the stripe is function of the orientation of the probe."-J.H. (What is meant here is the probability per unit length of probe, since by the definition of the problem, the probe in question always intercepts the stripe.) In the faulty derivation, the standard probe geometry was used which made the length of the intercept inversely proportional to the sine of the angle. In the "new" geometry the sine is replaced by the cosine. This explains the canceling of cosines in Eqn.(6.32). (The same thing happens in Eqn.(6.35) but it is hidden by the nature of the derivation.) In the "new" geometry, we see that a probe at an angle $\pi/2$ has a zero probability (per unit length) of intercepting the stripe since it is infinitely long and intercepts only once. Similarly in the standard geometry, a probe at an angle 0 behaves the same way. As already noted, this failure to allow for the probability of interception is a common source of error and has been the cause of several fallacious results in the literature. Fortunately, one can always check the derivation of a distribution by means of the general expression for $\overline{L}$; and, when the third moment exists, by Crofton's second theorem discussed in the previous chapter.

Integration of Eqn.(6.35) yields for the cumulative distribution,

$$F_t(L) = \frac{1}{L\sqrt{L^2 - t^2}}, t \le L < \infty\ . \tag{6.36}$$

The distribution can also be described in terms of the reduced variable, $\lambda = t/L$. This gives,

$$f(\lambda)=\frac{\lambda}{\sqrt{1-\lambda^2}} \tag{6.37}$$

and,

$$F(\lambda)=1-\sqrt{1-\lambda^2}\ , \tag{6.38}$$

where, $0 \le \lambda \le 1$. Notice that in terms of λ the distribution now proceeds from the largest intercept to the smallest.

3.2 Linear Intercept Distribution in an Aggregate of Stripes

The number per unit area is not usually a physically meaningful property for features such as infinitely long stripes. A more appropriate measure of the density of is $L_A(t)$ defined such that $L_A(t)dt$ is the length per unit area of the plane of stripes having width $t \pm dt/2$. Alternatively, one could use the areal fraction $A_A(t)$ similarly defined in terms of *t*. Since the two functions are related by,

$$A_A(t)=tL_A(t), \tag{6.39}$$

once one of these is known, the other can be immediately derived. So, we will consider only the estimation of $L_A(t)$.

Let $N_L(L,t)dLdt$ be the number of probe intercepts of length $L \pm dL/2$ per unit length of probe on stripes of width $t \pm dt/2$. In words, this equals the product of the expected number per unit probe length of intercepts with stripes within the specified range of widths times the probability that the resulting intercept falls within the specified range of lengths. We can use Eqn.(5.31) to give us the first factor if we specify that a probe line never ends within a stripe. In that case, if a probe line intersects the centerline of a stripe, it has an intercept with that stripe. Hence the first factor becomes $(2/\pi)L_A(t)$. The second factor is given by Eqn.(6.35), and

$$N_L(L,t)=\frac{2t^2L_A(t)}{\pi L^2\sqrt{L^2-t^2}}\,. \tag{6.40}$$

Integrating with respect to *t* gives the marginal distribution,

$$N_L(L)=\frac{2}{\pi L^2}\int_0^L \frac{t^2 L_A(t)}{\sqrt{L^2-t^2}}dt, \tag{6.41}$$

in which the upper limit is L since stripes wider than L cannot produce an intercept having that length. Except for the direction of integration and the order of variables under the radical, Eqn.(6.41) resembles Eqn.(6.12). It is another member of the family of Abel's type equations. Members of this family are listed by Polianin [Polianin and Manzhirov, 1998] who treats them as distinct forms. While they can all be solved by the method used to solve Eqn.(6.12) the form of which, by the way, is not listed by Polianin, only one of them is actually Abel's Equation. Equation (6.41) corresponds to Polianin's type 1.1.40. Abel's Equation of the first type, like Eqn.(6.12) and Eqn. (6.41) is a linear integral equation of the class termed *Volterra Equations of the first kind.* These are characterized by having the left hand side argument appear as a limit of the right hand side integral. What appears in the integrand besides the unknown function, $L_A(t)$ in Eqn.(6.41), is termed the *kernal* which also contains the argument of the left hand side.

Abel's Equation is written,

$$f(x)=\int_a^x \frac{y(t)dt}{\sqrt{x-t}}, \tag{6.42}$$

and has a solution,

$$y(x)=\frac{1}{\pi}\frac{d}{dx}\int_a^x \frac{f(t)dt}{\sqrt{x-t}}. \tag{6.43}$$

The general form applicable to Eqn. (6.41) is

$$f(x)=\int_a^x \frac{y(t)dt}{\sqrt{x^2-t^2}} \tag{6.44}$$

that has a solution,

$$y(x)=\frac{2}{\pi}\frac{d}{dx}\int_a^x \frac{t f(t)dt}{\sqrt{x^2-t^2}}. \tag{6.45}$$

Setting $f(t)=t^2 L_A(t)$ in Eqn. (6.44) and comparing with Eqn. (6.41) allows the desired inversion to be obtained from Eqn.(6.45),

$$L_A(t)=\frac{1}{t^2}\frac{d}{dt}\int_0^t \frac{\xi^3 N_L(\xi)d\xi}{\sqrt{t^2-\xi^2}} \quad . \tag{6.46}$$

As in the case of deriving Eqn. (6.20), We can differentiate the integral in Eqn. (6.46) by first integrating by parts to avoid the singularity. The right hand side of Eqn. (6.20) becomes

$$\frac{1}{t^2}\frac{d}{dt}\left[\left[-\xi^2 N_L(\xi)\left(t^2-\xi^2\right)^{\frac{1}{2}}\right]_0^t + \int_0^t \left(t^2-\xi^2\right)^{\frac{1}{2}} \frac{d}{d\xi}\left(\xi^2 N_L(\xi)\right)d\xi\right] .$$

The UV term (in the customary form for integration by parts), to the left of the integral sign, is evidently zero at its upper limit. By defining $N_L(0)$ equal to zero, the entire UV term drops out. Differentiating the integral then gives

$$L_A(t)=\frac{1}{t}\int_0^t \frac{\frac{d}{d\xi}\left(\xi^2 N_L(\xi)\right)}{\sqrt{t^2-\xi^2}}d\xi \ . \tag{6.47}$$

Thus we have found the distribution of stripe widths as a function of the distribution of interception lengths. The utility of such an expression is not often to be found in microscopy or other cases where the stripes themselves can be seen directly. However, were we to wish to analyze stripe domains in a polycrystalline nickel film using say a moving magnetic sensor, Eqn. (6.47) might prove to be helpful. It might also be useful to modeling the detectability of a bar code in an image.

3.3 Relationship between Stripe Width and Linear Intercept Moments

Consider the moments of intercept length. There is no upper bound on intercept length. As you increase the n in the moment μ_n, the result becomes increasingly dominated by the large values in the distribution. Since there is no upper bound on the lengths of intercepts, one would expect that μ_n would become undefined for n greater than some value. It is remarkable that even an average exists in this case. The value of n may of course be negative. In that case we would be finding the expected value of the reciprocal of intercept lengths raised to some power. It turns out that we can obtain a relationship between the negative moments of the intercept length distribution and the negative moments of the stripe width distribution. If the stripe width distribution is suitably bounded, as we would expect it to be in most real cases, positive moments would be expected to exist and characterize its distribution. The positive and negative moments of a distribution are related through the *moment generating function* or more generally its complex form, the characteristic function, discussed in Section 2.3.3. From this we may intuit that negative moments are a little like negative Fourier coefficients: they may be necessary to or redundant in the description of a function depending on the function and how it is portrayed in the transform space.

Multiplying Eqn.(6.41) by L^n and integrating with respect to L we obtain

$$\int_0^\infty L^n N_L(L)dL = \frac{2}{\pi}\int_0^\infty L^{n-2}\int_0^L \frac{\xi^2 L_A(\xi)d\xi}{\sqrt{L^2-\xi^2}}dL\,. \tag{6.48}$$

Changing the order of integration, Fig. 6.8, from that shown in (a) to that shown in (b)

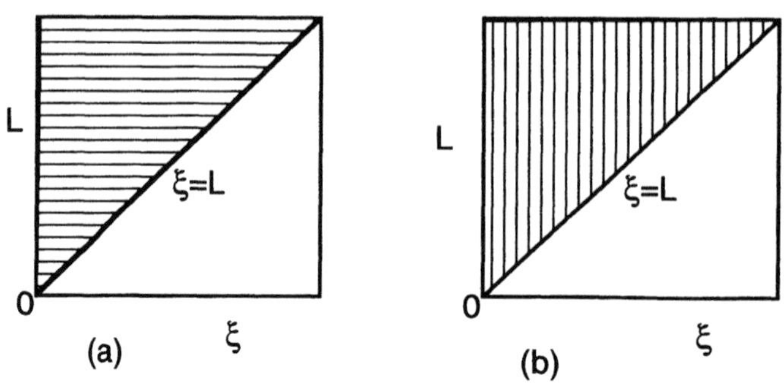

Figure 6.8: Regions of integration

results in

$$\int_0^\infty L^n N_L(L)dL = \frac{2}{\pi}\int_0^\infty \xi^2 L_A(\xi)\int_\xi^\infty \frac{L^{n-2}\,dL}{\sqrt{L^2-\xi^2}}d\xi\ .$$

Substituting $L = \xi/\sin\theta$ we obtain,

$$\int_0^\infty L^n N_L(L)dL = \frac{2}{\pi}\int_0^\infty \xi^n L_A(\xi)\int_0^{\pi/2} \sin^{1-n}\theta d\theta d\xi \ .$$

The sine appears to the power 1-n. This integral will not converge for any power of the sine less than 0. Accordingly the widest possible range of n is (-∞, 1]. The sine integral evaluates to a numerical factor yielding,

$$\int_0^\infty L^n N_L(L)dL = \frac{1}{\sqrt{\pi}}\frac{\Gamma\left(\frac{2-n}{2}\right)}{\Gamma\left(\frac{3-n}{2}\right)}^{\text{sup}\,t}\int_0^{} t^n L_A(t)dt \ . \tag{6.49}$$

We may "stretch the concept of a distribution function a little" and define $f(t)dt$ as the fraction of stripe length having a width $t \pm dt/2$. Then,

$$L_A f(t) = L_A(t), \tag{6.50}$$

and analogously

$$N_L f(L) = N_L(L). \tag{6.51}$$

Furthermore N_L and L_A are related by Buffon's equation which gives $L_A = \frac{\pi}{2} N_L$. Hence,

$$\mu_n'(L) = \frac{\sqrt{\pi}}{2}\frac{\Gamma\left(\frac{2-n}{2}\right)}{\Gamma\left(\frac{3-n}{2}\right)}\mu_n'(t), \tag{6.52}$$

where n ranges from negative infinity to positive unity. As mentioned, while positive moments greater than the mean would not be expected for the intercept length distribution, they would be expected for stripe width distribution. For a known distribution of widths, the relationship between its positive and negative moments can be deduced through application of the Mellin Transform, Section 2.2.3. This and other useful relationships between negative and positive moments are discussed by Cressie et al. [1981] and by Chao and Strawderman [1972].

4. INTERCEPTIONS OF A LINEAR PROBE WITH SPHERES

In this section we will be considering linear interceptions within three-dimensional spheres. In Chapter 4, we introduced a "Limited Axiom" to describe a random linear probe. From this axiom we deduce that random probe lines will be uniformly distributed with respect to any plane intersecting any sphere and any solid angle.

4.1 Linear Intercept Distribution in a Single Sphere

In this sub-section we will derive the function $f_D(L)$ where $f_D(L)dL$ is the probability that a random linear probe will produce an intercept of length $L \pm dL/2$ in a sphere of diameter D as shown in Fig. 6.9. The probe line is perpendicular to the hatched plane. In that plane the probe line intersects the radius, marked *D*/2 at the point indicated a distance ρ from the center of the sphere.

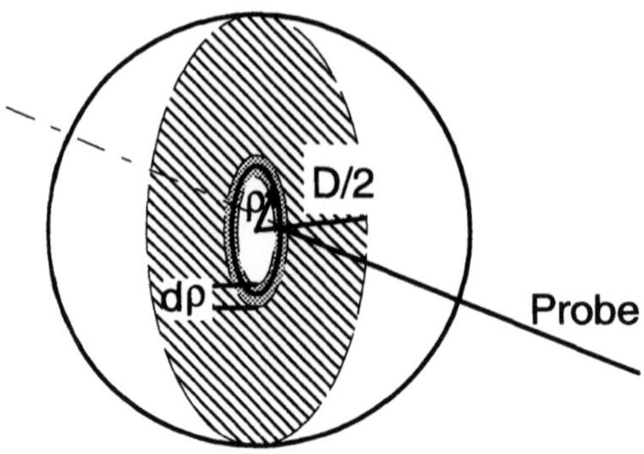

Figure 6.9: Linear probe geometry for a sphere

Since the sphere is spherically symmetrical, it will be noted that L only depends on the perpendicular distance ρ from the center of the sphere. Furthermore, since by the "Limited Axiom" the probe is uniformly distributed on the diametral plane normal to the probe, $f(\rho)d\rho$ is equal to the ratio of the area of the annulus of mean radius ρ and width $d\rho$ to the area of the diametral plane. (Compare with Problem 2.3.) And so,

$$f(\rho)d\rho = \frac{8\pi\rho d\rho}{\pi D^2} = \frac{8\rho d\rho}{D^2} \quad . \tag{6.53}$$

Now by the transformation rule much celebrated in the previous sections,

$$f_D(L) = f(\rho(L))|d\rho / dL| \; ; \tag{6.54}$$

and, since $\rho^2 = \frac{D^2}{4} - \frac{L^2}{4}$,

$$\frac{d\rho}{dL} = -\frac{L}{4\rho}. \tag{6.55}$$

Substituting from Eqns. (6.53) through Eqn, (6.55) we get

$$f_D(L) = \frac{2L}{D^2}. \tag{6.56}$$

The variable L is greater than or equal to zero but less than D. Finding the cumulative distribution is easy; it is

$$F_D(L) = \frac{L^2}{D^2}. \tag{6.57}$$

These distributions can also be written in terms of the reduced intercept length $\lambda = L / D$:

$$f(\lambda) = 2\lambda, \tag{6.58}$$

$$F(\lambda) = \lambda^2. \tag{6.59}$$

Here λ ranges between 0 and one.

4.2 Linear Intercept Distribution in an Aggregate of Spheres

We next consider the intercept distribution on a linear probe passing through an aggregate of spheres whose size distribution is defined by $N_V(D)$ where $N_V(D)dD$ is the number of spheres per unit volume with diameters in the range $D \pm dD/2$. We first determine the distribution $N_L(L, D)$ where $N_L(L, D)\, dLdD$ is the number per unit length of probe of intercepts in the range $L \pm dL/2$ that occur on spheres of diameter range $D \pm dD/2$. It follows from Eqn.(5.51), which relates the expected number of interceptions per unit probe length to the number of bodies per unit volume, that the expected number of interceptions per unit length of a probe

on spheres of diameter $D \pm dD/2$ is $(\pi/4)D^2 N_V(D)dD$, since the surface area of a sphere is πD^2. Combining this with the distribution of intercept lengths on a single sphere, Eqn.(6.56), yields for the joint distribution

$$N_L(L,D) = (\pi/2)LN_V(D). \tag{6.60}$$

Integrating we obtain what might qualify as the "marginal distribution" albeit non-normalized,

$$N_L(L) = \frac{\pi L}{2} \int_L^{\sup D} N_V(D)dD \,. \tag{6.61}$$

In this expression, the lower limit of integration is non-zero because the smallest sphere that can yield an intercept of length *L* has a diameter *D* = *L*. Since *L* does not appear in the integrand, this equation can be inverted to give the sphere diameter distribution from the intercept length distribution. Differentiation of both sides, cf. Eqn.(6.18), gives

$$N_V(D) = -\frac{2}{\pi}\frac{d}{dD}\left[\frac{N_L(D)}{D}\right] . \tag{6.62a}$$

Since this expression may seem slightly confusing, another way of stating Eqn.(6.62) is

$$N_V(D) = -\frac{2}{\pi} g(D), \quad \text{where } g(t) = \frac{d}{dt}\left[\frac{N_L(t)}{t}\right] . \tag{6.62b}$$

Furthermore, Eqn.(6.62) can also be written as above where,

$$g(L) = \frac{1}{L}\frac{d}{dL} N_L(L) - \frac{N_L(L)}{L^2} \,. \tag{6.63}$$

The foregoing derivation is similar to that given by Cahn and Fullman [1956]. Other derivations include those by Spektor [1950] and by Lord and Willis [1951].

Since we are already applying the language of (normalized) probability distribution functions to distributions such as appear in Eqn.(6.62), we may extend the analogy further by defining a "cumulative distribution" for sphere diameters. This follows from Eqn.(2.33) and Eqn.(6.61).

$$N_V\,F(D) = N_V - \frac{2N_L(D)}{\pi D}, \tag{6.64}$$

where the convention of Eqn.(6.62a) is followed, N_V is the total number of sphere centers per unit volume and *F*(*D*) is the true normalized cumulative distribution function for sphere diameters which we are using to avoid having to make up a new symbol for the product $N_V\,F(D)$. Note that Eqn.(6.64) was obtained by examining the limits in Eqn.(6.61) rather than by integrating Eqn.(6.62) since the latter has lost through differentiation a constant which results from the presence of the upper limit in Eqn. (6.61).

The final step is to determine N_V itself. By definition, as used to derive Eqn.(6.64), $N_V = \int_0^{\sup D} N_V(D)dD$. It thus follows from Eqn.(6.61) by dividing by *L* that

$$N_V = \frac{2}{\pi}\lim_{L\to 0}\frac{N_L(L)}{L}. \tag{6.65}$$

Evidently $N_L(L) \xrightarrow[L\to 0]{} 0$ if the limit is to exist. Applying L'Hopital's rule,

$$N_V = \frac{2}{\pi}\lim_{L\to 0}\frac{d}{dL}N_L(L). \tag{6.66}$$

Alternatively the same result can be obtained by differentiating Eqn.(6.61) directly. Thus, if only N_V is needed, it is not necessary to measure the complete distribution $N_L(L)$ but only values near zero. Since $N_L(L)$ and $N_V(D)$ are related by an integral, Eqn. (6.61), there is an integral transform relation between $N_L(L)$ and $N_V(D)$ when that integral is written so that it is taken over the entire real line. The ability to relate information about the whole of one function to that near zero of the other is often a property of integral transforms and hints at how problems of this type might be solved generally.

4.3 Relationship Between Sphere Diameter and Intercept Length Moments

Multiplying Eqn.(6.61) by L^{n-2} and integrating with respect to L gives us

$$\int_0^{\sup D} L^{n-2} N_L(L) dL = \frac{\pi}{2} \int_0^{\sup D} L^{n-1} \int_L^{\sup D} N_V(D) dD dL . \tag{6.67}$$

Changing the order of integration as diagramed in Fig. 6.10,

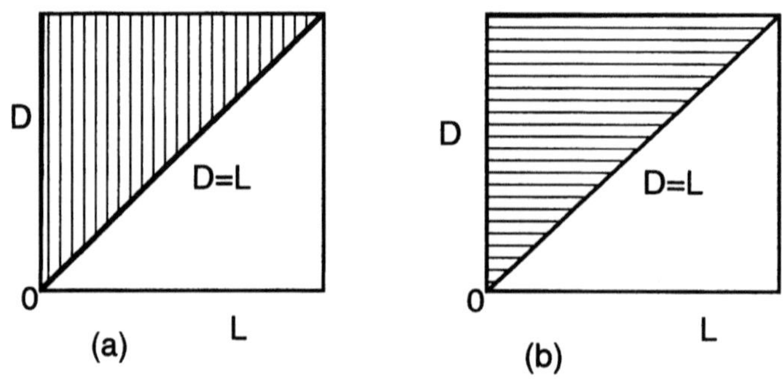

Figure 6.10: Regions of integration

$$\int_0^{\sup D} L^{n-2} N_L(L) dL = \frac{\pi}{2} \int_0^{\sup D} N_V(D) \int_0^{D} L^{n-1} dL dD . \tag{6.68}$$

Carrying out the integrations with respect to L and flipping things about gives us

$$N_V \mu_n'(D) = \frac{2n}{\pi} N_L \mu_{n-2}'(L) . \tag{6.69}$$

Before proceeding, we should examine how this was obtained. The above integration should work except when $n = 0$; because in that case, we would obtain a logarithm rather than a power of n. Equation (6.68) formally holds for negative moments provided the left hand side integral converges at the lower limit. Since $N_L = S_V / 4$ (one interception for each pair of intersections) and $S_V = \pi N_V \mu_2'(D)$, Eqn.(6.69) can also be written as

$$\mu_n'(D) = \frac{n}{2}\mu_2'(D)\mu_{n-2}'(L), \tag{6.70}$$

where $n \neq 0$. Because of this limitation on *n*, we cannot determine N_V from Eqn. (6.68). However, we have already shown it can be derived from the slope of $N_L(L)$ near zero, Eqn.(6.62).

5. INTERCEPTIONS OF A PLANAR PROBE WITH SPHERES

The interception of a planar probe with a sphere, Fig. 6.11, produces a circle. The diameter of this circular intercept can vary from zero to the diameter of the largest sphere in a group of spheres. There are two distributions of practical interest; one is with respect to the intercept diameters and the other is with respect to their areas. We will consider both.

5.1 Intercept Diameter Distribution

Comparing Fig. 6.1, a linear probe intersecting a circle, with Fig. 6.11, reveals that the intercept diameters generated by a planar probe in a sphere will be identical with that for a circle since both depend only on the perpendicular distance of the probe from the center. Consequently, all the results derived in Section 6.2 are directly applicable to the present case. Since a change in notation is involved, the principal results will be listed here for convenience in future reference.

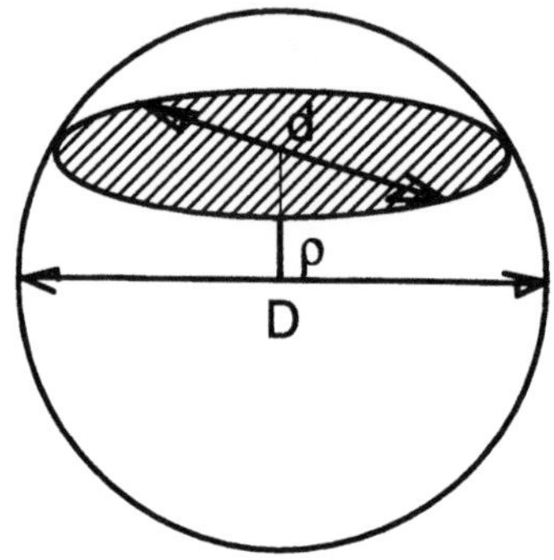

Figure 6.11: Planar intercept with a sphere

For the distribution of intercept diameters, d in a sphere of diameter D, we have from Eqn. (6.5)

$$f_D(d)=\frac{d}{D}\left(D^2-d^2\right)^{-\frac{1}{2}} . \tag{6.71}$$

As before, the equation only makes sense for $d < D$. The cumulative probability distribution is given by adaptation of Eqn.(6.8),

$$F_D(d)=1-\frac{1}{D}\left(D^2-d^2\right)^{\frac{1}{2}} . \tag{6.72}$$

In terms of the reduced variable λ, Eqns.(6.9) and (6.10) stay the same.

$$f(\lambda)=\frac{\lambda}{\sqrt{1-\lambda^2}} . \tag{6.73}$$

$$F(\lambda)=1-\sqrt{1-\lambda^2} . \tag{6.74}$$

Of course the definition of λ changes, $\lambda=d/D$. Its domain is still zero to one.

Defining $N_V(D)dD$ as the number of spheres per unit volume of diameter $D \pm dD/2$ and $N_A(d)\partial d$ as the number per unit area of circular interceptions of diameter $d \pm \partial d/2$, we have from Eqn.(5.91) giving the probability of interception and Eqn.(6.71),

$$N_A(d) = d \int_d^{\sup D} \frac{N_V(D)}{\sqrt{D^2 - d^2}} \partial D, \tag{6.75}$$

which is analogous to Eqn.(6.12) and still has the solutions given by Eqn.(6.20),

$$N_V(D) = -\frac{2}{\pi} \frac{\partial}{\partial D} \int_D^{\sup D} \frac{N_A(d)}{\sqrt{d^2 - D^2}} \partial d, \tag{6.76}$$

which is indeterminate if directly integrated but can be integrated by parts to give

$$N_V(D) = -\frac{2D}{\pi} \int_D^{\sup D} \left(d^2 - D^2\right)^{-\frac{1}{2}} \frac{\partial}{\partial d} \left[\frac{N_A(d)}{d}\right] \partial d. \tag{6.77}$$

For the relationships between the moments of the two distributions we have from Eqns.(6.24) and (6.26) for positive integer n,

$$N_V \mu_n'(D) = \frac{2}{\sqrt{\pi}} \frac{\Gamma\left(\frac{n}{2}+1\right)}{\Gamma\left(\frac{n+1}{2}\right)} N_A \mu_{n-1}'(d). \tag{6.78}$$

And,

$$\mu_n'(D) = \frac{2}{\sqrt{\pi}} \frac{\Gamma\left(\frac{n}{2}+1\right)}{\Gamma\left(\frac{n+1}{2}\right)} \overline{D} \mu_{n-1}'(d). \tag{6.79}$$

With $n = 0$, Eqn.(6.78) yields for the number of particles per unit volume

$$N_V = \frac{2}{\pi} \frac{N_A}{H(d)}, \tag{6.80}$$

where $H(d)$ is the harmonic mean of the intercept diameters. From Eqn.(6.79) we similarly obtain

$$\overline{D} = \frac{\pi}{2} H(d), \tag{6.81}$$

which in turn can be back substituted to evaluate Eqn.(6.79) when only the intercept distribution is known.

All of these preceding results have been given in some form previously by Wicksell [1925] who showed that some also apply to ellipsoidal particles. Wicksell later furthered the analysis for ellipsoidal particles [Wicksell, 1926]. However, because of the inevitable presence of elliptical integrals solved through numerical quadrature, these results, given in tabular form, are too cumbersome to reproduce here.

5.2 Intercept Area Distribution

Wicksell showed that the distribution function Eqn.(6.71) also applied to ellipsoidal particles independent of their eccentricity provided that suitable average diameters were used for D. Note that the distribution of the sizes of the intersections of an areal probe with any randomly oriented compact body has spherical symmetry, since the orientation is averaged out. This does not mean that the distribution itself is that of a sphere. But, when sizing particles that are nearly spherical, it may be preferable to use the distribution of area intercepts rather than the profile diameters, since the latter may not be uniquely defined but the average or effective diameter D may be meaningful. The following results are obtained by making the variable substitution $a = \pi d^2 / 4$ in the expressions for intercept diameters given in the preceding section. The reader should note that correct results can only be obtained by starting with the revised distribution function, Eqn.(6.82) and repeating all steps of the analysis -- see Problem 6.4.

For the distribution of intercept areas a in a sphere of diameter D, Eqn. (6.71) yields:

$$f_D(a) = \frac{2}{\sqrt{\pi}\, D \sqrt{\pi D^2 - 4a}} \; . \tag{6.82}$$

And, from Eqn.(6.72) we have for the cumulative distribution

$$F_D(a) = 1 - \frac{1}{D\sqrt{\pi}} \sqrt{\pi D^2 - 4a} \; . \tag{6.83}$$

We note that $0 \le a \le \pi D^2 / 4$. In terms of the reduced variable $\lambda = a / \sup a$

$$f(\lambda) = \frac{1}{2\sqrt{1-\lambda}} \tag{6.84}$$

and

$$F(\lambda)=1-\sqrt{1-\lambda}\,. \tag{6.85}$$

Corresponding to Eqn.(6.75) we have

$$N_A(a)=\frac{2}{\sqrt{\pi}}\int_{d(a)}^{\sup D}\frac{N_V(D)}{\sqrt{\pi D^2-4a}}dD, \tag{6.86}$$

where $d(a)=2\sqrt{\frac{a}{\pi}}$.

Equation (6.86) can be solved for $N_V(D)$. Equation (6.86) is another variation on Abel's equation but does not conform either the forms given as Eqns.(6.42) or (6.44). It is similar to Eqn.(6.12) but cannot be readily converted to that form either. Hence, we introduce two additional forms of Abel's equation giving us four forms in total; All are solved by the same method, but different multipliers and substitutions are used. For the record, generalizing on the solution of Eqn. (6.12) we obtain

$$f(x)=x\int_x^a\frac{y(t)}{\sqrt{t^2-x^2}}dt\,. \tag{6.87}$$

This equation has the solution

$$y(x)=-\frac{2}{\pi}\frac{d}{dx}\int_x^a\frac{f(t)}{\sqrt{t^2-x^2}}dt,$$

that becomes, when $f(a)=0$,

$$y(x)=-\frac{2x}{\pi}\int_x^a\left(t^2-x^2\right)^{-\frac{1}{2}}\frac{d}{dt}\left[\frac{f(t)}{t}\right]dt\,. \tag{6.88}$$

The presence of the factor x outside the integral in Eqn.(6.87) makes this an unsuitable form for solving Eqn.(6.86). By using $x\left(x^2-t^2\right)^{-\frac{1}{2}}$ as a multiplier rather than $\left(x^2-t^2\right)^{-\frac{1}{2}}$ we obtain for

$$f(x)=\int_x^a \frac{y(t)}{\sqrt{t^2-x^2}}dt, \tag{6.89}$$

$$y(t)=-\frac{2}{\pi}\frac{d}{dx}\int_x^a \frac{t f(t)}{\sqrt{t^2-x^2}}dt. \tag{6.90}$$

In the case that $f(a)=0$, this solution becomes

$$y(x)=-\frac{2x}{\pi}\int_x^a \frac{\frac{d}{dt}f(t)}{\sqrt{t^2-x^2}}dt. \tag{6.91}$$

This form is applicable to solving Eqn.(6.86). By substitution we obtain

$$N_V(D)=-D\sqrt{\pi}\int_{a(D)}^{a(\sup D)} \frac{\frac{d}{da}N_A(a)}{\sqrt{4a-\pi D^2}}da, \tag{6.92}$$

where $a(\sup D)=\pi(\sup D)^2/4$ and $a(D)=\pi D^2/4$. Taking all four forms of Abel's equation together, we should now be able to solve Abel's equation by inspection wherever we see it.

The relationship between moments is obtained formally by once again solving Abel's equation after first multiplying by $a^{\frac{n-1}{2}}$ and integrating. The results are these:

$$N_V\mu_n'(D)=\frac{2^n\Gamma\left(\frac{n+2}{2}\right)}{\pi^{\frac{n}{2}}\Gamma\left(\frac{n+1}{2}\right)}N_A\mu_{\frac{n-1}{2}}'(a), \tag{6.93}$$

and

$$\mu_n^{'}(D)=\frac{2^n\Gamma\left(\frac{n+2}{2}\right)}{\pi^{\frac{n}{2}}\Gamma\left(\frac{n+1}{2}\right)}\overline{D}\mu_{\frac{n-1}{2}}^{'}(a)\ . \tag{6.94}$$

In these expressions n must be zero or positive.

With $n=3$ these equations yield Delesse's Principle $A_A=V_V$ and $\overline{A}=\overline{V}/\overline{D}$, which is Eqn.(5.90) given only that the expectation of the cube of the diameter is $E(D^3)=6\overline{V}/\pi$. Setting $n=0$ in Eqn.(6.93) gives without any additional assumption,

$$N_V=\frac{N_A}{\sqrt{\pi}}\mu_{-1/2}^{'}(a)\ . \tag{6.95}$$

At this point, we need to reconsider where we are. Equations (6.92) through (6.95) strictly apply to spheres and therefore add nothing to what has already been said about them. For manual computations finding the diameter of the circular intercept is usually easier than finding the area since the former is an instant measurement with a caliper while the latter requires a planimeter. For machine calculation, the area is preferable to calculate since it amounts to counting the number of pixels inside the intercept. So, we have added a little to computer science in the way of a reminder.

Wicksell determined that the probability density function Eqn.(6.71) also applied to ellipsoidal features provided that d was the geometric mean of the principal elliptical diameters and that D was also the geometric mean of the principal diameters of the ellipsoid.

Table 6.1: Relations between spheres and ellipsoids

	Area	Volume
Circle<Sphere	$\pi d^2/4$	$\pi D^3/6$
Ellipse<Ellipsoid	$\pi d_1 d_2/4$	$\pi D_1 D_2 D_3/6$

Table 6.1 compares the volumes of spheres and ellipsoids with the areas of their intercepts to illustrate how the geometric mean diameter is that which preserves the appropriate content when inserted into the mensuration formula for the circle or sphere. For the ellipsoid, defining the effective

diameter this way gives $E\left(D^3\right) = 6\overline{V} / \pi$ as required above. These arguments strengthen the applicability of the area approach to sizing non-spherical particles. But, one must remember that the effective D in this case has little to do with the mean caliper diameter of the particle. So, while Eqns.(6.93) and (6.95) may be valid for ellipsoidal particles, Eqn.(6.94) is not unless we realize that we are averaging two different meanings of D and separate them accordingly.

6. INTERCEPTIONS OF A PLANAR PROBE WITH DISKS

Features in the form of circular two-dimensional disks may be present in a structure as a result of an interaction between other structural features. For example, a boundary between two spherical particles which have begun to coalesce -- as can be observed at the second stage of ceramic sintering -- takes the form of a disk. Alternatively, disks may be present in their own right as the idealization of disk-shaped particles whose diameters are large compared with their thickness. In this section we will be concerned only with the diameters of the disks and not their thickness, since, by definition, they have none. The thickness distribution of plate-like features will be considered later.

6.1 Intercept Distribution with a Single Disk

The interception of a plane probe with a disk is diagrammed in Fig. 6.12.

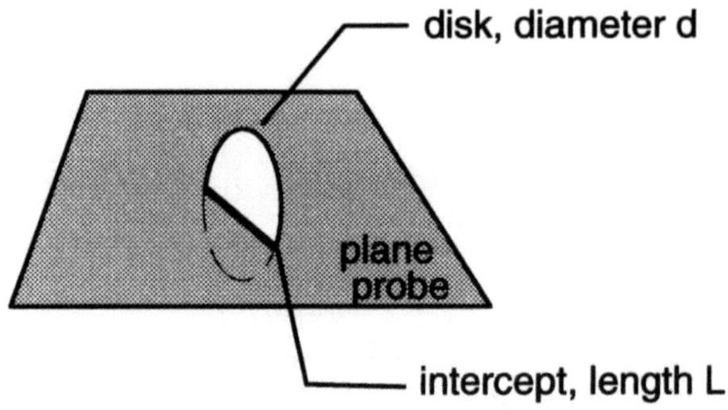

Figure 6.12: Interception of a plane probe with a disk

It is apparent from Fig. 6.12 that the distribution of intercept lengths L of a planar probe with a disk will be the same as the distribution of intercept lengths of a linear probe with a circle and will not depend on the relative orientation of the disk and plane. From Eqn.(6.5) the density function is

$$f(L)=\frac{L}{d}\left(d^2-L^2\right)^{-\frac{1}{2}}. \tag{6.96}$$

The cumulative distribution is likewise given in form by Eqn.(6.8),

$$F_d(L)=1-\frac{1}{d}\sqrt{d^2-L^2}. \tag{6.97}$$

6.2 Intercept Distribution in an Aggregate of Disks

Let $N_V(d)\partial d$ be the number of disks per unit volume with diameters $d \pm \partial d/2$. Then since the mean caliper diameter of a disk is $\pi d/4$ as shown in Problem 5.23, the expected number of interceptions with disks of diameter $d \pm \partial d/2$ per unit area of probe is $(\pi d/4)N_V(d)\partial d$. Noting that $N_A(L,d)=f(L)N_A(d)$, similar to Eqn.(6.11) we can write

$$N_A(L,d)=\frac{\pi L}{4}\frac{N_V(d)}{\sqrt{d^2-L^2}}. \tag{6.98}$$

The marginal distribution of intercepts in terms of L alone is similar to Eqn.(6.12),

$$N_A(L)=\frac{\pi L}{4}\int_L^{\sup d}\frac{N_V(d)}{\sqrt{d^2-L^2}}\partial d. \tag{6.99}$$

This equation is of the general form given as Eqn.(6.87). Consequently it has a solution,

$$N_V(d)=-\frac{8}{\pi^2}\frac{\partial}{\partial d}\int_d^{\sup d}\frac{N_A(L)}{\sqrt{L^2-d^2}}\partial L,$$

which becomes as before,

$$N_V(d) = -\frac{8d}{\pi^2} \int_d^{\sup d} \left(L^2 - d^2\right)^{-\frac{1}{2}} \frac{\partial}{\partial L}\left(\frac{N_A(L)}{L}\right) \partial L . \tag{6.100}$$

6.3 Relationship Between Disk Diameter and Intercept Length Moments

The moments of length and diameter for the disk are closely related to those of the circle following the same pattern as Eqns.(6.20) and (6.100).

The difference is the factor of 4/π, which appears when disks are considered. Again we may note that $N_A \mu_n'(d) = \int_0^{\sup d} d^n N_A(d) \partial d$ is the relationship that defines the moments of *d*. To arrive in this form, we multiply the expression for $N_A(L)$ by L^{n-1} and integrate as before. This process is simplified if we associate the factor π/4 with $N_V(d)$ while referring to Eqn.(6.21). This gives us

$$N_V \mu_n'(d) = \frac{8}{\pi^{\frac{3}{2}}} \frac{\Gamma\left(\frac{n+2}{2}\right)}{\Gamma\left(\frac{n+1}{2}\right)} N_A \mu_{n-1}'(L) . \tag{6.101}$$

For $n = 2$ this yields $\overline{L}_A = \frac{\pi}{4} S_V$ in agreement with Eqn.(5.99). Setting $n = 1$ we obtain $\overline{N}_A = \frac{\pi}{4} \overline{d} N_V$. Substituting this into Eqn.(6.101) gives us

$$\mu_n'(d) = \frac{2}{\sqrt{\pi}} \frac{\Gamma\left(\frac{n+2}{2}\right)}{\Gamma\left(\frac{n+1}{2}\right)} \overline{d} \mu_{n-1}'(L) . \tag{6.102}$$

The parameter *n* is restricted to positive numbers including zero in Eqns.(6.101) and (6.102).

Setting $n = 0$ in Eqn.(6.102) yields for the number of disks per unit volume:

$$N_V = \frac{8}{\pi^2} \frac{N_A}{H(L)} , \tag{6.103}$$

where $H(L)$ is the harmonic mean of the intercept lengths.

7. INTERCEPTIONS OF A LINEAR PROBE WITH PLATES

In this and in Section 6.8 we will be considering interceptions in flat plates whose lateral dimensions are large enough compared with their thickness for it to be justifiable to ignore interceptions which pass through the edges of the plates and consider only those which both enter and leave only through the faces. We can recall that for a random probe $N_L = S_V / 4$; hence the fraction of missed interceptions will approximately equal to the ratio of the surface areas of the edges to that of the faces under these circumstances. These results therefore may be expected to apply to lamellate structures and to those containing thin membranes provided that if any curvature is present, its radius is large compared to the thickness of the of the structure modeled as a plate.

7.1 Linear Intercept Distribution in a Single Plate

Consider a plate having an area A on a side and having a thickness W. This plate is for convenience oriented so that the customary z or polar axis is normal to the plate, Fig. 6.13.

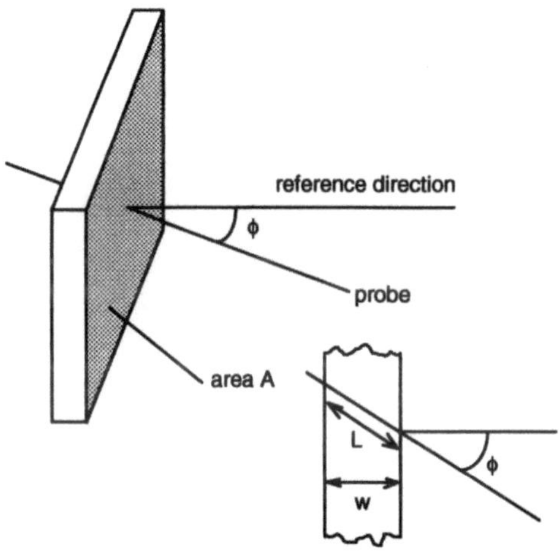

Figure 6.13: Linear interception with a plate

We will define the orientation of the probe by its angle ϕ with the normal to the plate. As a result, rotation of the probe, by angle θ, about the reference axis does not change the length of the interception. The location of the probe is defined as usual by the coordinates of its intersection with the reference plane normal to the probe as defined in Fig. 4.7. Following the convention of Fig. 4.9, we set the surface area equal to 2*A* for the purpose of writing the probability density function for the probe. Thus, from Eqn.(4.25) we have

$$f(x, y, \phi, \theta) = \frac{\sin\phi}{2\pi A},$$

where $0 \le \phi \le \pi$ and $0 \le \theta \le 2\pi$. Since the length of the interception is a function of ϕ only, we can integrate out x,y and θ. Integration with respect to x and y yields the projected area of the plate onto the reference plane which is $A|\cos\phi|$. Integration with respect to θ contributes a factor of 2π. Thus, $f(\phi) = \sin(\phi)|\cos(\phi)|$. Since we are only concerned with the range $0 \le \phi \le \pi/2$,

$$f(\phi) = 2\sin\phi\cos\phi \ . \tag{6.104}$$

Also, from the inset in Fig. 6.13

$$L = W / \cos\phi \ . \tag{6.105}$$

Hence, because there is a one-to-one mapping, we may find the distribution of L by transforming that of ϕ. In doing that it is convenient to write $\frac{d\phi}{dL} = \frac{\cos^2\phi}{W\sin\phi}$. Then in terms of $\phi(L)$ the new distribution is $f(\phi(L)) = f(\phi)\left|\frac{d\phi}{dL}\right| = \frac{2\cos^3\phi}{W}$. Explicitly this becomes

$$f_W(L) = \frac{2W^2}{L^3}, \tag{6.106}$$

where W is less than or equal to L which is less than infinity.

The lower limit on L is W since that is the shortest interception possible not counting the edges. Integrating Eqn.(6.106) gives the cumulative distribution,

$$F_W(L) = 1 - \frac{W^2}{L^2} \ . \tag{6.107}$$

In terms of the reduced variable $\lambda = W / L$ we have

$$f(\lambda) = 2\lambda \tag{6.108}$$

and

$$F(\lambda) = \lambda^2 \ , \tag{6.109}$$

where in both cases as before $0 \le \lambda \le 1$.

7.2 Linear Intercept Distribution in an Aggregate of Plates

An aggregate of plates may occur either through the aggregation of plate-like features or through the tiling of a surface such as a membrane. As a measure of the density of plates we will use the function $S_V(W)$ defined such that $S_V(W)dW$ is the total area, *counting only one side*, of plates of thickness $W \pm dW/2$ per unit volume. Alternatively one could use the function $V_V(W)$ similarly defined with respect to the volume fraction of the plates. Since the two distributions are obviously related by

$$V_V(W) = W S_V(W) \ , \tag{6.110}$$

we need only consider one of them.

Let $N_L(L,W)dW$ be the number of interceptions of length $L \pm dL/2$ with plates of thickness $W \pm dW/2$ per unit length of probe. Then, since by Eqns.(5.59) or (5.86) the total number of interceptions per unit length with plates of thickness $W \pm dW/2$ is

$\frac{S_V(W)}{2} dW$, we have

$$N_L(L,W) = \frac{S_V(W)}{2} f_W(L) \ . \tag{6.111}$$

So, on substitution from Eqn.(6.106) we have:

$$N_L(L,W) = \frac{W^2 S_V(W)}{L^3} \; . \tag{6.112}$$

Integrating out the W gives us

$$N_L(L) = \frac{1}{L^3}\int_0^L W^2 S_V(W) dW \; , \tag{6.113}$$

in which the upper limit on the integral is L because $L \geq W$. Multiplying both sides by L^3, differentiating and setting $W = L$,

$$S_V(W) = L\frac{d}{dL}N_L(L) + 3N_L(L) \; . \tag{6.114}$$

Or by using Eqn.(6.110), we may write this as

$$V_V(W) = L^2\frac{d}{dL}N_L(L) + 3L\,N_L(L) \, , \tag{6.115}$$

a result previously given by Cahn and Fullman [1956].

7.3 Relationship between Plate Thickness and Intercept Length Moments

Starting from Eqn.(6.113), multiplying both sides by L^n and following a procedure analogous to that used previously one obtains

$$(2-n)N_L\mu_n'(L) = \int_0^\infty W^n S_V(W) dW \; .$$

The meaning of the integral (on the right) depends on how the moment of W is to be interpreted. The form of the integral suggests that it represents the product with S_V the total one-sided surface area of plates per unit volume times a moment of W. And, so it is. This moment is averaged with respect to area. It gives

$$(2-n)N_L\mu_n'(L) = S_V\,\mu_n'(W) \, , \tag{6.116}$$

where $n \le 1$. Noting that $N_L = S_V / 2$,

$$\frac{2-n}{2}\mu_n'(L) = \mu_n'(W) \ . \tag{6.117}$$

Moments higher than the first do not necessarily exist for both L and W. Setting $n = 1$ gives

$$\overline{L}/2 = \overline{W} \ . \tag{6.118}$$

Setting $n = -1$ gives

$$H(W) = \frac{2}{3}H(L) \ . \tag{6.119}$$

When applied to a tiling of a membrane with plates, this states that the harmonic mean of the thickness of the membrane equals two-thirds the harmonic mean of the intercept distribution. This result appears to have been first noticed by Weibel and Knight [1964]. The harmonic mean of the thickness of a diffusion barrier is usually the appropriate average for transport calculations, e.g. gas transport in the lung.

8. INTERCEPTION OF A PLANAR PROBE WITH PLATES

As we have already seen, plates can be useful in themselves as model features and can also be used as tiles to represent surface layers, membranes and the like. When used as tiles, the significance of edge effects diminishes since the tiles abut upon each other reducing the amount of edge area. In the previous section the shape of the plate did not enter into the calculations. In the planar probe case, the shape enters into the calculation of the distribution only to drop out later on.

8.1 Planar Intercept Distribution in a Single Plate

The calculation of the probability density function for this case can be performed perhaps most easily for a circular plate. However, circular plates cannot tile a finite surface with anything less than an infinite number of tiles. Furthermore, the smallest tiles would have thickness W and area of differential order, anything but thin in the sense used here. Since, proving the validity of round-plate tiling would be a sidestep that some readers might be uncomfortable without, we will avoid the problem by doing the calculation with explicitly square plates and leave proving that the same results apply to round plates as exercise, Problem 6.6. The interception of a planar probe with a plate yields a stripe.

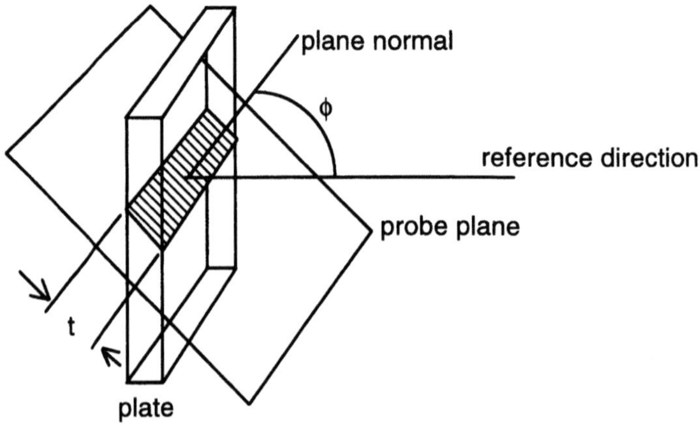

Figure 6.14: Planar interception with a plate

Figure 6.14 illustrates the interception geometry. The plate is intended to be square with sides of length a. The origin of the coordinate system is located at the center of the plate, which again is presumed to be thin. The thickness of the crosshatched interception, the width of the stripe, is t as shown. Again the thickness of the plate is W. The reference direction is the polar axis in the polar coordinate system used. The angle of the interception ϕ is that between the polar axis and the normal to the probe. Locating the origin in the center of the plate, while not necessary, makes calculation of the intercept distribution easier.

Since we have not calculated the mean caliper diameter for a square plate, we will now do so since it is required for the calculation of the distribution function. We do so using Eqn.(5.120) since a plate is a polyhedron: $D = a + W/2$. We have assumed that $W \ll a$. So, for the purpose of calculating the distribution, $D = a$.

The probability density for a planar probe in three dimensions is given by Eqn.(4.29). Substituting the value of mean caliper diameter just obtained, we get

$$f(\rho,\phi,\theta) = \frac{\sin\phi}{2\pi a} \tag{6.120}$$

for the plate. Looking at Fig. 6.14 we can infer that the intercept width t is independent of all variables save ϕ. Figure 6.15 makes this clearer.

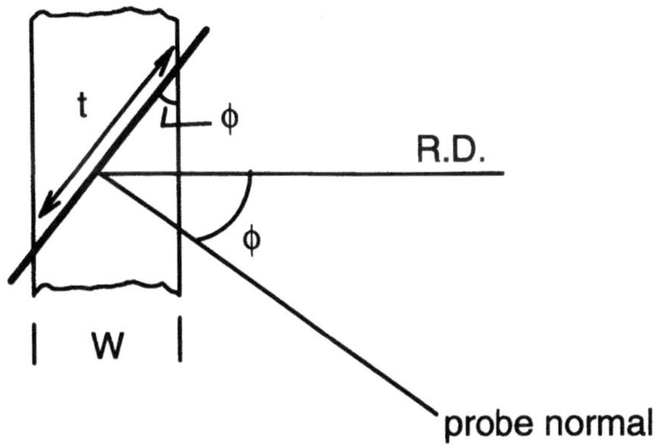

Figure 6.15: Interception width

The width of the interception is, by Fig. 6.15, $t = \dfrac{W}{\sin\phi}$. Accordingly,

we can integrate the other variables out of the distribution. The first step will be to integrate with respect to ρ. For any values of ϕ and θ, this integral will equal the height of the plate at angle θ projected onto the ρ-axis at angle ϕ. The first step becomes finding the height.

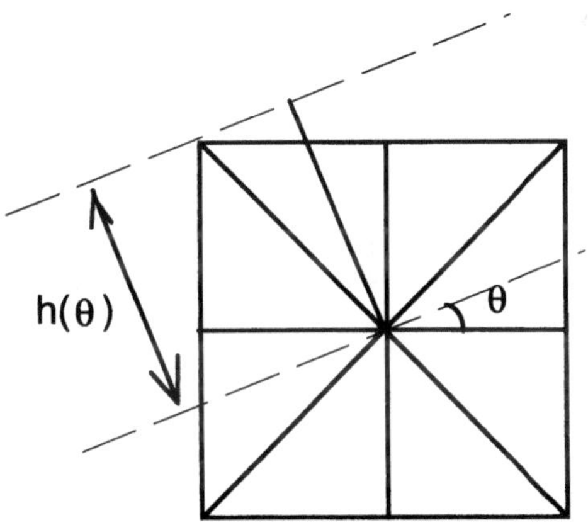

Figure 6.16: Height of plate as a function of longitudinal angle

We see that the square is divided into eight mirror-image sectors. In each the height varies between $a/2$ and $a/\sqrt{2}$. If we were to integrate this height with respect to θ over 2π radians, the resulting integral would be simply eight times that in any one sector. Hence we need only compute the height for the first sector. Since the length of the diagonal is $a/\sqrt{2}$, we see from

Fig.6.16 that $h(\theta)=\frac{a}{\sqrt{2}}\cos(\frac{\pi}{4}-\theta)=\frac{a}{2}(\cos\theta+\sin\theta)$ where $0\leq\theta\leq\pi/4$.

But, this height is not its projection onto the ρ-axis. Figure 6.17 illustrates the projection.

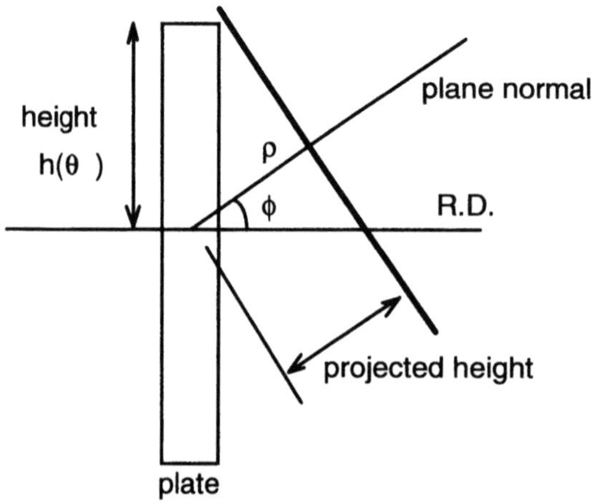

Figure 6.17: Projected height

From Fig. 6.17 we may deduce that the projected height, the range of interception for ρ is $h(\theta)\sin\phi$. Integrating this result then gives

$$\int_0^{2\pi}\int_{(\rho)} f(\rho,\phi,\theta)d\rho d\theta=\frac{8\sin^2\phi}{2\pi a}\int_0^{\frac{\pi}{4}}\frac{a}{2}[\sin\theta-\cos\theta]d\theta$$

leading to $f(\phi)=\frac{2}{\pi}\sin^2\phi$ where $0\leq\phi\leq\pi$.

The domain of ϕ can be reduced by exploiting symmetry. This gives

$$f(\phi)=\frac{4}{\pi}\sin^2\phi\ , \tag{6.121}$$

where $0\leq\phi\leq\pi/2$.

We can now transform the probability density Eqn.(6.121) written in terms of ϕ into one written in terms of our variable of choice t, since there is a one-to-one mapping. Therefore, $f_W(\phi(t)) = f(\phi)|d\phi / dt|$. Now,

$$t = \frac{W}{\sin\phi} \; ; \tag{6.122}$$

$$\frac{d\phi}{dt} = -\frac{\sin^2\phi}{W\cos\phi} \; . \tag{6.123}$$

Together these give,

$$f_W(t) = \frac{4}{\pi}\frac{W^3}{t^3\sqrt{t^2 - W^2}} \; . \tag{6.124}$$

Equation (6.124) does not contain a as a parameter. Is the shape of the plate somehow embedded in the expression? In Problem 6.6, the same distribution function is derived for the case of the circular plate. What properties have the two in common?

The cumulative distribution is derived by integrating Eqn.(6.124)

$$F_W(t) = \frac{2}{\pi}\left[\frac{W\sqrt{t^2 - W^2}}{t^2} - \sin^{-1}\left(\frac{W}{t}\right)\right] + 1 \, . \tag{6.125}$$

The one at the right hand end is an integration constant, which is added to avoid a negative result. The integration has taken place between W and the value of t since t cannot be less than W.

In terms of the reduced variable $\lambda = W / t$,

$$f(\lambda) = \frac{4\lambda^2}{\pi\sqrt{1-\lambda^2}} \; , \tag{6.126}$$

and

$$F(\lambda) = \frac{2}{\pi}\left[\sin^{-1}(\lambda) - \lambda\sqrt{1-\lambda^2}\right] , \tag{6.127}$$

where λ is limited to the interval [0,1]. No constant of integration appears in Eqn.(6.128) because of the backward nature of λ which decreases as t increases.

8.2 Planar Intercept Distribution in an Aggregate of Plates

Defining $L_A(t,W)dtdW$ as the length per unit area of the probe of intercepts of width $t \pm dt/2$ occurring on plates with thickness $W \pm dW/2$, and noting from Eqn.(5.99) that the total length of interceptions is $(\pi/4)S_V$, we obtain

$$L_A(t,W) = \frac{\pi}{4} S_V(t,W) = \frac{\pi}{4} S_V(W) f_W(t) \ . \tag{6.128}$$

In this expression, $S_V(W)$ is the (one-sided) area per unit volume of plates with thickness $W \pm dW/2$. Substituting from Eqn.(6.124) and integrating, we have

$$L_A(t) = \int_{\inf W \approx 0}^{t} \frac{W^3 S_V(W)}{t^3\sqrt{t^2 - W^2}} dW \ . \tag{6.129}$$

Since $t > W$, the lower limit is not strictly 0. This is a problem that we have encountered before, for the case of linear intercepts with stripes, and ignored. We can define $S_V(W)$ as zero for W < t. This will lead to a discontinuity in the derivative, which is not too important; we will examine it later. The result also applies to linear intercepts with stripes.

Multiplying both sides by t^3 and integrating we obtain an expression in the form of Eqn.(6.44). Noting this leads immediately to its solution,

$$S_V(W) = \frac{2}{\pi W^3} \frac{d}{dW} \int_0^W \frac{t^4 L_A(t)}{\sqrt{W^2 - t^2}} dt \ . \tag{6.130}$$

Here we can be more assured that $L_A(0)$ equals zero than we can that zero is indeed the proper lower limit. However, integrating by parts as before,

$$S_V(W) = \frac{2}{\pi W^2} \int_0^W \frac{\frac{d}{dt}\left[t^3 L_A(t)\right]}{\sqrt{W^2 - t^2}} dt \ . \tag{6.131}$$

We can now see the difficulty in defining the lower limit as zero: we are required to differentiate $L_A(t)$. This function is likely to go to zero abruptly at inf W. If it does, then the derivative is discontinuous. Is this important? Strictly speaking, it would not appear to be so since we are integrating. However, if what we are integrating is an approximating function, we might need along the way to be aware that higher derivatives of $t^3 L_A(t)$ may involve delta functions or not exist at all in the neighborhood of inf W. In short, we sweep this potential anomaly under the rug with some justification.

We have just again used a form of Abel's equation to find the distribution of features from the distribution of intercepts. Abel's equation appears frequently in such contexts. Were there more room in this book and more time in our lives (viz. John Hilliard's, Larry Lawson's and the reader's) we might use these techniques we have just practiced to obtain the intercept distributions of polygons, ellipses, cubes and ellipsoids in the greater detail originally wished for this work. But, by now the reader may have noted a certain repetitious quality to solving this class of problems. Should the reader really want to know the exact distribution for one of these shapes or for some other shape more exotic, I hope by now we have armed her with the tools for finding it herself. As a parting help, Table 6.2, lists where in this chapter particular forms of Abel-type equations can be found.

Table 6.2: Abel-type Equations

Form	Location	Comment
$\int_a^x \frac{f(t)}{\sqrt{x-t}}dt$	(6.42)	The genuine Abel's equation listed by Polianin
$\int_a^x \frac{f(t)}{\sqrt{x^2-t^2}}dt$	(6.44)	Variant Listed by Polianin
$\int_x^a \frac{f(t)}{\sqrt{t^2-x^2}}dt$	(6.89)	Not listed by Polianin
$\int_x^a \frac{xf(t)}{\sqrt{t^2-x^2}}dt$	(6.87)	Actually the same as Eqn.(6.42); illustrates substitution

8.3 Relationship between Plate Thickness and Planar Intercept Width Moments

Starting from Eqn.(6.129) and proceeding in the usual fashion, we obtain

$$\int_0^\infty t^n L_A(t)dt = \frac{\sqrt{\pi}}{2}\frac{\Gamma\left(\frac{3-n}{2}\right)}{\Gamma\left(\frac{4-n}{2}\right)}\int_0^{\sup W} W^n S_V(W)dW\,, \tag{6.132}$$

where n equals 2 or less as implied by the gamma function. If we define $f(W)dW$ as the fractional area of plates of thickness $W \pm dW/2$ and likewise, $f(t)dt$ as the fractional length of intercept stripes of width $t \pm dt/2$, then we can write

$$S_V \mu_n'(W) = \frac{2}{\sqrt{\pi}} \frac{\Gamma\left(\frac{4-n}{2}\right)}{\Gamma\left(\frac{3-n}{2}\right)} L_A \mu_n'(t). \tag{6.133}$$

Furthermore by setting $n = 0$, we confirm the shape-independent relationship $S_V = \frac{\pi}{4} \overline{L}_A$; hence, we can write in analogy with Eqn.(6.79)

$$\mu_n'(W) = \frac{\sqrt{\pi}}{2} \frac{\Gamma\left(\frac{4-n}{2}\right)}{\Gamma\left(\frac{3-n}{2}\right)} \mu_n'(t) . \tag{6.134}$$

9. INTERCEPTION OF LINEAR AND PLANAR PROBES WITH SHELLS

We have mentioned the idea of tiling surfaces. We might expect that we could tile any surface and thereby obtain the intercept distribution properties of a shell of that form as a limit for small tiles. We saw previously that we could obtain the Weibel and Knight formula, Eqn.(6.119) for the effective transport thickness of a membrane by reasoning along these lines. To what extent can we extend this concept? In order to see the limitations of this approach, we will briefly examine the intercept distributions for thin spherical shells.

Consider a section through the center of a thin spherical shell as illustrated in Fig. 6.18.

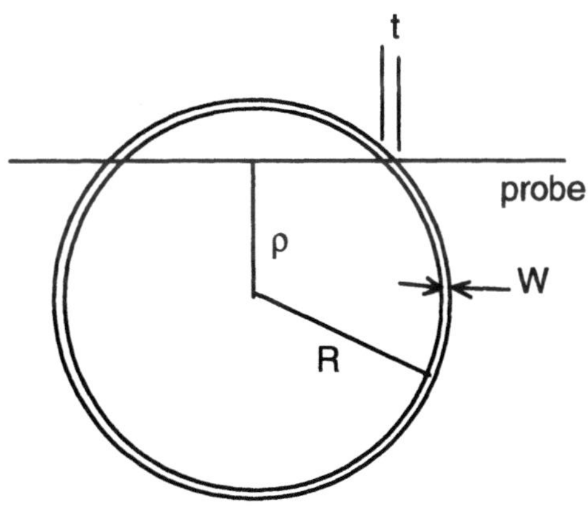

Figure 6.18: Intercept with a shell

The thickness of the shell is W. Its radius is R. The probe makes intercepts of width t. The width t depends only on ρ when the origin is located at the center of the sphere. If the probe is linear, there are, with probability one, two intercepts so long as the shell is sufficiently thin. If the probe is planar, the intercept consists of a single circular stripe.

9.1 Interceptions of a Linear Probe with a Spherical Shell

From Fig. 6.18, we can deduce that, so long as $\rho < (R\text{-}W)$, the length of the intercept is

$$t = \frac{W}{\sqrt{1 - \left(\frac{\rho}{R}\right)^2}} . \tag{6.135}$$

We next consider the probability density function for linear probes in three dimensions, Eqn.(4.25) and Fig. 4.7. This function is

$$f(x, y, \phi, \theta) = \frac{\sin(\phi)}{\pi S} \text{ where } 0 \le \phi < \pi ,\ 0 \le \theta < 2\pi \text{ and } \rho \in S_p .$$

By the symmetry of the problem, t does not depend on ϕ or θ. Integrating over ϕ and θ reduces the density function to

$$f(x, y) = \frac{4}{S} . \tag{6.136}$$

The projected surface S_p is simply a disk of radius R. So, $x = \rho \cos \omega$ and $y = \rho \sin \omega$ where ω is the angle the ρ-axis makes with the x-axis, cf. Fig. 4.7. As in the case of the stripe, we transform the distribution into one where one of the variables is the one of interest ρ and the other ω can be integrated out: $f(\rho,\omega) = f(x(\rho,\omega), y(\rho,\omega))|J|$. Evaluation of the Jacobean and integrating with respect to ω gives

$$f(\rho) = \frac{8\pi\rho}{S} . \tag{6.137}$$

Since we know t in terms of ρ, we can transform the distribution one more time to put it in terms of t. We obtain the density function

$$f(t) = \frac{2W^2}{t^3} \ . \tag{6.138}$$

But, this is the same as Eqn.(6.106) for the linear intercept distribution with a flat plate. It follows that all other linear probe results for the flat (or circular, see Problem 6.6) plates apply to a spherical shell. Were we to consider the shell as being made up of plates, these plates would be uniformly distributed and a linear probe would always intercept only two symmetrical plates at the same time.

9.2 Interceptions of a Planar Probe with a Spherical Shell

When examining the interception of a planar probe with a spherical shell, we do not need a new drawing. Figure 6. 18 will suffice. The width of the stripe intercept is as before given by Eqn.(6.135). Since the width of the intercept depends only on the variable ρ, the probability density function for the probe is that previously given as Eqn.(6.1), $f(\rho) = 1/R$. It follows from Eqn.(6.135) that in terms of t the distribution is

$$f(t) = \frac{W^2}{t^2\sqrt{t^2 - W^2}} . \tag{6.139}$$

This is not the same as that for a single plate in spite of the fact that the plates are uniformly distributed with respect to ρ and θ. Integrating gives the cumulative distribution

$$F(t) = \frac{\sqrt{t^2 - W^2}}{t^2} \ . \tag{6.140}$$

Following the methods developed already, we may examine the intercept distribution for an aggregate of spherical shells as a preliminary to examining the moments. We obtain

$$S_V(W)=\frac{8}{\pi^2}\int_0^W\frac{\frac{d}{dt}\left(t^2L_A(t)\right)}{W\sqrt{W^2-t^2}}dt\ . \tag{6.141}$$

Similarly we obtain

$$S_V\mu_n'(W)=\frac{8}{\pi^{\frac{3}{2}}}\frac{\Gamma\left(\frac{3-n}{2}\right)}{\Gamma\left(\frac{2-n}{2}\right)}L_A\mu_n'(t)\ . \tag{6.142}$$

And,

$$\mu_n'(W)=\frac{2}{\sqrt{\pi}}\frac{\Gamma\left(\frac{3-n}{2}\right)}{\Gamma\left(\frac{2-n}{2}\right)}\mu_n'(t)\ . \tag{6.143}$$

Although similar to Eqn.(6.134) for plates, Eqn.(6.143) is not the same. The substitution of Eqn.(6.134) as an estimator for the moments of intercept width for a spherical shell would give a biased result.

9.3 Tiling

Based on the results of the previous section for a planar probe with a spherical shell, reducing an arbitrary shell or aggregate of shells to an aggregate of tiles seems of limited value. However, the limitation is a result of a correctable bias caused be the shape of the sphere. We must realize that large values of t occur for intercepts near the poles and therefore would involve fewer tiles than would shorter intercepts nearer the equator. The length of the intercept on a sphere is proportional to the

number of tiles intercepted were the sphere tiled $L(\rho)=2\pi\sqrt{R^2-\rho^2}$, which also equals

$2\pi R\frac{W}{t}$. The average intercept length is $\overline{L}=\int_0^R L(\rho)f(\rho)d\rho=\frac{\pi^2R}{2}$. The ratio $\frac{L(\rho)}{\overline{L}}=\frac{4W}{\pi t}$. Multiplying Eqn. (6.139) by $\frac{L(\rho)}{\overline{L}}$

adjusts for the bias introduced by the spherical shape. The product is in fact Eqn.(6.124), the density function for a flat plate.

In order to understand this correction more easily, we emphasize that the probability density for intercept width for a shell is on a per intercept basis. When we to converted this to a per tile basis, we obtained the probability density for a single tile. Since the number of tiles intercepted is proportional to the length of the interception, weighting the shell probability density with the length of the corresponding intercepts and then dividing by a normalization factor $\overline{L}$ recovers the tile probability density. Reversing this process allows us to transform the probability density for isolated tiles into that for any shell for which we can determine the distribution of intercept lengths.

In general we define our probe probability density functions in terms of vectors $\boldsymbol{x}$ such as (x, y, ϕ, θ) or (ρ, ϕ, θ). The average length (or number, in the case of a linear probe) is thus

$$\overline{L} = \int_{(\mathbf{x})} L(\mathbf{x}) f(\mathbf{x}) d\mathbf{x} . \tag{6.144}$$

However, as we have seen, we can make the transformation

$L(\mathbf{x}) \to L(t)$. So, defining in terms of t, $\frac{L(t)}{\overline{L}} = w_L(t)$, we may generalize since $f_{W,\text{plate}}(t) = w_L(t) f_{W,\text{shell}}(t)$:

$$f_{W,\text{shell}}(t) = \frac{f_{W,\text{plate}}(t)}{w_{L,\text{shape}}(t)} . \tag{6.145}$$

This expression allows us to evaluate the planar intercept distributions of arbitrarily shaped shells and membranes.

In the case of the linear intercept distribution, there would be the

analogous expression $\frac{N(t)}{\overline{N}} = w_N(t)$, giving rise to a relation between

linear intercept density functions,

$$f_{W,\text{shell}}(t) = \frac{f_{W,\text{plate}}(t)}{w_{N,\text{shape}}(t)} . \tag{6.146}$$

However, as already indicated $N(t)$ equals two for all t in the particular shell under consideration and in any convex shell as well. Hence, $\overline{N}$ also equals two and $w_{N,\text{sphere}}(t) = 1$. In passing, we might notice that whether or not the plates in the tile aggregate are uniformly distributed with respect to ϕ and θ is of little direct importance.

10. INTEGRAL MEASURES OF STANDARD SHAPES

Using the methods developed in Chapter 5 along with mensuration formulae, the surface, volume and mean caliper diameter of a variety of standard shapes have been calculated for use with the methods of this chapter. These are given in Table 6.3.

Table 6.3: Properties of Shapes_for Particles of Various Shapes.

Sphere of radius r:	Cylinder of length l and radius r:
$D=2r$ $s=4\pi r^2=12.56637r^2$ $v=\left(\frac{4}{3}\right)\pi r^3=4.18879r^3$	$D=\left(\frac{1}{2}\right)[l+\pi r]$ $s=2\pi r[l+r]$ $v=\pi l r^2$
Hemisphere of radius r:	Thin Rod of length l:
$D=\left[1+\left(\frac{\pi}{4}\right)\right]r=1.78540r$ $s=3\pi r^2=9.42478r^2$ $v=\left(\frac{2}{3}\right)\pi r^3=2.09440r^3$	$D=\frac{l}{2}$
Cube of edge length a:	Thin Disc of radius r:
$D=\left(\frac{3}{2}\right)a$ $s=6a^2$ $v=a^3$	$D=\left(\frac{\pi}{2}\right)r=1.57080r$ $s=2\pi r^2=6.28319r^2$
Octahedron of edge length a:	Oblate Spheroid:
$D=a\left(\frac{6}{\pi}\right)\arccos\left(\frac{2}{3}\right)^{\frac{1}{2}}=1.17548a$ $s=a^2\,2\sqrt{3}=3.46410a^2$ $v=a^3\left(\frac{2}{9}\right)^{\frac{1}{2}}=0.47140a^3$	axes: $2a\geq 2b,\ \varepsilon=\left[1-\left(\frac{b}{a}\right)^2\right]^{\frac{1}{2}}$ $D=b+\left(\frac{a}{\varepsilon}\right)\arcsin\varepsilon$ $s=2\pi a^2+\pi\left(\frac{b^2}{\varepsilon}\right)\ell n\left[\frac{(1+\varepsilon)}{(1-\varepsilon)}\right]$ $v=\left(\frac{4}{3}\right)\pi a^2 b$

Dodecahedron of edge length a:

$$D=a\left(\frac{15}{\pi}\right)\arccos\left[\left(\frac{1}{2}\right)+\left(\frac{1}{10}\right)\sqrt{5}\right]^{\frac{1}{2}}=2.64312a$$

$$s=a^2\,3\left[5\left(5+2\sqrt{5}\right)\right]^{\frac{1}{2}}=20.64573a^3$$

$$v=\left(\frac{a^3}{4}\right)\left[15+7\sqrt{5}\right]=7.66312a^3$$

Icosahedron of edge length a:

$$D=a\left(\frac{15}{\pi}\right)\arccos\left[\frac{\left[1+\left(\frac{2}{5}\right)\sqrt{5}\right]}{\left[\left(\frac{3}{2}\right)+\left(\frac{3}{10}\right)\sqrt{5}\right]}\right]^{\frac{1}{2}}=1.74210a$$

$$s=a^2\,5\sqrt{3}=8.66025a^2$$

$$v=a^3\left(\frac{5}{12}\right)\left(3+\sqrt{5}\right)=2.18169a^3$$

Prolate Spheroid:

$$2a\geq 2b,\ \varepsilon=\left[\left(\frac{a}{b}\right)^2-1\right]^{\frac{1}{2}}$$

$$D=a+b\left(\frac{1}{\varepsilon}\right)\operatorname{arcsinh}\varepsilon$$

$$s=2\pi b^2+2\pi\left(\frac{a^2}{\varepsilon}\right)\arcsin\left(\frac{b\varepsilon}{a}\right)$$

$$v=\left(\frac{4}{3}\right)\pi ab^2$$

Parallelepiped
of edge lengths a, b, and c:

$$D=\left(\frac{1}{2}\right)(a+b+c)$$

$$s=2(ab+bc+ca)$$

$$v=abc$$

Tetrakaidecahedron of edge length a:	Tetrahedron of edge length a:
$D=3a$ $s=6\left(1+2\sqrt{3}\right)a^2=26.78461a^2$ $v=\left(8\sqrt{2}\right)a^3=11.31371a^3$	$D=a\left(\frac{3}{\pi}\right)arccos\left(\frac{1}{3}\right)^{\frac{1}{2}}=0.91226a$ $s=a^2\sqrt{3}=1.73205a^2$ $v=\left(\frac{a^3}{12}\right)\sqrt{2}=0.11785a^3$

11. BIBLIOGRAPHY

J.W. Cahn and R.L. Fullman, "On the use of lineal analysis for obtaining particle size distributions in opaque samples", *Transactions AIME (Metallurgical)*, **206** (1956) 610-612.

T. Chao and W.E. Strawderman, "Negative Moments of Positive Random Variables", *Journal of the American Statistical Association*, **67** (1972) 429-431.

R. Courant, *Differential and Integral Calculus*, Volume 2, Interscience, New York (1959).

N. Cressie, A.S. Davis, J.L. Folks and G.E. Policello II, "The Moment Generating Function and Negative Integer Moments", *The American Statistician*, **35** (1981) 148 - 150.

G.W. Lord and T.F. Willis, "Calculation of an air bubble size distribution from results of a Rosiwal transverse of aerated concrete", *ASTM Bulletin*, **177** (1951) 56-61.

A.D. Polianin and A.V. Manzhirov, *Handbook of Integral Equations*, CRC Press, Boca Raton, FL (1998).

A.G. Spektor, "Analysis of distribution of spherical particles in non-transparent structures", *Zavodskaja Laboratorija*, **16** (1950) 173-177.

E.R. Weibel, *Stereological Methods*, Vol 1., Academic Press, London (1979).

E.R. Weibel and B.W. Knight, "A morphometric study of the thickness of the air-blood barrier", *Journal of Cell Biology*, **21** (1964) 367-384.

S.D. Wicksell, "The Corpuscle Problem", *Biometrika*, **17** (1925) 84-99.

S.D. Wicksell, "The Corpuscle Problem. Second Memoir. Case of Ellipsoidal Corpuscles", *Biometrika*, 18 (1926) 151-172.

12. PROBLEMS

6.1) In Eqn.(6.7) the third moment about the origin was calculated and compared with the theoretical value given by Crofton's second theorem. Unfortunately no general relation exists for the second moment because it depends on the shape of the feature. However, knowing the variance of a distribution is important to assessing the relationship between the size of a sample and its expected accuracy in estimating the mean. (See Chapter 8) For a single feature having a fixed shape, knowledge of the mean intercept length determines all other moments of the intercept distribution. Show that the variance of the intercept length distribution for a circle of diameter d with a lineal probe is

$$\mu_2(L)=\left(\frac{32-3\pi^2}{48}\right)d^2. \tag{6.147P}$$

6.2) In Section 6.21, we placed the origin of the probe coordinate system inside the figure. Is this necessary? Show that for the geometry of Fig. 6.19P, the probability density function is

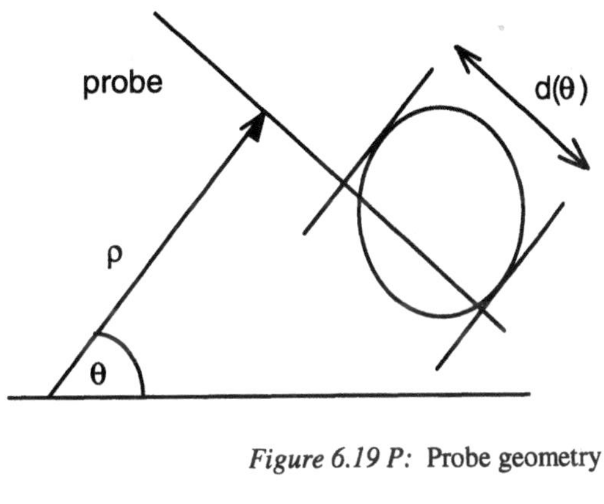

Figure 6.19 P: Probe geometry

$f(\rho,\theta)=1/c$ as in Eqn.(4.10) where c is the length of the perimeter.

6.3) A tetrakaidecahedron has 6 four-sided and 8 six-sided faces. Determine $\overline{S}$ for an aggregate of tetrakaidecehedra and compare this value with that given in Section 5.5.1.4 for the ideal case.

6.4) Compare the results of direct substitution of $a=\pi d^2/4$ into Eqns.(6.75, 6.76 and 6.77) with the corresponding equations in Section 6.5.2. The resulting error is systematic replacing a numerical factor with $\sqrt{a}$. Identify the numerical factor.

6.5) An exercise in the gamma function helps build the mind. Derive Delesse's Principle as a special case from Eqn. (6.93).

6.6) Show that Eqn.(6.124) is the probability density function for circular plates. Consider a circular plate of diameter R and thickness W. It has a mean caliper $D=\pi R/2$.

6.7) Derive Eqn.(6.56) for the linear intercept distribution in a sphere from Eqns.(6.71) and (6.12) by considering the following two-step process: (1) A planar probe intercepts an aggregate of spheres. (2) A linear probe is applied to the planar probe. A good strategy is to first convert Eqn.(6.12) into a form relating the two intercept probability density functions. In the process, Eqns.(5.19) and (6.16) will be needed.

6.8) The parthenogenic distribution Eqn.(2.132) is called such because if the distribution of sphere diameters follows this distribution, the circular intercepts with a planar probe will also follow this same distribution. This was first noticed by Wicksell [1925]. Direct substitution of Eqn.(2.132) into Eqn.(6.77) is straightforward but leads to a difficult integral. However, a theorem of probability states that if two distributions have identical moments, they are identical distributions -- at least with a probability of one. Hence, it is far easier to substitute the moments for the parthenogenic distribution into Eqn.(6.79).

6.9) Filaments or fibers especially in bundles are common features in sections. Figure 6.20P illustrates a small portion of a long fiber. From this figure we may see that when the probe is restricted to the standard range, $0 \le \phi < \pi$, interceptions will only occur for $\phi < \pi/2$. Hence there is no need to restrict the range. Identical interceptions occur for all values of θ.

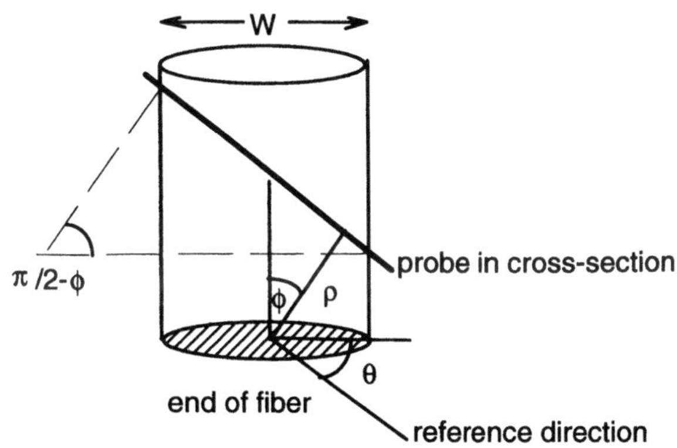

Figure 6.20P: A filament intersected by a planar probe

Interceptions are elliptical in shape. The minor axis, w_1, of the elliptical intercepts for a single fiber is always the diameter of the fiber, W. The major axes, w_2, of the intercepts is diagrammed in Fig. 6.20P. The distribution of major axes is closely related to the distribution of planar intercept widths in plates. Starting with Eqn.(4.29) and Table 6.3 show that the probability density function for the lengths of the major axes of intercepts with an single infinitely-long (or similarly slender) fiber is given by

$$f(w_2) = \frac{2W}{w_2^3} , \tag{6.148P}$$

which is the same as Eqn.(6.106). Show that the cumulative distribution function of fiber intercept major axes is given by Eqn.(6.107). Discuss the application of the formulas from Section 6.7.2 to an aggregate of fibers.

6.10) Lord and Willis [1951] devised a graphical method for estimating the size distribution of spheres from their linear intercepts, which captures the essence of several similar methods. Mostly because of end effects associated with estimating values near the extremes of the distribution, these methods have been criticized as inaccurate and biased. In evaluating these criticisms, one should consider the coarseness of these methods as practiced.

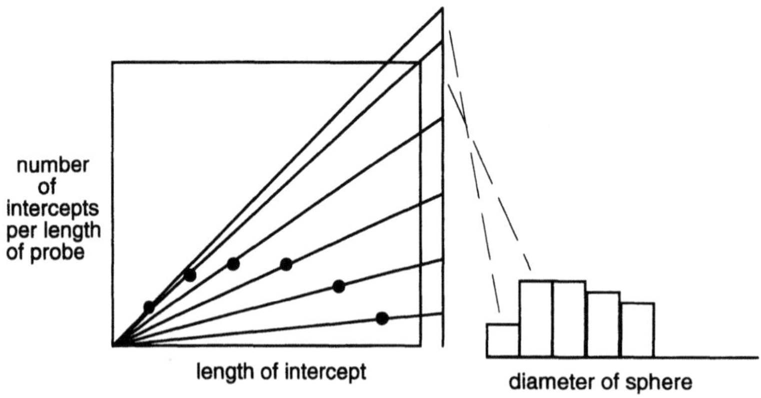

Figure 6.21P: Method of Lord and Willis

The method is illustrated in Fig. 6.21P and works as follows:

(1) The section is first scanned with linear probes and the lengths of the intercepts are recorded along with the total length of all the linear probes.

(2) The length data are then binned into bins of an appropriate length increment ΔL and plotted as a histogram as shown above.

(3) Lines passing through the origin are then used to project data points from the histogram onto a vertical reference line generating intercepts.

(4) The distances between adjacent intercepts are transferred to a second histogram which represents the distribution of sphere diameters.

(5) Appropriate scaling factors are applied to the derived histogram to give appropriate units.

a.) Show (using a similar triangles argument) that the process described corresponds to a numerical solution to Eqn.(6.62).

b.) On page 199 of his text, Weibel [1979] suggests that a raster of parallel test lines be used as the linear probe. He admonishes the reader that the spacing between parallel lines must be either greater than the diameter of the largest sphere or alternatively "very dense" such that most all spheres are multiply sampled. He states, "please note that no compromise between these two extreme line densities is possible". Discuss the significance of correlation in sampling with respect to this comment. Digital image analysis can readily provide abundant intercept data by thresholding or edge detection in which the pixel size is the ΔL. Would it be necessary to rotate the image and resample? If so, under what conditions would rotation of the image digitally (rather than rotation of the specimen and re-imaging) be likely to result in a biased distribution?

c.) In Fig. 6.21P, bins of decreasing interception length project to intercepts higher on the vertical reference line than those of bins representing longer interception lengths. On page 201 of the above-mentioned text, the author addresses the situation where this fails to be the case at the short end of the distribution. He suggests that this must be the result of imperfect counting and orders the reader to correct the distribution by increasing the height of such bins until they are tangent to the line passing through the highest intercept on the vertical reference. Try to justify this correction using the distribution functions. What is the underlying (but unstated) assumption about the distribution of sphere diameters that is needed and based on the property of a single projection line passing through multiple points on the length histogram?

6.11) Verify that the average intercept length in a plate calculated from Eqn.(6.106) agrees with the general relationship Eqn.(5.50) $\overline{L} = 4V / S$.

Chapter 7

RELATIONSHIPS FOR PROJECTED IMAGES

Transmission electron microscopists were among the first to be concerned about projection of a volume onto a plane since all their images are such projections. The applications of the subject have expanded to include a great deal of other projections especially those of astronomical bodies, particles of all types, leaves, plants, clouds and radiographic features. Projections of two-dimensional surfaces in a three dimensional space onto two-dimensional flats is a subject of interest in fractography, forestry and similar fields. Surprisingly little in terms of general results has appeared in this area although what is there is well worth discussion because of the importance of the applications. Closely related to projection is the problem determining N_V. Within stereology itself, the “disector” principle evolved has as a better way of solving the problem of measuring number per unit volume. The acceptance of the disector methodology within the medical regulatory community has denigrated the perceived value of traditional methods of inferring number densities such as the Schwartz-Saltykov method. This is unfortunate because disector methods are often hardware dependent and cannot be applied to pre-existing un-paired micrographs. The acclaim won by the disector method in neuroanatomy is due to the presence of nucleoli in nerve cells. These nucleoli are so small that they behave essentially as points making the counting of cells into a task of counting points per volume in a projected slice. More will be said on this.

1. RELATIONSHIPS FOR PARALLEL-PROJECTED VOLUMES (TEM AND OPTICAL MICROSCOPY)

In this section we will discuss the case where a volume is essentially shadow-cast onto a plane. The silhouettes of the features contained are parallel-projected onto this plane through rays originating "at infinity". We will not be directly concerned with focusing; the entire image will be presumed to be in focus. Focusing does play a role in some disector methods discussed later.

Historically, the first recognition of the role of slices of finite thickness was that their use could result in a distortion of the stereological parameters if that slice were misconstrued to be a plane intersection rather than a projection. This problem was first discussed by Holmes [1927] and was thereafter dubbed "the Holmes effect". The essence of the “effect” is that profile projections tend to be larger than plane intersections and as a result the insertion of the areas of profile projections into formulae derived for plane intersections results in error. One of the first approaches to the problem of finding volume fraction was the calculation of a correction for thickness introduced by Cahn [1959] expanding the earlier work of Holmes.

1.1 Introduction

We have so far in this book assumed that the measurements are made, in the case of optical microscopy, by reflected light or on an opaque material or, in the case of electron microscopy, by the use of replicas. However in the histological and some other sciences, specimens are frequently encountered that are not completely opaque; and, these are usually examined by "transmitted light". The transmission mode is also used in electron microscopy in order to improve resolution.

In the analysis of projected images that follows, we will consider two limiting cases. The first, which usually applies in optical microscopy, is when the slice thickness is thin compared with the scale of the structure. In this case we will find that the relationships developed for a two-dimensional section can be used if a correction is made for the finite thickness of the slice. At the other extreme when the slice is thick in comparison with the scale of the structure, which is usually the case with transmission electron microscopy, these relationships are not even approximately valid. There is also the additional complication of allowing for the overlapping of the images from structural features.

1.2 Thin Slices

Thin slices, although physically thin, do have a discernible volume. However overlap is not a major concern. What is paramount to the problem is that the centers of many of the features lie outside the slice. In this section, we will derive formulas, which are idealizations, to correct for finite thickness. Many of these apply only over limited ranges of thickness relative to some characteristic dimension of the microstructure. In the case of some very crowded microstructures they may not be applicable at all. The reader is encouraged to contemplate the applicability of these results and to wrestle with the problem of specifying the range of their utility.

1.2.1 Total properties in the zero thickness limit

Total properties are those discussed in Chapter 5 that are not dependent on the shape of the features. The word "total" is sometimes used in a different sense as in "total projection" [Underwood, 1970; Kendall and Moran, 1963] to mean "multiplied by the number of times of self-overlap". Thus, the *total projected area* of a convex body is twice the area projected. For example the area projected by a sphere of radius r is πr^2 while the total projected area of the same sphere is $2\pi r^2$ since the sphere overlaps itself once. While valuable for specialized purposes, the general use of total projected quantities tends to lead to confusion. In contrast with Underwood, in this chapter we will deal mostly with convex sets as features and will not use the concept of total projection as an integral part of the definitions of projected quantities.

1.2.1.1 Number per unit volume

Consider an aggregate of particles intercepted by a slice of thickness t. We will suppose that the "center", e.g. the center of mass, is assigned to each particle. The convention used for defining the center is immaterial providing it does not depend on the orientation or location of the particle. The expected number of particles with centers lying within the slice is tN_V per unit area of the slice. In addition there will be feature elements in the slice contributed by particles whose centers lie outside the slice.

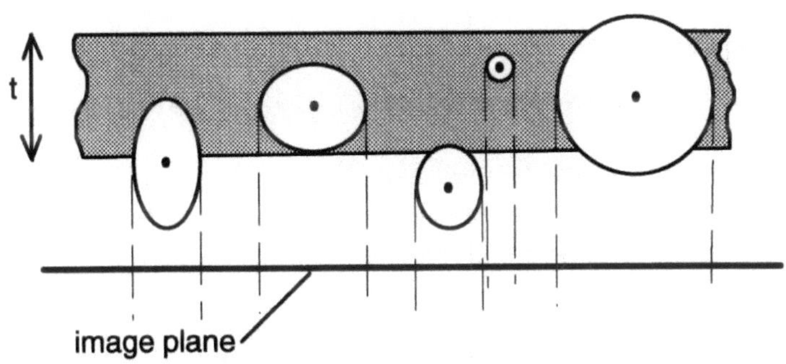

Figure 7.1: Thin Slice Projection

Locations of possible centers are illustrated in Fig 7.1. It follows from the relationship $N_A = N_V \overline{D}$, Eqn.(5.91), that the number of images per unit area from particles with centers above the top surface of the slice is

$N_V \overline{D}/2$. The factor of one half enters because on the average half the centers of intercepted particles will lie above the top surface and half below. For the same reason there will also be $N_V \overline{D}/2$ interceptions on the bottom plane from particles having centers lying below the bottom surface. Allowing for all particles intercepting the slice, the number $N_{A'}$ of projected images per unit area of the slice is:

$$N_{A'} = N_V t + N_V \overline{D} = N_V \left(t + \overline{D}\right) \tag{7.1}$$

as $t \to 0$. The $t \to 0$ restriction is required because we have not allowed for the overlap of particles.

It is evident that it is not possible to estimate N_V from measurements on a single slice unless $\overline{D}$ is known. However, if measurements can be made on a series of slices of different thickness, then a plot of $N_{A'}$ versus t yields both N_V and $\overline{D}$ since the slope is equal to N_V and the intercept is equal to $N_V \overline{D}$.

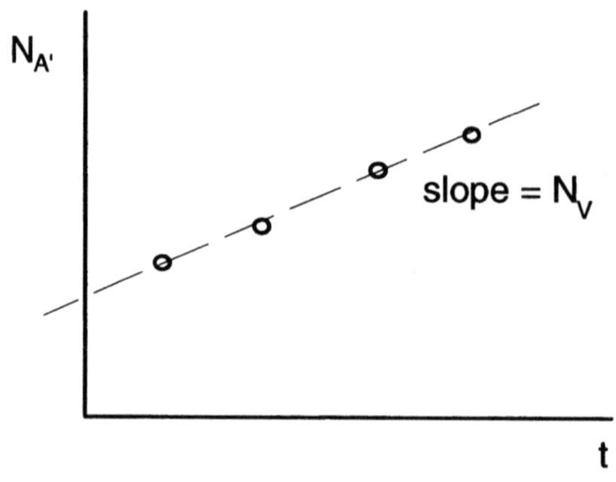

Figure 7.2: Plot of projected intercepts per area vs. thickness

Eqn.(7.1) has been used successfully. But, as $t \to 0$, the change in the number of particles intercepted per thickness increment tends to be Poisson-distributed introducing noise into the apparent slope. Consequently, thickness increments must be large enough to overcome the noise. But, this hints at a violation of the thin-slice requirement. In order to be useful, sample areas need to be large, the microstructure un-crowded and isotropic unlike the situation shown in Fig.7.3, where the slope tends to zero over the range in which overlap is negligible. Accordingly, we see that Eqn.(7.1) is only valid when $\overline{D}$ is much less that the spacing between particles.

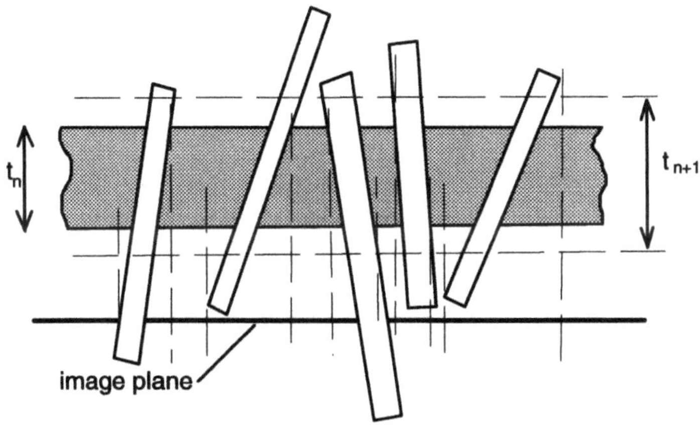

Figure 7.3: Inappropriate Geometry for Eqn.(7.1)

1.2.1.2 Area per unit volume

Since area per unit volume can in principle be determined from a single slice, its determination is less troublesome than that of number per unit volume. We first need to determine the average projected perimeter length of a single convex particle. As we have already noted, the mean caliper diameter of a particle can be regarded as its mean projected length and that this projection can be visualized as a two-step process in which the particle is first projected onto a plane and this projection is the projected onto a line within the plane. It thus follows that the mean projected height, previously defined as *d*, of the particle image is just equal to the mean caliper diameter *D* of the particle. But, from Eqn.(4.13) $d = c / \pi$, where *c* is the perimeter length of the particle image. Hence, if the orientation of the particles is isotropic,

$$\bar{c} = \pi D \ . \tag{7.2}$$

This can be applied to the shadows of particles in a slice only when the entire particle is contained in the slice or it has a center of symmetry that is contained within the slice.

We next consider an aggregate of particles of average mean caliper diameter $\overline{D}$ intercepted by a slice of thickness *t* where $t << \overline{D}$. We will assume that the particles are convex and in addition that they have centers of symmetry. If N_V is the number per unit volume of particles, then as already noted, the number per unit area of the slice whose centers fall within a slice of thickness *t* is tN_V. It follows from this and Eqn.(7.2) that the contribution of these particles to the projected image circumference per unit area of the projection plane is $tN_V \pi \overline{D}$. In addition there will be a contribution from particles which intersect the slice but whose centers lay outside the slice (cf. Fig. 7.1). From Eqn.(5.99) it is seen that the latter contribution is $(\pi / 4)S_V$. The total circumference length of particle images per unit area is therefore,

$$L_{A'} = N_V t \pi \overline{D} + \frac{\pi}{4} S_V \ . \tag{7.3}$$

This can be written,

$$S_V = \frac{\pi}{4} L_{A'} - 4N_V t \overline{D} \ , \tag{7.4}$$

where $t << \overline{D}$. Note that in Eqn.(7.3) there are again two contributions, upper and lower, added to form the term containing S_V. If $P_{L'}$ is the density of intersections of a linear probe with the

circumferences of particle images then by Eqn.(5.31) $L_{A'} = \frac{\pi}{2} \overline{P}_{L'}$.

Also, from Eqn.(7.1), $N_V \overline{D} = N_{A'} - tN_V$,
where as before, $N_{A'}$ is the number of particle images per unit area of the projection plane. With these substitutions, Eqn.(7.4) yields,

$$S_V = 2\overline{P}_{L'} - 4tN_{A'} + 4t^2 N_V \ . \tag{7.5}$$

Since t is usually much smaller than unity, t^2 is normally quite small indeed. So, to a first approximation,

$$S_V \approx 2\overline{P}_{L'} - 4tN_{A'} \ . \tag{7.6}$$

1.2.1.3 Volume fraction

We again assume that the particles are convex and do not overlap. We first determine an expression for the projected areal fraction of the particles. There are two types of particle surfaces that contribute to the projected area, cf. Fig. 7.1. The first are those formed by the interception of the particles with the top and bottom surfaces of the slice. These are single interceptions. The total intercepted area per unit area of the slice is $2A_A$; and, since these areas are parallel to the plane of projection, they project as their "true" values. The second type of particle surfaces are those lying within the slice and their total area per unit area of slice is tS_V, where t is the thickness of the slice and S_V the particle surface area per unit volume of the sample. It

follows from Eqn.(4.23b), $\langle dS_p \rangle = dS/2$, that allowing for a random

orientation of the interior surfaces their contribution to projected areal fraction is $tS_V/2$. The sum of both types of surface is $2A_A + tS_V/2$. Now since we have assumed that the particles are convex and do not overlap one another, each element of the projected area will come from two surfaces of each particle. Consequently, if we define $A_{A'}$ as the projected areal fraction,

$$A_{A'} = A_A + tS_V/4 \,. \tag{7.7}$$

Substituting $A_A = V_V$, Eqn.(5.93), and taking Eqn.(7.5) for S_V, we obtain,

$$V_V = A_{A'} - 2tP_{L'} + 4t^2 N_{A'} - 4t^3 N_V \,, \tag{7.8}$$

where $t << \overline{D}$. Or, if higher order terms are ignored,

$$V_V \approx A_{A'} - 2tP_{L'} \,. \tag{7.9}$$

Thus in a thin slice the volume fraction of particles can be estimated from a measurement of their projected areal fraction and the density of intersections of a linear test probe with the circumference of the particle images. This provides a correction to Eqn.(5.93) for finite section thickness. The derivation of Eqn.(7.7) follows the one given by Cahn and Nutting [1959] which is a generalization of the one given by Holmes [1927] of the special case where the particles are constant-size spheres.

Unfortunately, there may seem something cavalier about the division-by-two step in that there seems to be no clear rationale for dividing the S_V term by two. To illustrate this further, I now present a slightly different derivation given by Weibel [1979]. I do not want to waste the reader's time, but I do want to introduce the reader to the hermeneutics involved. So, this next derivation, which is much simpler, has distinctly different assumptions. The point I want to make is that derivations may be based on different sets of assumptions that are likely to co-exist. Often a result may be numerically true yet the supporting argument tenuous. To begin, let us look at the derivation just given.

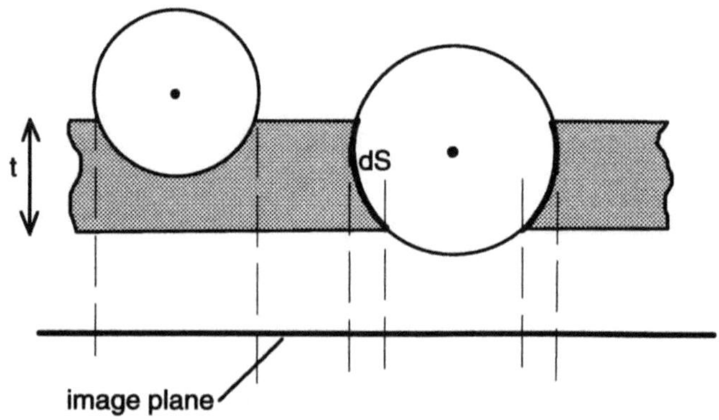

Figure 7.4: Intercepted Bodies

Presuming a center of symmetry, bodies that are intercepted either have their centers inside or outside the slice. In the case of the former, as on the left in Fig. 7.4, there is a single intercept which projects as its "actual" size in the sense of an intersection. Forgetting the center of symmetry requirement, it is also imaginable that a particle could have a single intercept and project one which is larger; but clearly the probability of this vanishes rapidly as $t \to 0$. We presume that for features which have only single intercepts $A_A = V_V$ when the total area of both sides of the slice is considered. On the right in that figure we see the double intercept case. In general, the projected intercept will either be the larger of the two, or as on the right in Fig. 7.4 may project as still larger than either. In deriving Eqn.(7.7) we attempted to make a correction for the additional area that resulted from the intersection between the 3-flat, which is the slice, and the surface of the body or feature. This area is marked with a heavy line and labelled as "dS" in Fig. 7.4. Using the Cauchy projection formula, Eqn.(4.23b), and summing over all the features centered within the slice, we obtained $tS_V/2$. The Cauchy formula assumes random projection of the areal element while near-normal projection seems disproportionately unlikely since the slice still appears relatively thick. This is one problem. A greater one is that while dividing by two to account for the two sides of the slice makes sense in the single-intercept case, there is less of justification for it in the two intercept case. In deriving Eqn.(4.24) dividing by two was justified because we were considering a closed convex body which had two sides one entirely in the shadow of the other. Looking at Fig. 7.4, we see that there may be some overlap adumbrating a portion of the surface *dS*. But, there is again no reason to presuppose the symmetry which underlies Eqn.(4.24). At best, we can hope for a cancellation of errors: the error from the lack of overlap might compensate the lost likelihood of near-normal projection.

The Weibel derivation is similar to the one just given. It essentially considers the two types of particles shown in Fig. 7.4 and adds to these a third type, particles contained entirely within the slice. The third type has the symmetry just mentioned and by Eqn.(4.24) an expected projected area one fourth that of its surface. The next argument is that those particles that have their surface only partially within the slice also project as one fourth their contained surface since the lost surface, outside the slice, is equal to that of those particles that have single intercepts and require no correction. Looking at the left side of Fig. 7.4, we see that at least for spheres this reasoning might be plausible since the probability that the particle is cut above the center must be equal to that of being cut below the center if in fact there is a center. This now justifies the division by two in the first derivation. The balance of the derivation does not need repeating.

Yet another way of looking at the problem is to consider the case where $t << \overline{D}$ that should certainly apply as $t \to 0$. The slice is then so thin that in the limit all interceptions are as they appear in Fig 7.5. The concern over the unlikelihood of near normal projection no longer seems relevant. We notice that if we could determine the area of each interception with the center plane (indicated by the center line in Fig. 7.5) then $A_A = V_V$. But, the projected area is dilated by the projection of a band of surface contained within the slice. We can estimate the width of this band using Eqn.(5.99), $L_A = (\pi/4)S_V$ and $A_A = tS_V$. The width before projection is $4t/\pi$. Using Eqn.(4.23b) the projected width is $2t/\pi$. From Fig. 7.5, we see that only half that width contributes to dilating the center plane area. Dividing by two and multiplying by Eqn.(5.99) gives the desired correction, $tS_V/4$. Hence $V_V = A_A - tS_V/4$ as previously determined.

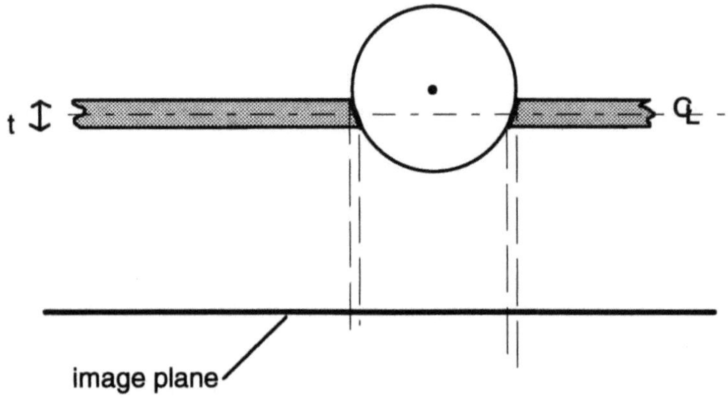

Figure 7.5: Centerline Area Estimation

1.2.2 Perimeter length per unit volume

The estimation of lineal features in a projected image differs from the estimation of N_V, S_V and V_V in that the effect of overlap is very much less important and in most cases can be neglected. As a consequence, most of the results derived in this section are applicable to relatively thick slices as well as to thin ones.

If the slice is very thin compared with the average length and radius of curvature of the lineal features, then it is justifiable to neglect the probability of a feature terminating in a slice. In this case, we can apply Eqn.(5.106), $P_A = L_V / 2$, to obtain

$$L_V = 2N_{A'} \text{ , as } t \to 0, \tag{7.10}$$

where $N_{A'}$ is the number of line segments per unit area of the projection through the slice.

If the thickness of the slice is large compared with the resolution (not depth of focus) of the microscope, then lineal features will project as line segments of resolvable length. In this case an alternative procedure is available for estimating L_V which eliminates the assumption that features do not terminate inside the slice. Consider a line of length l within a bounding body of volume V which is intercepted by a slice of volume V_i. Then it follows from the relationship $L_L = V_V$ (Eqn.(5.57) that the expected line length falling within the slice is

$$\bar{l}_i = \frac{l\, V_i}{V} \quad . \tag{7.11}$$

Since the derivation of Eqn.(5.57) did not require an average over the orientation of the linear probe, the expected line length in the slice is independent of the orientation of the line. (This is an important point, which the reader can check by considering a line inside a cubic box that is intercepted by a slice parallel to a face of the box. A line parallel to the slice will give a longer length of interception than one that is normal to the slice, but this is exactly offset by the difference in the probabilities of interception.) For a slice of thickness t and area A, $V_i = tA$ and thus from Eqn.(7.11),

$$\overline{L}_{iA} = tL_V \,, \tag{7.12}$$

where $\overline{L}_{iA}$ is the expected length of line intercepted per unit area of the slice. From Eqn.(5.142P) given in Problem 5.28, the expected ratio of the projected length to the true length of a line segment is $\pi/4$. Consequently we have for the projected line length per unit area,

$$L_{A'} = \frac{\pi}{4} L_A \,, \tag{7.13}$$

which in conjunction with Eqn.(7.12) gives

$$L_V = \frac{4}{\pi t} \overline{L}_{A'} \,. \tag{7.14}$$

The quantity $\overline{L}_{A'}$ can either be measured directly or, what is probably easier, estimated from the density $P_{L'}$, of intersections on a linear probe applied to the projected image. Substituting $L_{A'} = (\pi/2)\overline{P}_{L'}$ from Eqn.(5.31) we obtain

$$L_V = \frac{2}{t} \overline{P}_{L'} \,. \tag{7.15}$$

An alternative derivation is also available for this expression. The density of intersections on a linear probe applied to the projected image will be the same as that on a unit length of an intersecting planar probe that is normal to the plane of the slice. The area of this normal intersection per unit length is t. It follows from Eqn.(5.106) that the number of intersections $\overline{P}_L$ on a unit length is $tL_V/2$ in agreement with Eqn.(7.15).

1.3 Thick Slices and Overlap Correction

Apart from Cauchy's Theorem (Eqn. 4.23b) [Cauchy, 1850], the earliest work on the subject of the stereology of projected particle images appears to have been done by Vouck [1948] and Mack [1954]. Many articles have appeared since that time and many of these are listed by Underwood [1970]. Since overlap in the major concern in thick slice images, we will begin with Mack's treatment of overlap that was further developed by Cahn and Nutting [1959].

In estimating an overlap correction, it is necessary to assume that the particle images are Poisson-randomly[1] dispersed on the plane of projection. This is not quite equivalent to the assumption that the particles are randomly dispersed in three dimensions because at this point we will presume it physically impossible for the particles to interpenetrate one another.

1.3.1 Correction of $A_{A'}$ for overlap

Consider the addition of an image having an areal fraction of $\delta A_{A'}$ to a set of existing images having an areal fraction after allowing for overlaps of $A^o_{A'}$. (The superscript ^{o} will be used to denote properties corrected for overlap.) On the average, the fraction that will not overlap existing ones is $1-A^o_{A'}$. Hence the addition of the new image increases the projected areal fraction by $\delta A^o_{A'} = \delta A_{A'}(1-A^o_{A'})$. Since this applies only strictly in the limit, we rewrite this in terms of differentials, set $A^o_{A'}=0$ when $A_{A'}=0$ and solve the resulting ordinary differential equation to obtain the overlap correction,

$$A^o_{A'} = 1-\exp(-A_{A'}). \tag{7.16}$$

It will be noted that this expression is independent of the slice thickness. Of course $A_{A'}$ is not. Equation (7.16) is an approximation. Note that as $t \to 0$, $A_A \to V_V$ not to zero. However, it is also evident that as $t \to 0$, $A^o_{A'} \to A_{A'}$. The reason why Eqn.(7.16) fails at this limit is because of the assumption that it is "the images and not the particles which are randomly dispersed". This expression appears in both Macke's and Cahn and Nutting's treatments.

---

[1] The original treatment did not reference a Poisson process, but it is latent in the definition used of "random".

1.3.2 Correction of $L_{A'}$ for overlap

Let $L_{A'}$ be the perimeter length of particles per unit area of the projected image that would be seen in the absence of overlap and again $A^o_{A'}$ be the observed areal fraction of the particle images. Let us further suppose that to an existing array of images, we add one having a perimeter length of $\delta L_{A'}$ per unit area of the projection and an areal fraction of $\delta A_{A'}$. If it is assumed that the images are randomly dispersed, then the expected fraction of the perimeter length of the added image that will not be obscured by existing images is $1-A^o_{A'}$. The added image will also overlap a portion of the perimeter length of existing images. The loss of perimeter length due to this overlap is $L^o_{A'}\delta A_{A'}$. The net change in the perimeter length of the images is therefore,

$$\delta L^o_{A'} = (1 - A^o_{A'})\delta L_{A'} - L^o_{A'}\delta A_{A'} \; . \tag{7.17}$$

In order to integrate Eqn.(7.17) , we need a relationship between $\delta L_{A'}$ and $\delta A_{A'}$. We can find this by noting that for any defined family of shapes the expected total surface $S = k\overline{D}^2$, where k is a constant of the family and $\overline{D}$ is the mean mean (sic) caliper diameter. Further noting from Eqn.(4.13) that in projection, the expected perimeter of the shadow of a single feature is $\delta L' = \pi\overline{D}$ we find that

$$\overline{\delta A_{A'}} = k\left(\frac{\delta L_{A'}}{2\pi}\right)^2 , \tag{7.18}$$

where A' is the area of the projection. Substituting Eqn.(7.18) into Eqn.(7.17) and forming differentials shows that the second term in Eqn.(7.17) drops out as second-order. It then follows that

$$L^o_{A'} = L_{A'}\left(1 - A^o_{A'}\right) . \tag{7.19}$$

In obtaining Eqn.(7.19) there was a constant of integration. Since $L^o_{A'} \to L_{A'}$ as $A^o_{A'} \to 0$, this constant is zero. An alternative derivation is given by Underwood [1970].

1.4 Model-Based Treatment of Thick Slice Properties

Up to this point in the book we have tried or at least pretended to be free of the assumptions of specific models. In the earlier part of this chapter we pursued approximations, but their underlying assumptions were grounded in the experience of experimental practice. Results so obtained are often immediately useful but usually not very systematic. Since the last section of this book deals with stochastic geometry which is currently highly concerned with models as a way of achieving exactitude, we might as well continue our study of projected images by introducing an explicit model, the Poisson point process, in order to obtain systematic results for the properties of thick sections and begin our exposure to stereological models -- a subject treated more rigorously by Saxl [1989].

Once again we are indebted to R.E. Miles for the results that we are to follow [Miles, 1976]. We will consider a structure in three-dimensional space that is a union of random convex particles. We will consider first the distribution of their centers (centroids). These are distributed such that the following assumptions are met:

(1) The distribution is *simple*; there are no double points viz. two points located at the same place in space.

(2) The distribution is *stationary* and *isotropic* meaning that it is invariant respectively under translation and rotation. These are among the properties we required for a random probe.

(3) The scattering of the centers in space is *independent*; the location of one center has no effect on the location or even the existence of another. We can see that this requirement might conflict with that of assumption 1. However, from what we know already about the probability of coincidence of random points, we could amend by adding "with probability zero", if we had to. This assumption is often unrealistic since features usually cannot penetrate each other (this sets lower limits on center-to-center spacing) but frequently form in clusters. (Of course we may simplify by considering point processes made by convolving a localized process with an independent one.)

A Poisson process Φ with density λ defines the sort of process that meets the above assumptions. In such a process, the probability of finding a point in a volume element B with volume v is $P(\Phi(B)=1 \mid v \to 0)=v\lambda+o(v)$. Consistent with this statement is the probability of not finding a point, $P(\Phi(B)=0 \mid v \to 0)=1-v\lambda+o(v)$. These statements can be generalized by replacing the volume with the Lebesgue content of any Borel set in any number of dimensions. Referring back to Eqn.(2.62) we see that for small volumes the mean number of points in a small volume is λv. If we partition our space into a large number of very small elemental volumes or cells, the distribution of points will be given by a multi-dimensional version of

Eqn.(2.62), with one dimension for each cell, in the limit of infinitesimal subdivision of the space. If we specify any specific number of points to be found in that space, then that number implies some arrangement. Given such an arrangement, at the very least, the probability of the last point in such an arrangement would depend on the existence of the other points. The only case in which the probabilities of all the cells will be independent is the case where no point is found. This case is σ-addable and can be evaluated by integrating over the infinite number of cells involved. The result is called the *void probability* of the Poisson point process and is

$$\nu(V) = \exp(-\lambda V) , \tag{7.20}$$

where V is the volume of some subspace being examined. From assumption (2) above, we understand that the only relevant property of the subspace is its volume. Its location, shape and orientation are unimportant.

Let us now consider the distance from one point to its first or nearest neighbor. We evaluate this by considering a spherical volume. Since all volume elements are independent, we can choose to put a point in the center of the sphere without disturbing the rest of the distribution[2]. The volume is then $(4/3)\pi r^3$ in three dimensions (πr^2 in two dimensions). The probability of having found a point (at least one point) is simply one minus the probability of not having found one. Hence,

$$H(r) = 1 - \exp\left(-\lambda \tfrac{4}{3}\pi r^3\right) \tag{7.21}$$

is the probability of having found a point or what is called the *first contact distribution.* It is easily generalized to n-dimensional space through Eqn.(2.155).

1.4.1 Points and their projections

There are several similar measures of the mean distance between features. It is the average length of intercepts in the interphase. The mean free path between features in 3-space (indicated hereafter by the superscript 3) is related to the parameters observable on a plane section through Eqns.(5.57, 5.86 and 5.93) as

[2] This simple assumption can lead to complicated ramifications. These are treated in the theory of Palm distributions.

$$L_{mfp}^3 = \frac{1-A_A}{N_L} = \frac{4(1-V_V)}{S_V}, \tag{7.22}$$

where N_L is the number of intercepts per unit probe length. This is seen by noting that the number of intercepts per unit length with features must equal that with non-features since the two alternate. Further, one minus the areal fraction is simply the volume faction of non-features. Mean free path goes to infinity as the volume fraction goes to zero, since the probability of a line hitting a point is zero.

In terms of the Poisson process, we can obtain the mean path between nearest neighbors in three-space by differentiating Eqn.(7.21) to obtain the density function and then using this to find the first moment. The result is

$$r_{nn}^3 = \frac{\pi^{\frac{2}{3}} 2^{\frac{1}{3}} 3^{\frac{5}{6}}}{9\Gamma\left(\frac{2}{3}\right)} \lambda^{-\frac{1}{3}}. \tag{7.23}$$

The numerical factor is approximately equal to 0.5539. This formula is essentially that reported by Chandrasakhar [1943]. We note that under these circumstances $\lambda = N_V$, the number of points per unit volume.

We would like to determine the corresponding relationship for points projected onto a plane. The answer cannot be obtained from the expected length of a projected line, Eqn.(5.142P) because nearest neighbors in space need not be nearest neighbors in projection. However, the projection of a Poisson point process is another Poisson point process. Analogous with Eqn.(7.21), we consider a point process in the plane by replacing in that equation the volume of a sphere with the area of a circle. Differentiating as before and finding the first moment gives

$$r_{nn}^2 = \frac{1}{2} \lambda^{-\frac{1}{2}}, \tag{7.24}$$

where λ now refers to the intensity of the process in the plane, N_A or $N_{A'}$. In the case of projection $\lambda = N_V t$, where t is the thickness of the section. Equation (7.24) is then the mean distance between nearest neighbor points as they appear on the projection plane.

In passing, it may be worth noting some adjustments that can be made to these equations to estimate the mean central path when spheroidal-behaving[3] convex bodies are centered at the points and paths are near to the center-to-center line. This distance L^3_{mcp} is the average length between neighboring feature surfaces. If we just consider the distance along the centerline,

$L^3_{\text{centerline}} = r^3_{nn} - \overline{D}$. In the case where particle centers are far enough apart that the cosines of the angles between all paths connecting particles and their centerline paths nearly equal unity, we may substitute the average intercept length within the particle for $\overline{D}$. The resulting estimator for length we call the *mean central path*,

$$L^3_{mcp} = r^3_{nn} - \frac{4\overline{V}}{\overline{S}}, \tag{7.25}$$

following Eqn.(5.50). This expression has been used in estimating process zone sizes and the amount of plastic work in growing inter-particle voids [Lawson, 1991].

In the case of the two-dimensional projection the corresponding result is derived from Eqn.(5.17) with Eqns.(4.13 and 4.24),

$$L^2_{mcp} = r^2_{nn} - \frac{\overline{S}}{4\overline{D}}, \tag{7.26}$$

since in projection the mean caliper diameter of the projected image equals that of the body itself. Equation (7.26) applies only to complete projections, that is, projections of bodies entirely contained within the slice. Additional terms can be easily added to account for bodies intersected rather than projected. Figure 7.6 illustrates the path geometries of Eqns.(7.25 and 7.26).

[3]Such bodies have spherically-symmetric intercept distributions – usually due to random orientation.

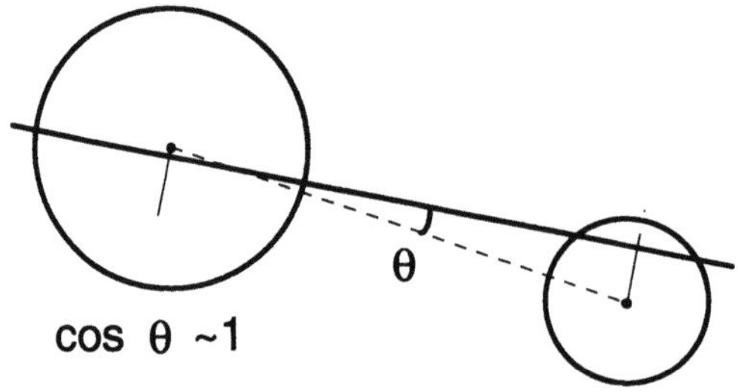

Figure 7.6: Mean Centerline Path

1.4.2 Bulk properties associated with unions of distributed convex bodies

We return to the important results presented by Miles for thick sections in projection microscopy [Miles, 1976]. Equation (7.20) gives us the probability that a given volume will contain no points. Subtracting this quantity from unity gives us the probability that the same volume will contain some points (at least one). If we associate the given volume on the average with the presence of a phase,

$$V_V = 1 - \exp\left(-\lambda \overline{V}\right). \tag{7.27}$$

In Eqn.(7.27), λ is the number of points per unit volume which is greater than or equal to the number of features per unit volume.

Since we have not yet begun to introduce the reader to the theory involved, and will only treat it in a very basic manner in Chapter 10, we refer the reader to the appendix of Miles's article for derivations. Intuitively, we might expect the surface area to be related to the volume by differentiation. Following Miles,

$$S_V = \lambda \overline{S} \exp\left(-\lambda \overline{V}\right), \tag{7.28}$$

which is the derivative of Eqn.(7.27). Since we are dealing with closed surfaces, mean curvature is given by

$$\mathrm{M}_V = \left(2\pi\lambda \overline{D} - \tfrac{1}{32}\pi^2\lambda^2\overline{S}^2\right)\exp\left(-\lambda \overline{V}\right). \tag{7.29}$$

This expression could prove useful in estimating the "ripening" of precipitate distributions. Miles also provides a formula for gaussian curvature.

Given these formulas relating the bulk behavior to the Poisson point process, Miles continues by finding their equivalents in projection analogous to the way in which Eqn.(7.23) becomes Eqn.(7.24) when projected. The projected areal fraction is the counterpart of the volume fraction. It is

$$A_{A'} = 1 - \exp\left(-\lambda\overline{V} - \frac{\lambda t\overline{S}}{4}\right) . \tag{7.30}$$

The projected length of interface curve between the set **X** of features and its complement is

$$L_{A'} = \pi\lambda\left(\frac{\overline{S}}{4} + t\overline{D}\right)\exp\left(-\lambda\overline{V} - \frac{\lambda t\overline{S}}{4}\right) . \tag{7.31}$$

The total curvature, as defined by Eqn.(5.112), divided by the area containing the curve was not specifically treated in Section 5.6 but is often given the symbol C_A. One reason it was overlooked was that

$$C_A = K_V \tag{7.32}$$

for an IUR section. So, the development proceeded quickly into the more general results for three dimensions. But, Eqn.(7.32), for sections, is inapplicable to projected images. Instead,

$$C_{A'} = \left[2\pi\lambda\left(\overline{D} + t\right) - 2\pi^2\lambda^2\left(\frac{\overline{S}}{8} + \frac{t\overline{D}}{2}\right)^2\right]\exp\left(-\lambda\overline{V} - \frac{\lambda t\overline{S}}{4}\right) . \tag{7.33}$$

In these formulas, it is necessary to know λ which is an artificial parameter. However, if t is known, then it is possible to solve Eqns.(7.30, 7.31 and 7.33) in sequence to obtain V_V , S_V and K_V . Alternatively, if two slices of differing but unknown thickness are available, the properties may be estimated using the following set of equations. Subscripts 1 and 2 will designate the slices from which the parameters were obtained. Copying directly from Miles:

$a = -\log(1 - A_{A'})$,

$$b = \frac{1}{\pi} \frac{L_{A'}}{1 - A_{A'}},$$

$$c = \frac{1}{2\pi} \frac{C_{A'}}{1 - A_{A'}} + \frac{\pi b^2}{4},$$

$$u = (a_2 - a_1)(c_1 a_2 - c_2 a_1) + \frac{(b_2 - b_1)(a_1 b_2 - a_2 b_1)}{(b_2 - b_1)^2 - (a_2 - a_1)(c_2 - c_1)},$$

$$v = \frac{c_1 b_2 - c_2 b_1}{(b_2 - b_1)^2 - (a_2 - a_1)(c_2 - c_1)},$$

$$\hat{V}_V = 1 - e^{-u},$$

$$\hat{S}_V = 4v(a_2 - a_1)e^{-u},$$

$$\hat{K}_V = \left[2\pi v(b_2 - b_1) - \tfrac{1}{2}\pi^2 v^2 (a_2 - a_1)^2\right]e^{-u}. \qquad (7.34a\text{-}h)$$

Carats indicate sample values. This may seem like a long collection of equations, but with computers being ubiquitous, the amount of work involved in using them is small. Additionally we have values of the intermediate parameters that may also be useful in themselves:

$$\hat{\lambda} = v(c_2 - c_1)$$

$$\hat{\bar{V}} = \frac{u}{v(c_2 - c_1)}$$

$$\hat{\bar{D}} = \frac{b_2 - b_1}{c_2 - c_1} \qquad (7.35a\text{-}c)$$

Still more information can be extracted. The interested reader is advised to read Miles's paper.

2. RELATIONSHIPS FOR PROJECTED SURFACES, AREA AND LENGTH

The topic of statistical inferences about projected surfaces is one of great current interest to geographers and biologists. It has evolved somewhat outside the field of stereology and yet it's tools can be formalized in terms of generalized probes. The quantities of interest include the ratio of area to projected area, the true areal fraction of some constituent, e.g. birch trees or wheat, the total area of leaves associated with some plot of land etc. The current techniques that have evolved to answer these questions strongly emphasize direct numerical integrations on a large scale and the use of color (spectrophotometry) as an indicator of the apparent numerical density of some species.

2.1 Introduction

Areas of interest pertaining to surfaces in which stereology may play a role include the characterizations of textures and patterns either using Fourier methods similar to the Fourier probes of Sections 5.4.2.3 and 5.4.4.3 or fractal techniques such as the multifractal spectrum. These techniques are, however, not "classical stereology". Results that derive from the provenance of classical stereology are those of El-Soudani [1978] and less classically but perhaps more precisely Janácek [2001]. Here we will confine are attentions to constructing definitions that generalize El-Soudani's approach. The reader who wants to pursue these further may find Janácek's article a starting point for actual computation. El-Soudani postulated that there is a relationship connecting two ratios: the ratio of the true surface area to the projected surface area and the ratio of the true length of a surface profile to its projected length.

El-Soudani derived a relationship for what he defines to be a purely random surface relating two ratios we will call R_S and R_P. The desired result would be an analogy to Rosiwal's principle allowing a line scan to predict a property of a three-dimensional surface. The first ratio is defined as $R_S = A / A'$, the ratio of true to projected area. The second is defined as $R_P = L / L'$, the ratio of true to projected length on line traced over the surface. Calculating the relation for a "purely random surface" is deceptively simple. We divide the surface into flat elements. We then take Cauchy's relationship for projected area, Eqn.(4.23b). Thus R_S is two for the "purely random surface" Were the line to meander randomly we might take Eqn.(5.142P) as giving a value of $4/\pi$ for R_P . However the line segment is constrained to a plane. Assuming that all orientations in the plane are equiprobable gives $R_P = \pi / 2$. This gives $R_S = (4/\pi)R_P$ or $R_S \sim R_P$. We could attempt to refine this approximation using the distributions given in Chapter 4, but would soon run into trouble since the assumptions about a "purely random surface" and that it gives a purely random profile when sectioned put a lot of stress on the concept of "random". However, for a given bounded surface $\overline{R}_S / \overline{R}_P$ must be a constant defining what we will call an "El-Soudani relationship". It is possible to construct a family of surfaces that obey a particular El-Soudani relationship.

The applications for which an El-Soudani-type relationship would be useful are typified by those of crop yield estimation on rough terrain and determination of surface annealing kinetics. In both cases, what we want to know is R_S. In principle we could find R_S if we knew the distribution of angles between the surface element normals and the normal the plane of reference. The sample of the terrain is typically too small to assume that all angles of projection are equally probable. Consequently, we need a different approach.

2.2 Discrete Methods for Estimating Surface Area

Consider a surface formed over a domain in the *x-y* plane. The coordinates of a point on the surface are (*x*, *y*, *z*) were we consider z to be a function of (*x*,*y*). If we consider a small element of surface over a domain element with sides δx and δy, the surface's projected area is simply the product $\delta x \delta y$. The area of the surface element itself is the magnitude of its normal in vector notation,

$$\delta \mathbf{a} = \mathbf{i}\left(-\frac{\partial z}{\partial x}\delta x \delta y\right) - \mathbf{j}\left(\frac{\partial z}{\partial y}\delta x \delta y\right) + \mathbf{k}(\delta x \delta y) . \tag{7.36}$$

Since the content of the projected area divides out, it immediately follows that for this surface element the ratio of true area to projected area is

$$R_{\delta a} = \sqrt{1 + \left(\frac{\partial z}{\partial x}\right)^2 + \left(\frac{\partial z}{\partial y}\right)^2} . \tag{7.37}$$

The overall ratio of true surface area to projected surface area (weighted by the projected surface since this is what is visible) is thus

$$R_S = \mu'_{\frac{1}{2}}\left(1 + \left(\frac{\partial z}{\partial x}\right)^2 + \left(\frac{\partial z}{\partial y}\right)^2\right) . \tag{7.38}$$

Eqn.(7.38) could be estimated on a point sampling basis. "How?" you might ask. Answers include photogrammetry as discussed in Chapter 1 and remote sensing techniques such as LIDAR. As a side note, I (L.L.) would like to point out that although we live in a time when massive computational resources are available to practically everyone in the form of personal computer workstations in homes and public libraries, brute-force approaches based on positioning a large number of nodes and numerically integrating are not always as successful as one might wish and cannot obviate the need for the type of reasoning on which stereology is based. For example, news reports of the time stated an estimate of the volume of rubble from the destruction of the World Trade Center by numerical integration of LIDAR data on a workstation. The calculation, intended to guide city officials allocating manpower, severely underestimated the amount of rubble to be removed.

Continuing with the problem of relating surface area to profile length, we can examine the product of the length of the two sides of the true surface element in Eqn.(7.36) since these individually would contribute to the profile length ratio and we would like to understand the relationship between the profile length ratio and surface area ratio we have discussed. The length of the element which projects onto a line parallel to the x-axis is

$\delta x\sqrt{1+(\partial z/\partial x)^2}$. That which projects onto a line parallel to the y-axis is of the same form and their product is

$$\delta x \delta y\sqrt{1+(\partial z/\partial x)^2+(\partial z/\partial y)^2+(\partial z/\partial x)(\partial z/\partial y)}.$$

Comparison with Eqn.(7.37) shows that this expression has an extra term relative to the true area. This extra term's existence points out that although the true area element projects as a square patch, the angles at the corners of the true area patch are in general not 90 degrees. This complicates the problem of relating R_S to R_P.

Let us examine the problem of characterizing a surface analytically. The length of a 2-D graph is given by

$$L=\int_a^b\sqrt{1+\left(\frac{dy}{dx}\right)^2}\,dx\,. \tag{7.39}$$

What sorts of functions for dy/dx lead to tractable answers in closed form for values of L? A survey of elementary functions reveals that linear, quadratic and logarithmic functions are good choices individually.

Trigonometric, hyperbolic, exponential, gaussian and polynomials of degree greater than 2 are among those that do not lead to tractable results. Sums of shifted linear and quadratic functions are reducible and remain linear and quadratic functions respectively. Sums of shifted logarithms lead to “messy” results. Consequently, analytical treatments of surface length and area problems are scarce. The simple type of general result obtained in Chapter 5 for sections of bodies do not seem to be available for projected surfaces.

The lack of analytical solutions and the ready availability of digital computers have focused attention on modeling surfaces using splines. There is an enormous literature on possible tessellations of surfaces with different forms of tiles, which in turn represent spline functions [Aziz and Bhat, 1990, Miller, 1997]. Because of the problems associated with integration just mentioned and the often present need to use the same tiles to fill a volume; flat-surface tiles are most often used to represent surfaces. For simplicity, we will consider an extension of the rectangular layout discussed. The tiles have square projections onto the reference surface, the surface the observer views.

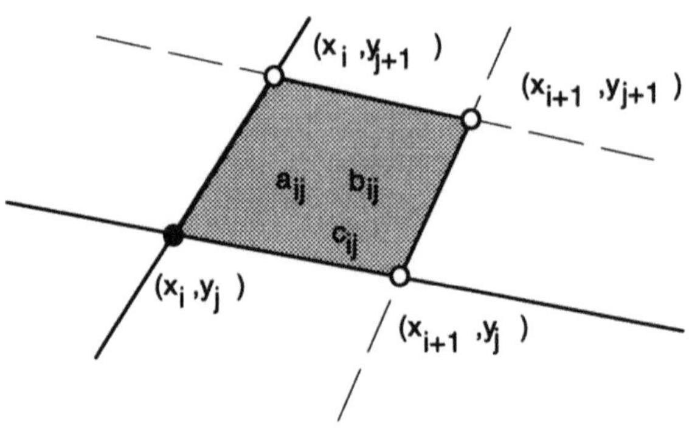

Figure 7.7: Example of a Surface Tile

Figure 7.7 illustrates a simple square-in-projection tile. The lower left node is the reference node. The sides of the tile in projection are δ in length. The projected area is thus δ^2. The z-coordinates have been omitted from the nodes since they are functions of the indices. The surface of a tile is given by

$$z = a_{ij}(x - x_i) + b_{ij}(y - y_i) + c_{ij} : (x, y) \in B_{ij} \quad . \tag{7.40}$$

Since the subscripted coordinates are the coordinates of the reference node,

$$\frac{\partial z}{\partial x} = a_{ij}, \frac{\partial z}{\partial y} = b_{ij} \Big| (x, y) \in B_{ij} \ , \tag{7.41}$$

where B_{ij} is the Borel set in the domain representing the particular tile. From Eqn.(7.38),

$$R_S = E_\delta\left(\frac{A}{A'}\right) = \mu'_{\frac{1}{2}}\left(1 + a_{ij}^2 + b_{ij}^2\right). \tag{7.42}$$

Since, in a computer modeling, all the coefficients are at hand, the area ratio could be calculated without sampling. We subscript the expectation with δ because the area ratio is a function of δ. As δ becomes smaller, the area ratio tends to increase.

Similarly the length ratio for a line parallel to the x-axis is given by

$$R_P = E_\delta\left(\frac{L_x}{L'}\right) = \mu'_{\frac{1}{2}}\left(1 + a_{ij}^2\right). \tag{7.43}$$

If the surface is isotropic, the a_{ij} will be identically distributed to the b_{ij}. From this it is possible to relate root-mean-square estimates of roughness for two and three dimensions, but this not the same as relating the projected area ratios. The apparent lack of analytical solutions to the problem of reducing L and dy/dx in Eqn.(7.39) to related series does not mean that a series approach is not possible for relating surface to area. Rather, analyticity may be sacrificed. Consider sums of rectangular boxcars as illustrated in Fig. 7.8 as forming a 2-D example of a construction in an arbitrary number of dimensions.

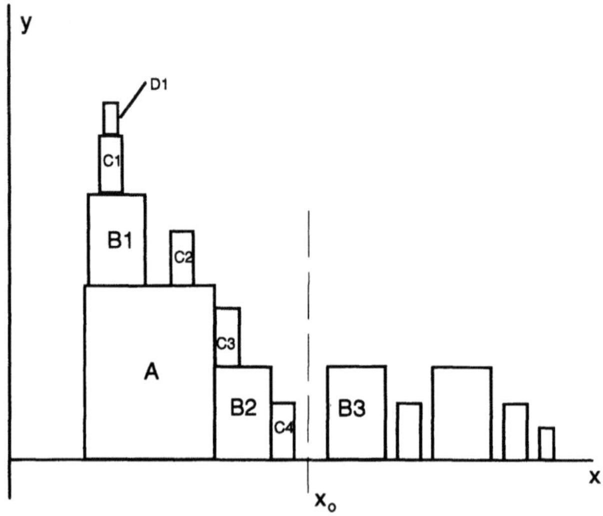

Figure 7.8: Bump Hierarchy

Each boxcar represents a function of x. The pattern shown in Fig. 7.8 is the sum of the boxcar functions. The exposed perimeter of each boxcar consists of one top and two sides. In isolation a boxcar covers a length equal to its bottom, which because it is a boxcar, is also equal to that of its top. Consider the profile of the boxcar sum over the domain $[0, x_0]$. We have distributed the boxcars in order of size. In the absence of any boxcars, the profile length is x_0. We add the largest boxcar, A. If $T(\ )$ denotes the measure of a boxcar's top and $S(\ )$ that of its sides, then the profile with the first order boxcar added is $P = x_0 + (2S(\mathrm{A}) + T(\mathrm{A})) - T(\mathrm{A})$ where the second (grouped) term is the profile of the first order boxcar and the last term is the amount of initial profile lost through coverage by adding the first order boxcar. We now add the second order boxcars.

$$P_B = [x_0 + (2S(\mathrm{A}) + T(\mathrm{A})) - T(\mathrm{A})] + 2S(\mathrm{B1}) + T(\mathrm{B1}) + 2S(\mathrm{B2}) + T(\mathrm{B2})\ldots$$
$$\ldots - T(\mathrm{B1}) - T(\mathrm{B2}) - 2S(\mathrm{B2})$$

.

Since we have not included any third-order boxcars, we have subscripted the profile with a "B" to indicate the highest order over which it applies. Clearly the general profile length is in the form of a recursive sum:

$$P_{H+1} = P_H + \sum_{(i)} L_{H+1}(i) - \sum_{(K)} \sum_{(j(K))} \sum_{(i(H+1))} L(i(H+1) \mid j(K)), \qquad (7.44)$$

where $L(i(H+1)|i(K))$ is the interface contact length between the i-th member of the set of order H+1 boxcars and the j-th member of the set of order K boxcars where $K \leq H$. The simple pattern exhibited by Eqn.(7.44) will not give a very faithful representation because it uses straight sides. Hence it cannot represent both surface and volume simultaneously. However, it could readily be extended to other "bump" functions besides boxcars although the accounting would become more involved. The essence of a good bump function is that it takes finite values only over a compact convex Borel (for emphasis) set and is zero everywhere else. We note that in order to form a hierarchy, it should not be allowed to hang a higher order function off the edge of a lower order one. That is, higher order functions shall not straddle lower order ones. (The same effect is produced "legally" by using two separate higher order functions each belonging to a different lower-order one.) The x-axis line itself is the zero order of functions. Trapezoidal bump functions could also be added to give a more faithful representation; in the case of these, the straddle prohibition extends to the zones formed by the projection of the sloping sides onto the domain. Mixing shapes is desirable, since the highest order function can more easily take a curvilinear form.

This series treatment of surfaces amounts to a sequence of decreasingly complete tilings, each overlaid upon another. Variations on this theme are nearly endless. They can model the structure of arbitrary surfaces or processes such as sequential vegetative overgrowths. “But, where is the stereology?” you might ask. The stereology is in the fitting process. Linear probes parallel to the direction of projection passing through the surface and the reference plane intercept a distance equal to the height of the surface. Since the available methods for realizing such probes are all discrete, there is always a discrete number of points sampled. Hence the process of fitting the surface is always a statistical one. For most stereologists, the topic of surfaces amounts to a digression and the topic of fitting functions to sampled points is treated well enough in books on numerical methods. Hence, we turn this pursuit of this subject over at this point to the reader.

2.3 An Example of an El-Soudani Relation in Closed Form

The ratios of true to projected surface and true to projected profile depend on the specific distribution and even on specific parameters within that distribution. It is not easy to construct a test case in closed form. However, we will present such a case now using a somewhat artificial distribution. The purpose of this exercise is to illustrate the process of calculating the relationship between R_S and R_P. In practice, this process, when needed to characterize a type of surface, would be best done numerically.

To begin we hypothesize that our model distribution is isotropic with respect to rotations in the *x*-*y* plane; the projection plane is as above given by $y = 0$. Consequently a_{ij} and b_{ij} are both taken from the same iid distribution. We next postulate that this distribution is such that a_{ij}^2 is uniformly distributed over the interval $[0, M^2]$. This is necessary to the computation; it leads to a density function of the form

$$f(a_{ij}) = 2a_{ij} / M^2 \Big|_{0 \le a_{ij} \le M} . \tag{7.45}$$

Since the most probable values of a_{ij} are those near M, this suggests possibly some sort of roughly faceted structure. Under these circumstances we can find an expression for, neglecting the subscripts and treating a and b as random variables, the moments of $1+a^2+b^2$. This expression was found using the symbolic mathematical package MAPLE™ integrating separately with respect to the two variables. Even with the simplifying assumptions, the full result is too long to reproduce here. However, from this general expression, one can find a relation for the ratio of the 1/2-th to the first moment (a special property of this particular distribution). Thus, knowing the average of $1+a^2+b^2$ allows calculation of the area ratio for this particular distribution. Since it is the squares that are uniformly distributed, the average is simply M^2+1. Accordingly,

$$R_S = \frac{4}{5}\frac{M^2+1}{3M^6+3M^4} \times \ldots \\ \left[\left(4M^2+4M^4+1\right)\sqrt{2M^2+1} - \left(2+2M^4+4M^2\right)\sqrt{M^2+1} + 1\right] . \tag{7.46a}$$

Analogously, a formula for the length ratio can also be derived

$$R_P = \frac{4}{3}\frac{\left(1+\frac{1}{2}M^2\right)\left[\left(M^2+1\right)^{\frac{3}{2}} - 1\right]}{\left(M^2+1\right)^2 - 1} \tag{7.46b}$$

Figure 7.8 plots the ratio of R_S to R_P as a function of M^2. The plot starts at unity and

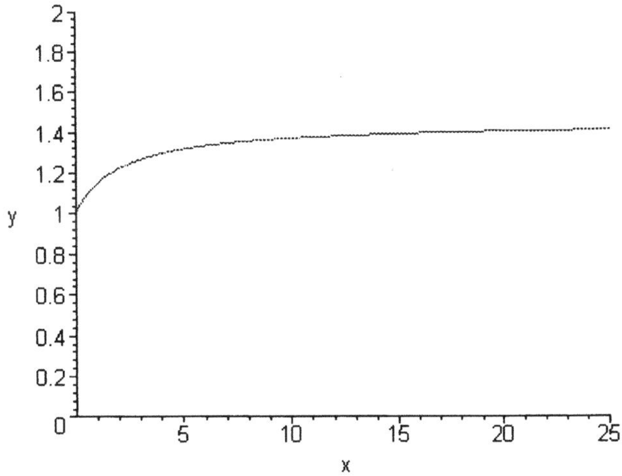

Figure 7.8: Ratio of R_S to R_P

rises to about 1.45 at $M^2 = 25$, the plot's end. Since the choice of distribution was made restrictive in order to make computation possible, these specific results may have practical use only to some highly enlightened sand paper manufacturer. But, it is interesting to observe that in this case El-Soudani's hypothesis appears to approximately hold so long as M does not grow too big.

3. NUMBER PER UNIT VOLUME

Since a plane can intersect a point only with probability equal to zero, some form of projection process is needed to determine N_V albeit subtle in many cases. The problem of determining number per unit volume, especially when it is necessary to break this quantity into size groupings, is knotty because it involves the often-unknown topology of the features being counted. Topological parameters, e.g. the connectivity number, are typically difficult at best to determine by stereological methods. The available stereological methods for counting topologically simple features, such as convex bodies, tend to be difficult and inexact. Hardware based approaches such as those using microfocus x-ray tubes to project a sample volume onto a plane image are gaining in popularity. They have great appeal, since they are easily understood and give easily visualized results. Methods based on single plane sections are definitely not in vogue at this time. They have been severely criticized for their potential biases and inaccuracies and have been supplanted by multiple-section methods. [Gunderson et al, 1988]. Evaluating number per unit volume can be viewed as sweeping a test plane through a volume, which, discounting overlap, is the same as projecting the volume. The optical disector (to be discussed) is a projection technique of this sort. The Schwartz-Saltykov technique is not and might have been better included in Chapter 6. However, it is treated here because it competes with projection methods.

We have already reviewed the mathematical basis for the methods of determining number per unit volume from plane sections. Common sense tells us that $N_V = V_V / \overline{V}$. In Chapter 5 we derived the relation $N_V = \overline{N}_A / \overline{D}$, Eqn.(5.91). We examined the basis for all existing single-section methods for opaque materials in Chapter 6 when we solved variations of Abel's equation for different shapes. In Section 7.1.4 we identified the Poisson process intensity λ with N_V for the case of a distribution of non-interpenetrating convex bodies and saw how λ could be determined either through having a transparent section of known thickness or through multiple transparent sections of unknown thickness. So, we would seem to have covered the subject already. What more can we add?

In the following we will review the most successful of the traditional opaque single-section methods and thereafter introduce the disector method. We are studying the former not for its immediate profitability but to learn the pattern of problem solving which it conveys. We remind the reader that the purpose of this work is not to create another instruction manual for how to do stereology. Instead, our purpose is to help the reader understand the stereological literature and expose the reader to stereological results that may be useful in constructing theories and models.

3.1 Quasi-spherical Particle Size Distributions and the Schwartz-Saltykov Method

In Chapter 6 we derived a number of results for spherical and spheroidal particles using continuum approximations. Continuum approximations usually confer analyticity, which in turn makes available a powerful mathematical apparatus for their solution. Wicksell [1925] treated the problem of plane sections through opaque objects containing quasi-spherical particles analytically and his results form the basis of the treatment given in Chapter 6. But, this treatment does not provide any practical way of estimating number and size distributions. The difficulty is that solutions to problems of the sort discussed in Chapter 6 require differentiation of the intercept size distribution. Experimentally obtained intercept size distributions are not analytically smooth; they are discrete and subject to a variety of noises including the Poisson noise associated with any counting process. Scheil [1931] was the first to address this problem through a method of binning the data. Simple in principle, his method, which closely followed the analytical solution, involved estimation of intermediate variables. This led to an error propagation problem. Others authors introduced an assortment of improved methods [Underwood, 1970]. Of these methods, the last and best scheme is the Schwartz-Saltykov method [Saltykov,1967; Underwood, 1970] is the sum of several publications by the named authors.

The Schwartz-Saltykov method reduces the differentiation problem to one of matrix algebra. This approach is, in outline: The particle sizes and intercept sizes are divided into an equal number of bins for each so that the relationships between the various particle sizes and intercept sizes can be summarized in a square matrix. The indices of the elements of this matrix correspond to a scaling of the dimensions such that the indices themselves actually represent the sizes in some system of units. Another feature of the method is that although the observables, the intercept sizes, are binned and treated as continuous, the particle sizes are assumed to be in a discrete distribution having only specific-sized particles. This assumption seems weird and leads to bias but has stood the test of time. The reader is encouraged to think about this assumption and consult Saltykov's paper for further elucidation of this question.

The first step is to identify the largest intercept and call it D_{max}. This intercept is presumed to arise from an equatorial intersection between the plane of polish and the largest particle in the distribution. This is actually an unlikely assumption. But, so long as there is a large enough sample of intercepts from particles occupying a sufficiently narrow size range relative to the number of intercepts recorded, for all practical purposes the error in the size of the largest particle will not affect the usefulness of the results. ("...You hope", said a fellow schoolmate of mine. $_{LL}$)

The second step is to choose the number of bins in the intercept size distribution. This number of bins should be greater than "seven or eight", said Underwood. On the basis of standard statistical practice, less than five bins might be problematic. But, at the same time, normal theory expects a minimum of five variates per bin. Since particle distributions are not normal, even more variates per bin may be required at a minimum. The interested reader might apply Chebyshev's inequality to the problem of how many bins and how many variates. Once the number of bins is established, a quantity Δ is defined which is the basic size unit for both particles and intercepts. If there are k bins, $\Delta = D_{max} / k$. We identify the bins using the subscript i. Each bin contains intercepts in the interval $((i-1)\Delta, i\Delta]$. We assign subscript j to the diameter of particles. Particles in this case are spheres; they have diameters $j\Delta$ and no others.

We now consider how intercepts arise on a plane of polish. Figure 7.9 illustrates the two possible positions for the particle center.

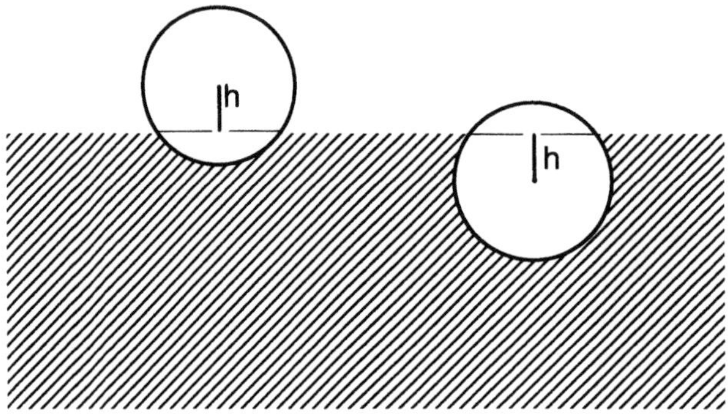

Figure 7.9: Intercepts of the same size from particles above and below the plane of polish

Since, if the probe plane is random, there is no reason to suppose that there are more intercepts from particles above the plane than below it, we will concentrate on the latter case only and double the result.

Spheres have two desirable properties for developing this type of approach. First is that their maximum diameter is also their mean caliper diameter. Second is that they have a simple density function. Following the note at the beginning of Section 6.5, the density function is a constant equal to the reciprocal of the of the radius, Eqn.(6.1). These properties are important in the next steps.

The third step is to relate the intercept size bins to the sphere diameters.

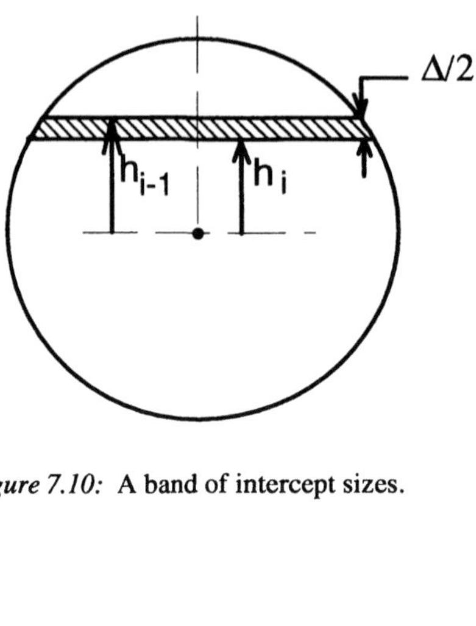

Figure 7.10: A band of intercept sizes.

Figure 7.10 shows one classification or bin of intercept sizes, as they appear on a sphere capable of yielding these intercepts. Each classification of intercept sizes has an interval width of Δ; increasing index corresponds to increasing diameter but decreasing distance from the center. By means of our scaling, the fact that a sphere can produce no intercept larger than its diameter leads to the restriction $j \geq i$. Hence, if we call the sphere diameter in Fig.7.10 $j\Delta$. It follows that

$$h_i = \frac{\Delta}{2}\sqrt{j^2 - i^2} \ , \tag{7.47}$$

and by substitution the same applies to h_{i-1}.

The fourth step is to relate the number of intercepts of size classification i to the number of spheres of size j per volume:

$$N_A(i, j) = P(i \,|\!\uparrow j) N_A(j) \ , \tag{7.48}$$

where $P(i \,|\!\uparrow j)$ is the probability that an intercept of classification i results given the plane intercepts a sphere of size j and $N_A(j)$ is the number of intercepts per unit area with spheres of size j. Using Eqn.(5.91) which incorporates the multiplicity illustrated in Fig. 7.9, we obtain a very general relation worth noting,

$$N_A(i, j) = N_V(j) D_j \int_{h_{i-1}}^{h_i} f(h) dh, \tag{7.49}$$

where D_j is the mean caliper diameter of bodies of size j and $f(h)$ is the density function for intercepts (given in terms of a convenient parameter). We could make the parameter d and use Eqn.(6.71) but it is easier to make the parameter h as defined above and use Eqn.(6.1), since D_j then divides out. We obtain,

$$N_A(i, j) = 2(h_{i-1} - h_i) N_V(j). \tag{7.50}$$

If we substitute Eqn.(7.47) into Eqn.(7.50) we obtain,

$$\begin{aligned} N_A(i, j) &= N_V(j) \Delta \alpha_{ij}, \\ \alpha_{ij} &= \sqrt{j^2 - (i-1)^2} - \sqrt{j^2 - i^2} \ . \end{aligned} \tag{7.51a,b}$$

The fifth step is constructing a set of equations for $N_A(i)$ in terms of $N_V(j)$ and solve.

$$N_A(i) = \sum_{(j \geq i)} N_A(i, j) = \Delta \sum_{(j \geq i)} N_V(j)\alpha_{ij} \ . \tag{7.52}$$

Equations (7.52) give us the intercepts as a function of the particle diameters. Putting this in matrix form we obtain the solution,

$$\mathbf{N_V} = \frac{1}{\Delta}[\alpha_{ij}]^{-1}\mathbf{N_A} \ . \tag{7.53}$$

Because in the case of spheres $[\alpha_{ij}]$ is a lower triangular matrix, the solution of Eqn.(7.53) can take a form easily suited to manual computation. This solution is worked out in [Underwood, 1970]. Clearly through Eqns.(7.49 and 7.53) the Schwartz-Saltykov method can be applied to non-spherical shapes so long as there is a single parameter which characterizes the size of an intercept and the size of a body. If multiple variables are required to characterize these sizes, than multivariate extensions of the theory can be evolved along the same lines.

Many processes that generate particles, such as milling, tend to produce size distributions that are lognormal. This is awkward to handle in the format just given. The conversion of the Schwartz-Saltykov method to a logarithmic scale results in the Johnson-Saltykov method. This is discussed in the reference by Underwood.

3.2 The Disector Method

In deriving the Schwartz-Saltykov method, we mentioned two sources of bias: the origin of the distribution is based on assuming that the size of the largest intercept is that of the largest particle (almost certainly untrue) which tends to shift the distribution averages toward smaller sizes, and the assumption that particles occur only in discrete sizes. For discussions of the method's shortcomings, see Gundersen et al., which lists references on bias.) In spite of it faults, the Schwartz-Saltykov method remains the best for estimating particle size distributions where only a single plane section exists. More accurate results are obtained using a multi-section probe called a *disector.*

The disector is a pair of parallel sectioning planes enclosing a volume. It is assumed that each particle or feature is traceable from one plane to the next to the extent necessary to identify and match its intercepts on both sample planes and that the interplanar spacing is smaller than the mean caliper diameter of the smallest particle to be measured. Matching from one section to the other can be easy to accomplish with stained sections of biological materials. In fact the *optical disector* uses a microscope lens with a short depth of focus to isolate parallel planes of known spacing in a transparent specimen. The disector uses an exclusion counting principle to insure that features are counted in a way which does not depend on their height: We may recall the derivation for finding N_V using thin sections, Section 7.1.2.1.1. We found the slope of the number of projected intercepts as a function of section thickness. At a minimum, this slope could be found from two parallel sections as the difference in the number of projected intercepts divided by the separation between the sections. Making the features traceable by assumption, overcomes the need for projection as well as concern about possible overlap.

The disector is a variant of the method of Section 7.1.2.1.1. To derive the disector we will first consider a mono-disperse system. From Eqn.(5.91) we know that the number of intercepts per unit area will be DN_V. The fact that we are attempting a stereological measurement means we have assumed this distribution to be stationary and hence invariant to translation. If we consider the underlying point process associated with the centers (centroids) of the features being counted, we realize that sweeping a unit area plane through a distance h will sweep a volume containing $N_V h$ points. If we consider the set of centers associated with intercepts on a single planar probe, when we sweep this plane (a short distance) along a direction normal to its surface through a distance h, the expectation of the total number of intercepts will not change. However because of this sweeping we will have lost $N_V h$ associated centers from the original set of centers, and gained the same amount of new centers. So, if we can restrict ourselves to the set of intercepts with a given reference plane (of unit area) while examining a parallel plane (also of unit area) displaced from the first by a distance h, called the *lookup plane*, it should contain the original number diminished by $N_V h$, of original intercepts in the limit as $h \to 0$. If we can trace the features from the reference probe plane to the lookup plane, then by Eqn.(5.91),

$$N_{\text{reference}} - N_{\text{lookup}} = AN_V D - \left(AN_V D - AN_V h\right) , \tag{7.54}$$

where A is the area of the probe plane. It follows immediately from Eqn.(7.54) that

$$N_V = \frac{N_{\text{reference}} - N_{\text{lookup}}}{Ah} . \tag{7.55}$$

We note that the factor D drops out. Hence we can apply Eqn.(7.55) to polydisperse distributions. Since the disector is a variant of the method of Fig. 7.2, we can appreciate that noise is inherent in the restriction to thin intersection spacings and better results may be had from a sequence of such sections as suggested by Fig. 7.2. Also, the "inappropriate geometry" shown in Fig. 7.3 applies to the disector for reasons of noise. On the other hand, the disector method, because of its simplicity in concept is capable of great generalization and has had many daughters. It holds a key position in medical stereology. The reader is encouraged to consult the reference by Gundersen et al. for a bibliography on this subject.

The D-weighting of intercepts (the D is what is doing the projecting) haunts all methods involving plane sections rather than volume samples (viz. slices). In Schwartz-Saltykov-type methods, the decreasing probability of interception leads to increasing noise in the small-diameter end of the distribution. In both the Schwartz-Saltykov and the disector methods, it sets a lower-threshold beneath which features cannot be counted. In the former, there is a presumed smallest feature (but all intercepts are counted) in the latter; intercepts below a certain size are ignored. Methods involving plane sections are therefore ill suited to many polydisperse systems in which the small particles play a dominant, role e.g. sintering of powder compacts.

4. STAR MEASURES

Star measures are something that has recently evolved but does not clearly fit into any particular chapter of this work. Yet the use of star measures is extremely popular and their literature is extensive. Because they do not depend on shape, they might to belong in Chapter 5. But because, although unbiased in a sense, they tend to be poor (noisy) estimators. Since they may be applied a projected image, Section 10.4, they are included here. Star measures evolved from work as early as 1974 (see Chapter 10) but first appear as a feature of mathematical morphology [Serra, 1982]. The principal star measure is the *star volume,* which is defined as the apparent volume of something as seen from one of its internal points. Consider a void in a material. The void could be a convex space in which case an observer at any point within could see the entire volume of the void. Alternatively, the void could be part of a system of twisting tunnels. In that case the star volume of such a void would be the volume of tunnel that an observer at a random point within could see on the average. Star measures are based on the intersections of a linear probe and can be constructed in any dimensionality. When used to characterize a distribution of pores or particles, they may be volume-weighted (which comes "naturally") or surface weighted [Reed and

Howard, 1998; Gittes, 1990] We will only discuss the volume-weighted star volume.

To begin, we examine a star measure of area often used to calculate volume. Consider the problem of finding the volume of a cam lobe. Cam lobes have a curved perimeter but a constant thickness in the direction parallel to the axis of the camshaft. So, it suffices to find the area of the cross-section of the cam in a plane perpendicular to the axis of the camshaft. This problem, which should be familiar to the reader, appears in many elementary calculus books. It is solved using the integral,

$$A = \frac{1}{2} \int_0^{2\pi} r^2(\theta) d\theta \ , \tag{7.56}$$

by breaking up the total area into elementary triangles where the radius r is measured from the "center" of the cam. But, the center of the cam can be located almost anywhere the cam designer wants it to be. If the profile of the cam is convex, any point can be used as the center. The cam will function differently, but the area will stay the same. If the cam is not convex, the center for integration must be one from which all points on the periphery can be reached by straight lines drawn from the center. If we write

$$\bar{r}^2 = \frac{1}{2\pi} \int_0^{2\pi} r^2(\theta) d\theta \ , \tag{7.57}$$

Then the area can be written,

$$A = \frac{2\pi}{2} \bar{r}^2 = \pi \int_0^{\infty} r^2 dF(r) \ . \tag{7.58}$$

The integral on the right is the Stieltjes integral for the second moment of r. The reason the two was left in both the numerator and denominator was to show a pattern. We can see at a glance that this holds for a circle. We pursue this pattern when we examine the equivalents of these equations in three dimensions.

In three dimensions, Eqn.(7.56) can be written to find the volume as,

$$v_3 = \frac{1}{3} \int_{(4\pi)} r^3 d\Omega \ , \tag{7.59}$$

where solid angle is now used and the integral is taken over 4π steradians. By analogy with Eqn.(7.58),

$$v_3 = \frac{4\pi}{3}\bar{r}^3 = \frac{4\pi}{3}\int_0^\infty r^3 dF(r) \,. \tag{7.60}$$

It is again immediately obvious that this applies to a sphere. And, for any convex body, the location of the origin from which r is drawn is arbitrary. In fact,

$$v_n = \frac{S_n}{n}\int_0^\infty r^n dF(r), \tag{7.61}$$

where S_n is the Lebesgue content of the surface of a unit sphere in n dimensions. Again, the location of the origin is immaterial for a convex body. Similar dimensional generalizations will be discussed in Chapter 9. So far all we have presented is a series of integration formulas.

The integration formulas presented can be adapted to stereology through the definition of star volume and some added assumptions. Because we can write Eqn.(7.60) as a Stieltjes integral, we can begin to think about a distribution of radii rather than radius as a continuously varying function in spherical coordinates. To generalize the application of Eqn.(7.60), we adopt the definition of star volume previously given in order to dispense with the assumption of convexity. We now need an assumption of isotropy either of the bodies themselves or of their sampling. Given these two assumptions, Eqn.(7.60) defines an unbiased volume estimator for a body or of a distribution of bodies. This estimator is a star volume and has whatever weighting F has.

The star volume defined by Eqn.(7.60) is inconvenient to evaluate because we need to find a three-dimensional distribution of radii. In practice we could use the isotropy assumption: Were we looking at orthogonal projections of the body or bodies, and were these bodies isotropic, we could justify substituting the two dimensional distribution of radii, found by locating random points within the profiles and drawing radial lines to their peripheries. This can be done. But it is not a very efficient estimator since much work is required. Hence, we would like to recast Eqn.(7.60) in terms of the linear intercept distribution and random sections.

We can relate the linear intercept distribution to the distribution of radii from a random point by considering the distribution of points along an intercept [Cruz-Orive, 1987]. Given any intercept of a linear probe with the body in question, a random point along that intercept yields two radii which sum to the length of the intercept l. Looking at either one of them we can easily find the conditional expectation of the cube,

$$E\left(r^3 \mid l\right) = \frac{1}{l}\int_0^l \varsigma^3 d\varsigma = \frac{l^3}{4} . \tag{7.62}$$

Inserting this result into Eqn.(7.60) gives what is a conventional formula for star volume[4],

$$v^* = \frac{\pi}{3} E(l^3) , \tag{7.63}$$

where the intercept lengths are obtained through an IUR sampling process. We could also insert Eqn.(7.62) into Eqn.(7.61) to obtain a dimensionally-generalized form.

What happens when an intersection rather than a projection is used? Spheres are the only bodies isotropic in themselves. The linear intercept density for a sphere of diameter D is given by Eqn.(6.56) as $2l/D^2$. Integrating we find,

$$E\left(l^3\right) = \frac{2}{D^2}\int_0^D l^4 dl = \frac{2D^3}{5} . \tag{7.64}$$

If the result of Eqn.(7.64) is inserted into Eqn.(7.63) the "star volume" obtained is 20% smaller than the true volume. An additional possible reason for the difference is that for Eqn.(7.63) to be applicable, all the intercepts averaged must pass through a single point. Yet, it has not been uncommon for workers to use only a single intercept per feature since this corresponds more closely to using a random linear probe. Furthermore, were one to use multiple linear intercepts on the profiles of planar sections of spheres, the error would increase. For these reasons, star volume has been used mainly as an index of relative pore volume in such difficult-to assess specimens as cancellous bone. On the other hand, for spheroidal features it seems an ideal technique to apply to projected profiles of thick sections in transparent materials or to apply through x-ray methods such as laminography, the x-ray equivalent of the optical disector, to opaque specimens.

---

[4] This is given by Serra [1982] and Gunderson et.al. [1988]. Other formulas exist for different problem geometries.

5. BIBLIOGRAPHY

N. Aziz and S.Bhat, "On the Computation of Integral Properties of Objects", *Advances in Engineering Software,* **12**:4 (1990) 174-180.

J.W. Cahn, “A Quantitative Correction for the Holmes Effect”, *American Minerologist,* **44** (1959) 435-437.

J.W. Cahn and J. Nutting, "Transmission quantitative metallography", *Transactions of the American Institute of Mining, Metallurgical and Petroleum Engineers,* **215** (1959) 526-528.

A. Cauchy, "Mémoire sur le rectification des courbes et la quadrature des surfaces courbes", *Mémoires de l'Academie des Sciences,* **22** (1850) 3 (pages 167-177 in *Oeuvres complètes,* Series 1, Vol. 2 (1908)).

S. Chandrasekhar, “Stochastic Problems in Physics and Astronomy” *Reviews of Modern Physics,* **15** (1943) 1-89.

L.M. Cruz-Orive, "Particle number can be estimated using a disector of unknown thickness: the selector", *Journal of Microscopy,* **145** (1987) 121-142.

S. El-Soudani, "Profilometric Analysis of Fractures", *Metallography,* **11** (1978) 247-336.

F. Gittes, “Estimating particle volume and number from random sections by sampling profile boundaries”, **158** (1990) 1-18.

A. Holmes, *Petrographic Methods and Calculations* (Murby, London 1927).

H. Gundersen, P. Bagger, T. Bendtsen, S. Evans, L. Korbo, N. Marcussen, A. Moller, K. Nielsen, J. Nyengaard, B. Pakkenberg, F. Sorensen, A. Versterby and M. West, "The new stereological tools: Disector, fractionator, nucleator and point-sampled intercepts and their use in pathological research and diagnosis, Review article", *Acta Pathologica, Microbiologica et Immunologica Scandinavica,* **96** (1988) 857-881.

J. Janácek, “Estimating length and surface area by systematic projections”, *Image Analysis and Stereology,* **20** (Suppl 1) (2001) 89-94.

M.G. Kendall and P.A.P. Moran, *Geometrical Probability* (Chas. Griffin, London, 1963).

L. Lawson, "Fracture Toughness and Dimple Formation in a Whisker-reinforced 6061 Aluminum Matrix Composite", *Scripta Metallurgica at Materialia,* **25** (1991) 1553-1556.

C. Mack, "The Expected Number of Clumps when Convex Laminae are Placed at Random and with Random Orientation on a Plane Area", *Proceedings of the Cambridge Philosophical Society*, **50** (1954) 581-585..

R.E. Miles, "Estimating Aggregate and Overall Characteristics from Thick Sections by Transmission Microscopy", *Journal of Microscopy*, **107** (1976) 227-233.

E. Miller, *An Analysis of Surface Area Estimates of Binary Volumes Under Three Tilings* (M.S. Thesis, Massachusetts Institute of Technology, 1997).

M. Reed and C. Howard, "Surface-weighted star volume: Concept and Estimation", *Journal of Microscopy,* **190** (1998) 350-356.

I. Saxl, *Stereology of Objects with Internal Structure* (Elsevier, Amsterdam, 1989).

S. Saltykov, "The determination of the size distribution of particles in an opaque material from a measurement of the size distribution of their sections", in *Stereology*, Proceedings of the Second International Congress for Stereology, H. Elias, ed. (Springer Verlag, New York, 1967) 163-173.

E. Scheil, "The calculation of the number and distribution of sizes of spherical crystals in opaque substances by means of the circles obtained with a plane section", *Zeitschrift für anorganische und allgemeine Chemie*, **201** (1931) 259-264.

J. Serra, *Image Analysis and Mathematical Morphology* (Academic Press, London, 1982) Chapter 10.

S. Torquato, T. Truskett and P.G. Debenedetti, "Is Random Close Packing of Spheres Well Defined?", *Physical Review Letters*, **84** (2000) 2064-2067.

E. Underwood, *Quantitative Stereology* (Addison-Wesley, Reading, MA, 1970).

V. Vouk, "Projected Area of Convex Bodies", *Nature*, **162** (1948) 330-331.

S. Wicksell, "The Corpuscle Problem", Biometrika, **17** (1925) 84-99.

6. PROBLEMS

7.1) Show that Eqn.(7.44) applies to negative-going bumps. An application of negative-going bumps would be the modeling of flat bottom holes in a surface. If one were use the BET adsorption isotherm method in the presence or absence of a larger molecule than the test molecule and find either a) the presence of the larger adsorbate does not significantly affect the apparent surface area or b) significantly reduces the apparent surface area, how would this change the hierarchal model? John Hilliard held that the

BET isotherm method constitutes a form of stereological probe. If this is true, what sort of stereological probe is it in terms of a random measure?

7.2) Show that for a flat surface the ratio of R_S to R_P is inherently one. Construct a Taylor's series approach to argue that all nearly flat surfaces obey El-Soudani's hypothesis.

7.3) Consider a distribution of spheres classified by size in the range 1 through 5. We may call $N_A(i)$ the total number density of intercepts of size i with a test plane. We call $N_A(i,j)$ the number density of intercepts of size i resulting from spheres of size j. Since size 5 is the largest size, $N_A(5)=N_A(5,5)$ only. Since $N_A(5,5)$ is known, $N_A(4,5)$ can be calculated though $N_V(5)$. Since $N_A(4)=N_A(4,5)+N_A(4,4)$, by the same process we can calculate $N_V(4)$ and so on. This is the pencil and paper approach for the Schwartz-Saltykov method. Show that it is equivalent to solving Eqns.(7.52) by Gauss-Jordan elimination.

7.4) Consider a Poisson point process in three-dimensional space in which spheres are centered at each point. These spheres are generated by picking points at random and inscribing a sphere among the adjacent points (and previously-generated spheres) such that the inscribed sphere occupies the entire open space available without interpenetrating other points or spheres. The maximum attainable volume fraction of such spheres is about 0.3. Noting that the mean sphere radius is half the mean inter-point spacing, construct a first-order estimator for the volume fraction of this process. Cave! Although the mean radius is half the mean interpoint spacing, the mean cube of the radius is not one-eighth the mean interpoint spacing. Consequently finding the mean cube of the interpoint spacing and dividing by eight is a worse approximation than simply cubing the means. Since the mean radius is half the mean inter-point spacing, we can imagine a second process in which we inscribe spheres of radius equal to half the distance to the nearest neighbor point when the nearest neighbor point has no sphere already inscribed. For nearest neighbors with inscribed spheres, we inscribe the sphere to contact it as in the first process. Again the mean radius remains half the mean inter-point spacing. But, the volume fraction of inscribed spheres is significantly lower. For close packed spheres of the same size the volume fraction is 0.74, but random close packing (obtained by physically jamming the structure) gives at best about 0.64 for a volume fraction. Yet, still higher volume fractions are readily achieved by jamming spheres or similar shapes having a gradation of radii. Asphalt paving is an example. It usually has over 85% stone by volume. For a discussion of the ambiguity of random packing see Torquato et al. [2000].

Chapter 8

STEREOLOGICAL SAMPLING AND STATISTICS

This chapter examines sampling distributions for stereological parameters. Most of the results are obtained indirectly. Several useful estimators are derived. Since at this time numerical methods provide the best answers to problems of this class, an introduction to them is given.

1. INTRODUCTION TO SAMPLING STATISTICS

In considering the practical application of the results derived in the previous chapters we will be concerned with such matters as the estimation of sampling errors, optimization of analyses and the numerical reduction of experimental data. In this chapter (which is a counterpart to Chapter 2) we will develop results that will be needed for these purposes. Over the years, statisticians have developed a great deal of theory concerning sampling. Parameters of the population, such as the mean or the standard deviation, are often estimated either on one hand as lying with some predetermined certainty within a *confidence interval* or on the other as the values with the *greatest likelihood* of being correct. These and other approaches are chosen depending on circumstance. A confidence interval tells you more about the certainty of value than a greatest likelihood estimate. Under ideal circumstances, not only an estimate of a parameter is obtained but along with it an actual function describing the probability of it being correct to within some error. Obviously, this latter case is best. However, there is a tradeoff. More assumptions are needed to provide this additional information and some of these assumptions may be wrong. After all, statistics as a subject treats problems of uncertain knowledge. Theory should follow experiment, not the other way around.

The sampling theory for stereological parameters is based on general sampling theory, but it is more specialized. We will be approaching the subject primarily through examining the uncertainty associated with the several elementary stereological probes that we have studied. As we pursue these results, we should keep in mind that their application in digital imaging technology differs somewhat from their application to the manual techniques. In the latter case, a limitation on the number of data points usually dominates the error estimation. Digital technology typically gives a great number of data points, and other sources or error are likely to be significant.

1.1 Notation

We will follow the convention of distinguishing between a population value of a statistic and that estimated from a sample by using a Greek letter for the former and a the corresponding lower case Roman letter for the latter. A statistic is defined as a function of a random variable that does not depend on any of the unknown parameters. That is, it is a function of the distribution alone and not on how that distribution is written nor on the process from which it was derived. Statistics such as the mean and standard deviation are often used to parameterize a distribution. For example, the mean of a population is written as μ. The average of a sample would then be m following in this convention. More commonly the variance of a population is written σ^2 while the variance of a population estimated from a sample is most often written s^2. The use of m for the mean is unusual, so the more common notation using a bar will be used; the mean of a sample of the variable X will be written $\overline{x}$. In the previous sentence the capital letter X was used to denote a random variable itself while the lower case x was used to denote a *variate*, that it a specific value taken by that variable in a specific instance. This is commonly used notation. One of the authors (J.H.) established a second use for a capital letter: the capital letter in the case of a statistic is used to denote an unbiased estimator of that statistic (for emphasis). When it is possible that confusion would arise, a statistic may be written as a function of the random variable to which it applies, e.g. $\sigma^2(x)$. For convenience a lower-case letter is sometimes used to designate the random variable in that case without any special meaning. Since n is now required to indicate the sample size, the letter k will be used for an index on moments in place of n.

1.2 Sampling

In Chapter 5 we derived a large number of expressions relating the expected value of some measured property to a structural parameter. For example, re-arranging Eqn.(5.59) we have

$$S_V = 2\overline{P}_L, \tag{8.1}$$

which enables us to estimate the boundary area per unit volume from the density of intersections of on a linear probe. It is now time to examine the meaning of the equality sign in such expressions. No one would be so naive as to assume that, even in the absence of experimental errors, an analysis is going to yield an *exact* value of S_V in the region in which the measurements were made, let alone in the specimen as a whole. Instead, Eqn.(8.1) and similar expressions have to be interpreted as follows.

Taking the specimen as the bounding body, we assume it is sectioned in such a manner as to generate a "random" section as defined in Chapter 4. A "random" linear probe is then applied to the section with its perimeter as the bounding figure, and the number P of intersections and the total length L of the probe are determined. The specimen is then re-assembled (this is obviously a hypothetical experiment) and the process repeated n times. If on the i-th replication, the values P_i and L_i are the values measured, then on the basis of Eqn.(8.1) we can state

$$S_V = \lim_{i\to\infty} \frac{2\sum_{i=1}^{n} P_i}{\sum_{i-1}^{n} L_i}, \tag{8.2}$$

where S_V is the boundary area of the specimen divided by its volume. It is thus apparent that Eqn.(8.1) holds as an equality only at the limit of an infinite number of replications and provides no information on the reliability of the estimate obtained from a finite number of replications. This is the matter that will be one of our primary concerns from no on in this chapter.

The sampling procedure just described would be almost impossible to apply in practice. Fortunately, the structure in many specimens is sufficiently homogeneous and isotropic to justify confining measurements to a single arbitrarily chosen section. When this is not the case, procedures are available that will minimize the sampling required. There is nothing sacrosanct about random sampling. Namely, when anisotropy is present, it is often preferable to select a non-randomly oriented section and modify the relationships accordingly. For example, if a specimen contained boundaries all of which were parallel nearly to a given direction, then one would apply a circular linear probe to a section normal to that direction of orientation and use $S_V = \overline{P}_L$, omitting the factor of two, to estimate S_V. Also as we shall see, it is often simpler to determine the size distribution of non-spherical particles having a common alignment than it is for randomly oriented particles.

Finally, it hardly seems necessary to point out that the specimen being analyzed is usually a sample from some larger object that in turn is a sample drawn from a set of similar objects and so on. There is thus a whole hierarchy of sampling problems. Since these are so diverse it is impossible to offer any generalizations for their solution other than that the investigator should be aware of them.

1.3 Independence

We will frequently be encountering results that are valid only if the random variables involved are independent of each other. For example, the standard error of the sum of n measurements decreases as $1/\sqrt{n}$ *if* the measurements are independent. Formally, independence is defined by the conditions for which Eqn.(2.48) is valid. However, it is often not feasible to apply this test and one has to resort to more intuitive considerations. So, let us consider some examples of independence and non-independence (correlation) that relate to stereological measurements.

As we have seen, one method of estimating the volume fraction V_V of a constituent is by point counting. This is usually performed by applying a grid of points successively to non-overlapping regions of the image and counting the fraction of points falling in the constituent. It will be shown that the variance in the estimate of V_V is inversely proportional to the total number of points applied if certain conditions are satisfied. Now consider an analysis in which after each count, the grid is not moved but instead is re-applied to exactly the same points of the image. Barring observational errors, each application would yield the same estimate of V_V, and thus the average of the results of n applications would be no more reliable than that yielded by a single application. Now, let us suppose that the grid is moved between applications but by a distance that is very small compared with the scale of the structure. In this case the probability that the status of any given point will change during movement of the grid is small so that there will be a high degree of correlation (or non-independence) between the successive counts. Because of this correlation, the variance for a given total number of points applied (in n applications of the grid) will be higher than it would have been had the grid been moved a large enough distance between each application to nullify the correlation. This nullifying distance is usually conceptualized as some multiple of the *correlation length* of the structure being examined.

Similar considerations apply when parallel linear probes (or scans) are used. If the probes are closely spaced there will again be a high degree of correlation in the measurements and this will increase the variance. Yet, as discussed in Problem 6.10 part b, when the scan lines become so numerous and close together as to approach a Riemann sum over the probability space, the significant of correlation may decrease. This is a consideration that deserves further examination.

Lack of independence may also occur when analyzing structures containing a high degree of regularity or periodicity. A common example in metallurgy of such a structure is a constituent of steel known as pearlite. Pearlite consists of alternating lamellae of the two iron phases, ferrite and cementite. It occurs in colonies within which the lamellae have approximately the same orientation. Let us suppose that we wished to determine the average true spacing of the lamellae from a measurement of the apparent spacings on a planar probe. Because the apparent spacing depends on the orientation of the lamellae, measurements within any colony will be correlated. Therefore, it would not be justifiable to estimate the variance from the total number of measurements without correcting for this effect.

2. AVERAGE AND VARIANCE OF A FUNCTION OF RANDOM VARIABLES

Many structural parameters are functions of one or more experimentally determined variables. We therefore need to know how to estimate the expected value and variance of a parameter from the expected value and variances of the measured properties. We will first consider the case of linear functions for which exact expressions are available and then the more general case (but one which can only be solved approximately) of non-linear functions.

2.1 Linear Functions

Consider the linear function $X = c_1 x_1 + c_2 x_2$ where the x's are random variables and the c's are constants. If $f(x_1, x_2)$ is the joint density function for the x's, then by definition

$$\overline{X} = \int\limits_{(x_1)}\int\limits_{(x_2)} (c_1 x_1 + c_2 x_2) f(x_1, x_2) dx_1 dx_2 ,$$

which can be written

$$\overline{X} = c_1 \int\limits_{(x_1)}\int\limits_{(x_2)} x_1 f(x_1, x_2) dx_1 dx_2 + c_2 \int\limits_{(x_1)}\int\limits_{(x_2)} x_2 f(x_1, x_2) dx_1 dx_2 .$$

Carrying out the integration of the first term with respect to x_2 and the second with respect to x_1, the expression may be written in terms of the marginal distributions. Utilizing Eqn.(2.47), we obtain

$$\overline{X} = c_1 \int\limits_{(x_1)} x_1 f(x_1) dx_1 + c_2 \int\limits_{(x_1)} x_2 f(x_2) dx_2 = c_1 \overline{x}_1 + c_2 \overline{x}_2 . \tag{8.3}$$

It is evident that for the general case $X = \sum\limits_{(i)} c_i x_i$, we have

$$\overline{X} = \sum_{(i)} c_i \overline{x}_i . \tag{8.4}$$

Next we consider the variance for the sum of two variables. From Eqn.(2.40) $\sigma^2(X) = \mu_2'(X) - \overline{X}^2$, we have

$$\begin{aligned}\sigma^2(X) &= \int\limits_{(x_1)}\int\limits_{(x_2)} (c_1 x_1 + c_2 x_2)^2 f(x_1, x_2) dx_1 dx_2 - \overline{X}^2 \\ &= c_1^2 \int\limits_{(x_1)}\int\limits_{(x_2)} x_1^2 f(x_1, x_2) dx_1 dx_2 + c_2^2 \int\limits_{(x_1)}\int\limits_{(x_2)} x_2^2 f(x_1, x_2) dx_1 dx_2 + \\ &\quad 2c_1 c_2 \int\limits_{(x_1)}\int\limits_{(x_2)} x_1 x_2 f(x_1, x_2) dx_1 dx_2 - \overline{X}^2 .\end{aligned}$$

Evaluating the two integrals and substituting $\overline{X}$ from Eqn.(8.3)

$$\sigma^2(X)= c_1^2\left[\mu_2'(x_1)-\bar{x}_1^2\right]+c_2^2\left[\mu_2'(x_2)-\bar{x}_2^2\right]-2c_1c_2\left[E(x_1x_2)-x_1x_2\right], \tag{8.5}$$

where $E(\cdot)$ denotes the expected value of the operand. The last square bracket term is the covariance of x_1 and x_2. Equation (8.5) can be rewritten

$$\sigma^2(X)=c_1^2\mathrm{Var}(x_1)+c_2^2\mathrm{Var}(x_2)+2c_1c_2\mathrm{Cov}(x_1,x_2). \tag{8.6}$$

If x_1 and x_2 are independent of each other, than as hinted in Section 2.3.4 the covariance is zero. In that case

$$\sigma^2(X)=c_1^2\sigma^2(x_1)+c_2^2\sigma^2(x_2). \tag{8.7}$$

We can generalize on Eqn.(8.6) using a special notation. If the use of a prime in conjunction with a moment designates "about the origin", than what would $\mu_1(x)$ indicate? It would appear to be always zero. If we have two variables we can write $\mu_{11}(x_i,x_j)$ to mean the first product moment in which the first variable is taken to the first power about its mean and the second variable is also taken to the first power about its mean. This is another way of writing the covariance of x_i with x_j. The difference is that it allows us to set $i=j$ and obtain the variance without violating a convention. It follows that in the multivariable case,

$$\sigma^2(X)=\sum_{(i)}\sum_{(j)}c_ic_j\mu_{11}(x_i,x_j), \tag{8.8}$$

where $X=\sum_{(i)}c_ix_i$.

The summation in Eqn (8.8) includes the permutation of the subscripts when $i\neq j$, so that these terms will appear twice as required.

An important application of Eqn.(8.7) is in the estimation of the variance of the mean of a set of n observations. By definition

$\bar{x}=\frac{1}{n}\sum x_i$. Thus, if the measurements are independent

$$\sigma^2(\bar{x}) = \frac{1}{n^2}\sum \sigma^2(x_i). \tag{8.9}$$

Furthermore, since these measurements may be presumed to be made on a stationary population, the variance of that population is the same for all measurements. Consequently,

$$\sigma^2(\bar{x}) = n\sigma^2(x)/n^2 = \frac{\sigma^2(x)}{n}, \tag{8.10}$$

or, in terms of the standard deviation,

$$\sigma(\bar{x}) = \frac{\sigma(x)}{\sqrt{n}}. \tag{8.11}$$

The inverse dependence of the standard deviation of the mean on the square root of the number of observations is a well-known phenomenon, but it must be emphasized that it is strictly valid only if the results are independent. That has been the purpose of re-deriving it here.

Before leaving the subject of linear functions, some remark is needed about constants. After all a general linear function such as is given by Eqn.(8.4) usually has a constant added to it. Some readers might feel that the topic was slighted if the constant was not mentioned. Let me reassure these readers that the constant drops out in the case of the variance since the variance is a moment taken about the mean and constants do not contribute any variation.

2.2 Non-Linear Functions

Let X' be a function $g(x)$ of the random variable X. Then by definition

$$\bar{x}' = \int g(x) f(x) dx\ , \tag{8.12}$$

and

$$\sigma^2(x') = \int [g(x)]^2 f(x) dx - \bar{x}'^2\ . \tag{8.13}$$

We have already considered the case where $g(x)$ is linear. For a non-linear function, the quantities $\bar{x}'$ and $\sigma^2(x')$ can only be evaluated exactly if $f(x)$

is known. In many instances, only $\bar{x}'$ and $\sigma^2(x')$ are available. We will therefore derive approximate expressions for this case. It will be assumed that $g(x)$ is analytic and therefore can be expanded in a Taylor's series about the average of x. Thus,

$$\begin{aligned} g(x) &= g(\bar{x}) + (x-\bar{x})g^{(1)}(\bar{x}) + \tfrac{1}{2}(x-\bar{x})^2 g^{(2)}(\bar{x}) + \ldots \\ &= g(\bar{x}) + \sum_{r=1}^{\infty} \frac{1}{r!} g^{(r)}(\bar{x})\ (x-\bar{x})^r , \end{aligned} \tag{8.14}$$

in which $g^{(r)}(\cdot)$ is the r-th derivative of g. Substituting this for $g(x)$ in Eqn.(8.12) we find,

$$\bar{x}' = g(\bar{x}) + \sum_{r=2}^{\infty} \frac{1}{r!} g^{(r)}(\bar{x}) \mu_r(x) . \tag{8.15}$$

The lower limit $r = 2$ results from the vanishing of the term containing $x-\bar{x}$ to the first power when integrated. If only the first term in the summation were retained,

$$\bar{x}' \approx g(\bar{x}) + \frac{1}{2}\frac{d^2 g}{dx^2}\sigma^2(x) . \tag{8.16}$$

As one would expect, the zeroth-order approximation $\bar{x}' = g(\bar{x})$ improves as the variance of x decreases.

We next consider the variance of X'. Substituting from Eqns.(8.14) and (8.15) into Eqn.(8.13) and retaining only those terms involving the second derivative of g or less we obtain,

$$\sigma^2(x') \approx \left(\frac{dg}{dx}\right)^2 \sigma^2(x) + \frac{d^2 g}{dx^2}\frac{dg}{dx}\mu_3(x) + \left(\frac{d^2 g}{dx^2}\right)^2 \left(\frac{\mu_4(x) - \sigma^4(x)}{4}\right) . \tag{8.17}$$

(Recall that the derivatives in Eqns.(8.16, 8.17) are evaluated at $\bar{x}$.) If we have reason to believe that the second derivative of g at $\bar{x}$ is significantly less than the first derivative, we can posit a "zeroth-order" order approximation, involving the first derivative,

$$\sigma^2(x') \approx \left(\frac{dg}{dx}\right)^2 \sigma^2(x) . \tag{8.18}$$

The extension of this approach to the multivariable case is easy provided an approximation of the form of Eqn.(8.18) is valid.

Let $X' = g(X_1, X_2,..X_n) = g(\mathbf{X})$, then

$$\bar{x}' \approx g(\bar{x}_1, \bar{x}_2,..\bar{x}_n) + \tfrac{1}{2}\sum_{(i)}\sum_{(j)} \frac{\partial^2 g}{\partial x_i \partial x_j}\mu_{11}(x_i, x_j). \tag{8.19}$$

Similarly an approximation for the standard deviation can be constructed along the same lines,

$$\sigma^2(x') \approx \sum_{(i)}\sum_{(j)} \frac{\partial g}{\partial x_i}\frac{\partial g}{\partial x_j}\mu_{11}(x_i, x_j), \tag{8.20}$$

where the derivatives in both Eqn. (8.19) and (8.20) are evaluated at the mean value of their arguments in keeping with the remark following Eqn.(8.17) because the Taylor expansion of g is taken about the means. The function g is presumed to be smooth enough that these derivatives exist.

In case X is of the form $C\prod x_i^{p_i}$ where C and the exponents are constants, we have

$$\left.\frac{\partial g}{\partial x_i}\right|_{\mathbf{x}=\bar{\mathbf{x}}} = \frac{p_i}{\bar{x}_i} C \prod_{(i)} \bar{x}_i^{p_i} \tag{8.21}$$

in which the averages appear because they are the components of $\bar{\mathbf{x}}$ about which the expansion is taken. Substituting this into Eqn.(8.20) gives

$$\sigma^2(x') \approx g^2(\bar{\mathbf{x}}) \sum_{(i)}\sum_{(j)} \frac{p_i p_j}{\bar{x}_i \bar{x}_j}\mu_{11}(x_i, x_j) . \tag{8.22}$$

Finally, Eqn.(8.20) can be generalized to give the covariance between two transformed variables, X, Y, where X is as previously defined and Y is $h(y_1, y_2 .. y_m)$,

$$\mu_{11}(x', y') \approx \sum_{(i)} \sum_{(j)} \frac{\partial g}{\partial x_i} \frac{\partial h}{\partial y_j} \mu_{11}(x_i, y_j) \ . \tag{8.23}$$

This result is given by Kendall and Stuart [1977] as their Eqn.(10.13).

3. VARIANCES IN THE ESTIMATION OF THE MOMENTS OF A DISTRIBUTION

Several of the structural parameters that we will be considering involve the moments of some measured distribution; for example, intercept lengths in a constituent. In the case of large samples, moments about the origin are computed from

$$m_r'(x) = \frac{1}{n} \sum_{(i)} x_i^r \ , \tag{8.24}$$

and those about the mean from[1]

$$m_r(x) = \frac{1}{n} \sum_{(i)} (x_i - \bar{x})^r \ . \tag{8.25}$$

As with any other measurement, these estimates are subject to sampling error. The reader who wonders why n-1 does not appear instead of n in the expression for $r = 2$ may recall that that substitution of n-1 removes a small-sample bias relative to the true moment of the population. It is a posterior correction. It is included as part of the definition of the variance in many applications. In other words, these large-sample estimators are not necessarily the greatest likelihood estimators in the case of small samples, but biases vanish in the limit as n goes to infinity.

For these estimators, variances and covariances can be calculated using the methods just developed. A number of these are given in Kendall and Stuart, Chapter 10 [1977]:

---

[1] These definitions follow Kendall and Stuart but are not standard.

$$\sigma^2\left(m_r'\right)=\frac{\mu_{2r}'-\left(\mu_r'\right)^2}{n} \tag{8.26}$$

This and the following are "exact" expressions.

$$\mu_{11}\left(m'_r, m'_q\right)=\frac{\mu_{q+r}'-\mu_q'\mu_r'}{n} \tag{8.27}$$

Approximate expressions for the variances of estimators of moments about the mean can also be calculated and are given by Kendall and Stuart.

$$\sigma^2(m_r)\approx\frac{1}{n}\left(\mu_{2r}-\mu_r^2+r^2\mu_2\mu_{r-1}^2-2r\mu_{r-1}\mu_{r+1}\right). \tag{8.28}$$

$$\mu_{11}\left(m_r, m_q\right)\approx \frac{1}{n}\left(\mu_{r+q}-\mu_r\mu_q+rq\mu_2\mu_{r-1}\mu_{q-1}-r\mu_{r-1}\mu_{q+1}-q\mu_{r+1}\mu_{q-1}\right). \tag{8.29}$$

Example 8.1

Equation (8.28) estimates the variance of a moment about the mean. We may use it to estimate the variance of a large-sample variance estimator m_2 for a normal population. From Eqn.(8.28) we obtain directly,

$$\sigma^2(m_2)\cong\frac{1}{n}\left(\mu_4-\mu_2^2+4\mu_2\mu_1^2-4\mu_1\mu_3\right).$$

Substituting the values for the moments from a normal population, this becomes,

$$\sigma^2(m_2)\cong\frac{1}{n}\left(2\sigma^4+4\sigma^2\mu_1^2\right)=\frac{2\sigma^4}{n},$$

since $\mu_1=0$. The sampling distribution of the variance of a normally distributed population is chi-squared.

And, the variable $p = \frac{\nu m_2}{\sigma^2}$ is chi-square distributed with ν degrees of freedom. Solving for p we see that m_2 is a linear transformation of p. Since the variance of the chi-squared distribution is 2ν, it follows that the variance of m_2 is $\sigma^2(m_2) = \frac{2\sigma^4}{\nu}$. Since $\nu = n - 1$, the two formulas agree for large n. This reminds us that the estimator m_r is a large sample estimator, and for the case of r = 2 differs from the conventionally defined sample variance.

An estimator (given by J.H.) for the covariance of a central moment estimator with that of a moment about the mean is,

$$\left|\mu_{11}\left(m_r, m_q'\right)\right| \approx \frac{1}{n}\left|\mu_{r+q} - \mu_r\mu_q - r\mu_{r-1}\mu_{q+1}\right| . \tag{8.30}$$

We will need this estimator, but because this expression contains both a moment about the mean and one about the origin, it is neither exact nor can convergence be attained as readily as in Eqn.(8.29). Equation (8.30) can estimate only the magnitude but not the sign of the covariance and even that not as well as Eqn.(8.29). The covariances of finite samples of m_i or m'_i do not approach any limit rapidly. And, of course the above formulas hold only so long as the moments involved exist. In the approximations, terms on the order of n^{-2} and higher negative powers are omitted. It will be noted that Eqn.(8.10) previously given for the variance of the mean, is a special case of Eqn.(8.26) with $r = 1$.

Because these estimators will be important in subsequent work, they deserve further examination. We begin with Eqn.(8.27) which is Kendall and Stuart's Eqn.(10.6.) The derivation is straightforward. Among other terms, we obtain one product of sums, $\sum x^q \cdot \sum x^r$. We expand this product of sums as a sum of products and obtain $\sum x^{q+r} + \sum_{j \neq k} x_j^q x_k^r$.

The first term represents diagonal terms in the expansion of the initial product. It is tempting to view these diagonal terms as accounting for all correlation between the x's. If such were the case, then the x's in the second term would be independent, in fact "iid". In that case, the second term becomes $n(n-1)\mu_q'\mu_r'$ while the first term becomes $n\mu_{q+r}'$. But, this iid assumption is not strictly true. We can take a simple example, the product of two sample means. This corresponds to setting $q=r=1$.

As in the above, $\overline{xx} = \frac{\sum x}{n}\frac{\sum x}{n} = \frac{\sum x^2 + \sum_{i \neq j} x_i x_j}{n^2}$. But, using the above expansion containing $n(n-1)\mu_q'\mu_r'$, we obtain for the product of the two means, $\frac{E(x^2)}{n} - \frac{\bar{x}^2}{n} + \overline{xx}$.

This contains two more terms than we would like to have and can only be true in the limit if n is extremely large or if the variance of x is extremely small. Alternatively an error term must be subtracted when using the iid assumption for sums of the $i \neq j$ type to eliminate these unwanted terms. This error term is examined in Problem 8.1. Since it does not appear to have been addressed in the derivation of Eqn.(8.27), its existence raises a question of just how exact that equation is.

The accuracy of Eqn.(8.27) was evaluated numerically and appears to be excellent. The procedure, which was also used to check Eqns.(8.29) and (8.30) is now described briefly. A random number generator was used to generate tables of 10,000 random numbers distributed either with a normal (symmetric) or a Weibull (non-symmetric) distribution. From these tables with known parameters, samples of either ten or 1000 variates were drawn. The moments, (e.g. m_q' and m_r') were calculated for 200 pairs of these samples. From a table of these 400 moments, the observed covariances between moments were calculated. For each pair of q and r-values, five runs (each with a new table of 10000 numbers) were performed which yielded minimum, maximum and median values for the sample covariances. These were compared with the estimator. For Eqn.(8.27), the results are best stated in terms of the median values because these proved to be close to the estimator, Eqn.(8.27). Values of q and r between unity and five were explored. Coefficients of variation, the ratio of the standard deviation to the mean, of 0.1 and 1.0 were used in the input data. In the great majority of cases the magnitude of the error was less than 5%. However as q and r increase there is a tendency for the error to increase. The extremes of error were seen with the extreme pair of moments tested, 5 and 3 but were not regarded as statistically significant.

Unlike Eqn.(8.27), the apparent accuracy of the inexact estimators was less than excellent. For example, Eqn.(8.29) with $q = r = 4$ was tested with the Weibull distribution at a coefficient of variation of 0.1. The error of the estimator relative to the median of five runs was -18% with $n = 10$ and -78% with $n = 1000$. However, the latter estimator was still only a few percent less than the lowest value in the five Monte Carlo runs. For this particular formula the error seemed to increase with n. This trend was not observed with the other two formulas tested. The accuracy of Eqn.(8.30) was poorer than that of Eqn.(8.29).

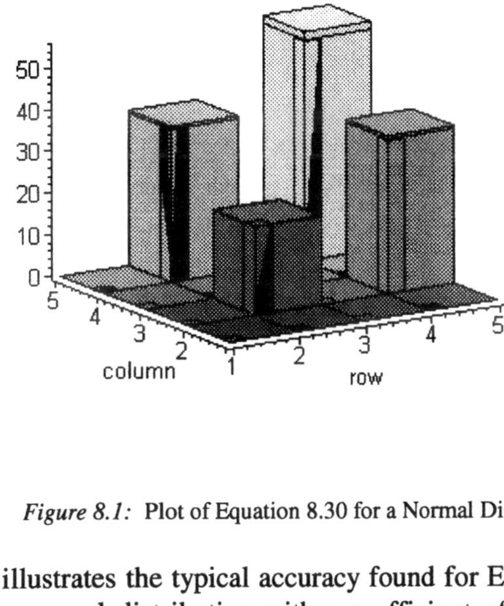

Figure 8.1: Plot of Equation 8.30 for a Normal Distribution

Figure 8.1 illustrates the typical accuracy found for Eqn.(8.30). It shows the result for a normal distribution with a coefficient of variation of unity and $n = 1000$. The vertical axis is the natural logarithm of the covariance. The covariance for a pair of moments is represented by a bar at the row and column corresponding to the indices of the respective moment. The band on the top of the bar represents the range of the covariance as calculated numerically for five runs of Monte Carlo data. The inset portion of the bar represents the estimator. Although, considering all test runs, the accuracy of the estimator with respect to the median of five runs is only within an order of magnitude; the dispersion of the runs is wide enough that the estimator is within the range or near to it in nearly all cases.

Example 8.2

As an illustration of the use of the foregoing results we will determine the sampling variance of $R = s(x)/\hat{x}$, which is the *coefficient of variation.* We will also be encountering this statistic as a structural parameter. For example, if the x_i are the intercept lengths between particles, then R is a measure of the spatial dispersion of the particles. (How does R change for particles as they cluster?)

Comparing the definition of the coefficient of variation, given above, with the product form for X associated with Eqn.(8.21), we can put the coefficient of variation into the product form by setting $C = 1$, $x_1 = s(x)$, $x_2 = \hat{x}$, $p_1 = 1$ and $p_2 = -1$. Omitting the argument on s, and substituting Eqns.(8.20, 8.21) we obtain the variance of R. Then from Eqn.(8.19) by neglecting the second order derivative term we obtain,

$$\frac{\sigma^2(R)}{\overline{R}^2} = \frac{\sigma^2(s)}{s^2} + \frac{\sigma^2(\hat{x})}{\hat{x}^2} - 2\frac{\mu_{11}(\hat{x}, s)}{\hat{x}s}. \tag{8.31}$$

Now s is defined as $\sqrt{m_2}$, and that is approximately $\sqrt{\mu_2}$. Furthermore $\mu_1 = 0$. Assembling the terms of Eqn.(8.31), we then have from Eqn.(8.28) followed by Eqn.(8.18),

$$\sigma^2(s) \approx \frac{\sigma^2(m_2)}{4s^2} \approx \frac{\mu_4 - \mu_2^2}{4n\mu_2}. \tag{8.32}$$

Also, from Eqn.(8.26),

$$\sigma(\hat{x})^2 = \frac{\mu_2' - (\mu_1')^2}{n} = \frac{\mu_2}{n}, \tag{8.33}$$

and from Eqn.(8.30) followed Eqn.(8.23),

$$\mu_{11}(s, \hat{x}) \approx \frac{\mu_3}{2n\sqrt{\mu_2}}. \tag{8.34}$$

Substituting these quantities into Eqn.(8.31) and using the population values for s and $ we obtain,

$$\frac{\sigma^2(R)}{\overline{R}^2} \approx \frac{1}{n}\left(\frac{\mu_4 - \mu_2^2}{4\mu_2^2} + \frac{\mu^2}{(\mu')^2} - \frac{\mu^3}{\mu_2\mu_1'}\right). \tag{8.35}$$

---

4. VARIANCE OF CORRELATED SUMS

When variables are identically and independently distributed (iid) the variance of a sum of them is simply the sum of the variances. Often variables are not iid. Even in a point count analysis the variates, viz. the values of the property being measured at the selected locations, may be neither independently nor identically distributed. Losing the property of being identically distributed need not be a great loss so long as no assumption about the form of the distribution is required. Furthermore, there are central limit and Lindbergh-Feller theorems, which show that such sums tend to be normally distributed so long as the distributions of the individual variates are not too wild. The loss of independence is somewhat more significant. But, dealing with that loss is not difficult: The variance of the sum of two variables, x and y, is

$$\mathrm{Var}(x+y) = \mathrm{Var}(x) + \mathrm{Var}(y) + 2\mathrm{Cov}(x,y). \tag{8.36}$$

This is a formula standard in textbooks. Obviously $\mathrm{Var}(x+y+z) = \mathrm{Var}((x+y)+z)$ by the linearity of Eqn.(8.36). The result is

$$\begin{aligned}&\mathrm{Var}(x+y+z) = \\ &\mathrm{Var}(x) + \mathrm{Var}(y) + \mathrm{Var}(z) + 2\mathrm{Cov}(x,y) + 2\mathrm{Cov}(x,z) + 2\mathrm{Cov}(y,z)\end{aligned}. \tag{8.37}$$

We may recognize that $\mathrm{Var}(x) = \mathrm{Cov}(x,x)$. Consequently,

$$\mathrm{Var}\left(\sum_{(i)} x_i\right) = \sum_{(i)}\sum_{(j)} \mathrm{Cov}(x_i, x_j)\ , \tag{8.38}$$

where i and j are allowed to range freely through all possible permutations. This result will prove useful and can be written in a variety of integral forms depending on the dimensionality and purpose. In Eqn.(8.38) the x's are variates. To extend this equation we can make the x's functions of relative location.

Since integrals over random functions are sums, we could write $x_i \rightarrow \lambda(x_i)$ changing the meaning of x_i from a variate to a specific relative location on the real line. The variate is now $\lambda(x_i)$. We can formally replace the l.h.s with the variance of an integral over a convex subset X of the real line of which v is the measure. Eqn.(8.38) becomes,

$$\mathrm{Var}\left(\int_X \lambda(x)dx\right) = \int_X \int_X Cov(\lambda(\xi), \lambda(\eta))\, d\xi\, d\eta$$

and can be rewritten,

$$\mathrm{Var}\left(\int_X \lambda(x)dx\right) = \int_X \int_X \lambda(\xi)\lambda(\eta)d\xi d\eta - v^2(x)\bar{\lambda}^2 \; . \tag{8.39}$$

Note that $\bar{\lambda} = \frac{1}{v}\int_{(x)} \lambda(x)dx$.

5. ERROR IN THE ESTIMATION OF RECIPROCAL VOLUME RATIOS, V_V, S_V AND L_V

The subject of the efficiency of various stereological procedures is far from closed. Even what "efficiency" means is not universally agreed upon. Generally we will take efficiency as an intuitive quantity relating the sampling error in an estimate to the number of samples required. Here we will examine perspectives on the efficiency of various estimators for several stereological parameters with the intent of reaching at least an intuitive generalization about efficiency.

5.1 Introduction

In the following sections we will consider the estimation of three of the most commonly determined structural properties, V_V, S_V and L_V. The relationships of these quantities to measurements on a section or a projection have already been developed. Our primary concern here is to determine which of the alternative procedures are the most efficient, which conditions would optimize an analysis and which method to use to estimate of sampling errors. We will see that the estimation of sampling errors is dependent on the path of the calculation that in turn depends on the actual structure of what is being sampled. We could in fact imagine flow graphs for the various possible courses of analysis and using the methods introduced in Section 8.3 determine a variance for the sampling distribution based on models for the variances of the components in the calculation. Since the result would be path dependent, we should then have average over all possible paths in some suitable way. There is nothing very appealing in such a procedure since it promises little insight in return for much work. On the other hand, more elegant treatments require simplifying assumptions. The Hilliard-Cahn estimators are developed by following a "typical" path.

5.2 Historical development

It follows from Eqns.(5.93, 5.57 and 5.8) that apart from a direct volume-based analysis (for example grinding up the sample and perform either a chemical or an x-ray powder diffraction analysis.) there are three different methods of determining the volume fraction of a constituent from measurements on a section: from two-dimensional data, one-dimensional data and zero-dimensional data respectively. The first of these to come into actual use was areal analysis (two-dimensional data) by Delesse [1848]. This entails measuring and then summing the areas of interception of a constituent with a planar probe. Delesse did this by first tracing the structure onto paper, then cutting up the tracing and weighing the pieces. Modifications of this method utilizing micrographs have been used intermittently up to the present despite the observation of Huntington and Desch [1908] that this procedure is much more time consuming than using a planimeter or its modern day equivalent.

Lineal analysis was apparently first used by Rosiwal [1898] who measured individual intercept lengths on a linear probe and then summed them at the end of the analysis. Later, microscope stages were developed that allowed the distances traversed in various constituents to be integrated. In one such stage described by Hunt [1924] the specimen can be traversed under the cross hairs of a microscope by any one of several independently acting micrometers, each of which is assigned to a particular constituent. The differences in the readings of the micrometers before and after the analysis yield the total distance traversed in each constituent. Other devices include those developed by Dollar [1937] and Hurlbut [1939] in which the stage is driven via a flexible shaft by a variable speed electric motor. The latter was also connected to a set of Veeder-Root™ (mechanical) counters and actuating keys. The drive mechanism could be started by depressing any one of the keys, which also caused the drive shaft to engage the corresponding counter. Distance was then counted in discrete increments. An additional counter also kept a tally of the total distance traversed. These same ideas in a more sophisticated form were incorporated into a device developed by Smith [1960], which allowed the stage to rotate to obtain a circular traverse. Here an optical encoder was used to generate electrical pulses. Depending on the constituent under the cross hairs, the pulses were switched into any one of several counters.

The use of mechanical counters is becoming obsolete. More recently, there has been an explosion in the number of proprietary software methods based on digitized images. They are to numerous to enumerate here. Typically these combine lineal and areal analysis in the limit as the second approximates the first. The number of pixels in each scan line within a given constituent is counted. An edge-finding algorithm adjusted through the application of dilation, erosion, closure and similar image algebraic methods often decides what constitutes a constituent. These make linear and areal methods into point counting.

Point counting methods did not come into use until many years after areal and lineal methods. The first application to microstructural analysis appears to have been made by Thompson [1930]. He superimposed a two-dimensional lattice of points on the image either by the use of a ruled reticule in the eyepiece of a microscope or by the simultaneous projection of the point lattice and a lanternslide of the structure. Glagolev [1934] introduced a one-dimensional point count performed by advancing the specimen in discrete steps under the cross hairs of a microscope. Similarly, a simple modification of the standard stage was devised by Chayes [1949], which involved the addition of click stops on the traversing screws so that the specimen could be advanced in constant increments.

Scanning instruments of various sorts have been introduced including an early one by the author [Hilliard, 1976]. If the scans are closely spaced, the analysis becomes an areal one. If the scans are widely spaced, the analysis tends toward a linear one especially if the scan lines are suitably rotated. The two analyses have somewhat different statistical properties. The in-between case was mentioned in Problem 6.10 part 5(b) in which we referenced Weibel's comment that scan lines *must* be either closely spaced or widely spaced but never in between. The reason for this is that the closely spaced approach assumes a different set of uncertainties than the widely spaced one. In the latter case, the scan lines are thought of as statistically independent so that their correlations can be neglected. In the former case, the areas sampled are assumed statistically independent so that their correlations might be neglected. In the case of point count methods, the points are assumed to be uncorrelated. It may be posited that in reality, all methods actually reduce to point count methods because there is always a limiting resolution even if we must resort to Heisenberg's uncertainty principle to find it. What differs is the amount of correlation between the points.

As we have already seen, there are three different procedures available for estimating volume fractions and all three have been used. The question therefore arises as to which, if any, is to be preferred. In making a choice, the first thing we have to note is that it is meaningless to ask which is more *accurate*. In the absence of experimental errors all three will yield what will tend toward an exact measurement in the limit as the number of replications goes to infinity. Instead, the choice has to be made on the basis of their relative efficiencies, the most efficient procedure being defined as that requiring the least effort for a result of given reliability, viz confidence interval and level of significance.

In principle, the relative efficiencies could be determined experimentally. One method of analysis would be repeated a number of times on the same specimen and the results analyzed to establish the reproducibility as represented by say the standard deviation. The same operation would then be carried out using another method of analysis, thus providing a comparison between the two methods in terms of the effort required to attain a given standard deviation. Amazingly, such comparisons have actually been performed. But, unfortunately the results are in conflict. For example Cheyes [1949] has reported that a particular form of point counting required less than half the time required for lineal analysis with a Hurlbut counter. On the other hand, Howard and Cohen [1947] had claimed an analysis with a Hurlbut counter was at least ten times quicker than point counting for the same accuracy. As will be shown later, the source of the discrepancy is that the efficiency of an analysis depends on several factors that were not explored in these evaluations. Consequently, these experimental estimates of efficiencies hold only for the specific conditions under which the analyses were performed. They provide a poor guide to the efficiencies realizable under optimum conditions. We will therefore explore an alternative approach developed by Hilliard and Cahn [1961]. This involves a theoretical estimate of the standard deviation of the various procedures in terms of the number of experimental measurements and certain parameters determined by the form of the structure. It amounts to determining the variances associated with typical procedures and emphasizes the variance contributed from procedural aspects.

A quite different view of the problem was given by Miles [1980]. He derived formal integrals for the variances of plane section estimators of V_V and related more general scalar properties. These were further expanded by Saxl [1989]. These integrals calculate variances based on the intrinsic variability of the material rather than the method. In a continuum, these integrals do not generally reduce to relationships between known quantities, as would be the case with ordinary point-count sampling statistics. Furthermore, there is a pervasive assumption that plane sections pass through the object in question in its entirety rather than through only a sample of it.

The first of Miles's integrals expresses the variability of the mean quantity, λ, of a function of position (ξ, η) over an area of intersection. He calls this the *mean grade* when the value is averaged over the entire volume. Taken over a section, we may think of it as a sectional mean grade. The variance of this sectional mean grade is

$$\mathrm{Var_A}\left(\bar{\lambda}_A\right)=$$

$$E\left\{\frac{1}{A(X \cap T)^2}\int_{X\cap T}\int\int_{X\cap T}\int\lambda(\xi,\eta)\lambda(\xi',\eta')d\xi d\eta d\xi' d\eta'\right\}-\Lambda_V{}^2, \quad (8.40)$$

where E denotes expectation, X the body being sampled and T the test plane. The similarity between this expression and that of Eqn.(8.39) is notable. Eqn.(8.40) is Eqn.(8.39) in two dimensions where $\lambda(x)$ has been replaced by $\lambda(\xi,\eta)/A(X \cap T)$, since we are computing the variance of an average rather than that of a sum. Equation (8.40) thus reduces to Eqn.(8.38) if the specimen is regarded as a set of discrete points. The bar over the lambda in Eqn.(8.40) does not appear in Miles's paper; it is added for emphasis to remind the casual reader that we are not deriving the variance of lambda on some surface but rather the variance of the integral average of lambda over a family of surfaces. Here Λ_V denotes the average per volume and X is not a set in the real line but the three dimensional specimen itself. The subset being interrogated is $X \cap T$ where T is the probe. The coordinates of points on the intersection are ξ and η. The area of the intersection is $A(X \cap T)$. Equation (8.40) can be further expanded to eliminate the expectation by using the distribution function for an A-weighted probe. This serves to lengthen the expression materially. The interested reader should consult Miles's paper. Equation (8.40) can be converted from one for a planar probe to one for a linear probe. Since the linear probe gives shorter expressions, we will pursue that shortly as a beginning. The basic difficulty of evaluating Eqn.(8.40) lies in finding the expectation. Substituting the planar probe's distribution function into it does not appear to facilitate its solution. Unlike the integrals which gave rise to the formulas of Chapter 5, Eqn.(8.40) does not simplify easily. As we shall see, reverting to the discrete sum approach does allow an approximate solution, which we will discuss.

Miles's second integral relationship concerns the estimating the global grade variance throughout a specimen from the *sectional grade variance* defined as

$$\sigma_{\lambda}^{2}(X \cap T)= \iint_{X\cap T}\{\lambda(\xi,\eta)-\lambda_{A}(X \cap T)\}^{2} / A(X \cap T)\ . \qquad (8.41)$$

Miles shows that such an estimate is negatively biased, as common sense would suggest. He also shows that this bias diminishes as the specimen becomes more "homogeneous". Kriging, to be discussed in Chapter 10, provides another way to define the global variance especially of large specimens.

For comparison, consider a material made up of a finite number of discrete points. Any mixture of phases can be reduced to a two-phase mixture problem by identifying the phase of interest as one phase and the others as the "other phase". In a two-phase mixture problem, λ takes the value of one for points in the phase of interest and zero for point in all others. For any fixed number of points the global grade variance is given by the binomial variance, $f(1-f)$ where f is the volume fraction of the phase of interest. Hence the same value also applies to the sectional variance except that f varies from section to section.

The comparative intractability of the Miles integrals combined with the realization that the intrinsic variability treated by these integrals is frequently negligible in comparison to variances from other causes has altered the focus of research toward sources of extrinsic variability. Gundersen and Osterby [1981] treated the optimization of stereological studies in biology from the standpoint that the major source of variation was between biological specimen subjects rather than within the stereology itself. A similar emphasis on finding the controlling variance of a stereological measurement in factors outside of stereology proper was also placed by Uylings, van Eden and Hofman, [1986] and Regeur and Pakkenberg [1989]. This list is certainly not exhaustive. Procedural details probably contribute the major part of the total variance in most practical stereology. But, however small it may be, understanding the intrinsic sampling variance is still important to understanding stereology. From the above papers we can enumerate four categories of variance: first, intrinsic variance which is the sampling variance of the stereology formulae themselves; second, the variance which results from the spatial limitation of the sample relative to the whole specimen being tested, third the variance related to lost information from the coarse discretization of the stereological data (see [Serra, 1982]); and fourth, variance contributed through technique and choice of specimens -- especially when it is desired to characterize not only a specimen but some more general entity such as the "brains of schizophrenics". Because this is a book about the principles of stereology, we will focus on the first source of variance. We tip our hats to the second by introducing a new term, "subarea". The *subarea* is the portion of the specimen on which the stereological sampling is performed. Recognizing that the subarea may be different from the specimen as a whole is at least a step toward accounting for the variance of the second category.

5.3 Estimation of Volume Fraction by Analytical Techniques

Volume fraction is perhaps the most often estimated structural property. Hilliard and Cahn [1961] introduced the following analytical approximations. Considering that the contribution of the stereological process to the total variance is usually small, the important task is to show when it can be neglected. Monte Carlo methods will be discussed following the analytical ones.

5.3.1 Variance of a random point count

The simplest case to analyze is that of a random point count. Randomness often presupposes independence, which complicates analyses increasingly as the dimensionality of the sample increases. Point counts are zero-dimensional, so the complication is minimal.

Consider a point count performed on a subarea in which the true areal fraction of some constituent we call α is A_A. Since we presuppose independence, the total number of points falling within α profiles will follow a binomial distribution. From the results given in Section 2.4.1, we have for the expected number of points $\overline{P}_\alpha$ and their variance in terms of the total number P of test points applied,

$$\overline{P}_\alpha = PA_A \ , \tag{8.42}$$

$$\text{and } \sigma^2\left(\hat{P}_\alpha\right) = PA_A\left(1 - A_A\right). \tag{8.43}$$

The latter equation can be rewritten in terms of the point fraction $P_P = P_\alpha / P$. Using Eqn.(8.6) for re-scaling of the variance,

$$\sigma^2\left(\hat{P}_P\right)=\frac{P_P\left(1-P_P\right)}{P}. \tag{8.44}$$

The expected value of the point fraction is the volume fraction and some authors have used these expressions for the variance of the volume fraction estimated by point counting. But clearly this is not valid. According to Eqn.(8.44) the variance goes to zero as $P \to \infty$. However, no matter how high the point density goes, the minimum variance can never be less than that of an areal analysis performed on the same subarea. In other words, we must allow for the variation in areal fractions from one subarea to another. As an approximation we can assume that sections are uncorrelated so the variances are additive. In the case of independent areal sections, we obtain,

$$\sigma^2\left(\hat{V}_V\right)=\frac{P_P\left(1-P_P\right)}{P}+\sigma^2\left(A_A\right). \tag{8.45}$$

This is similar to a more detailed expression derived by Hilliard and Cahn [1961] excepting for an additional term of order $\sigma^2\left(A_A\right)/P$, which can be regarded as negligible when at least several hundred test points are applied to the section. Evaluation of the second term requires evaluation of Eqn.(8.40) or its generalization for the sections involved. We have assumed independence in the point count and between sections in deriving Eqn.(8.45) and for that matter in deriving Eqn.(8.40). Clearly then, Eqn.(8.45) would not apply to point counts on closely-spaced serial sections.

5.3.2 Variance of a systematic point count

In the previous section we derived the variance of a point count on the assumption that the points were randomly distributed on the section. Prior to digital image analysis, such a procedure was avoided in practice because of the experimental difficulties of generating a random set of points. Instead, a grid of points was superimposed on the structure. In statistics, this corresponds to a systematic sampling of the population and is known in most cases yield a lower sampling variance than random sampling for a given size sample. Of course as noted in Section 8.1.3, there may be exceptions if the structure has a high degree of regularity. The reason for the lower sampling variance is that a systematic sampling precludes certain extreme results that can occur in random sampling. For example, in a random point count, it is possible although admittedly improbable for all the points to fall within the profile of a single feature. This and similar unlikely outcomes are allowed for in the variance of a binomial distribution.

A systematic point count can be viewed as a discretization of the surface. Let us define a function over the surface as the indicator function (see Appendix) that takes the value of one when the point falls in the alpha phase and zero when it falls elsewhere. We are seeking to find the average of this function over the area, since that is the areal fraction. We note a useful property of the indicator function; it is equal to its own square. It follows from Parseval's theorem of Fourier analysis (see for instance [Brigham, 1974]) and the fact the indicator is real and equals its own square that

$$A_A \equiv \frac{1}{A} \int_{(A)} \varepsilon_\alpha(\mathbf{x}) dA = \frac{1}{A} \int_{-\infty}^{\infty} \int_{-\infty}^{\infty} \left| F[\varepsilon_\alpha(\mathbf{x})] \right|^2 d\mathbf{f} , \tag{8.46}$$

where ε_α is the indicator function for the alpha phase and $\mathbf{f}$ is the frequency vector. We see from Eqn.(8.46) that we must integrate over the entire frequency space of the Fourier transform of ε_α. In the case of a sampled space, such as a systematic point count, the right hand side of Eqn.(8.46) can be replaced by a sum over the discrete Fourier transform components. Such a sum is valid only when the sampling is conducted so as to meet the Nyquist sampling theorem, which says that the spatial frequency of samples (*Nyquist limit*) must be at least double the highest spatial frequency of the function being sampled. Not meeting this criterion gives rise to two types of error in our case. The first is not so serious and results from step function nature of ε_α which requires an infinite range of frequencies to reproduce. The second is called *aliasing* and is a moiré effect, which occurs when there is periodicity in the structure at frequencies above the Nyquist limit. It is particularly severe when the sampling frequency and structural frequency are near each other. The reader has probably seen the effects of aliasing as ripples in digitized photographs. If the reader has tried to re-digitize a halftone photograph, the reader is aware of how frustrating the problem can be and has probably observed that sampling must be either at a frequency (dots per inch for a photograph) well above any periodicity in the image itself or well below it in order to obtain an acceptable result.

For sampling conforming to the latter case, Hilliard and Cahn obtained an estimate for the variance which is the same as Eqns.(8.44 and 8.45) for random sampling on account of both the random and the systematic samplings being dilute by assumption. The reader should note that random sampling in principle includes all spatial frequencies.

5.3.3 Variance of a lineal analysis

Using Eqn.(8.39) and the definition of the indicator function, we can state a one-dimensional variant of Eqn.(8.40):

$$\sigma^2(\bar{L}_L) = E_L\left\{ L(X_2 \cap T_1)^{-2} \int\limits_{X \cap T} \int\limits_{X \cap T} \varepsilon(x)\varepsilon(x')dx'\,dx \right\} - \mathrm{E}_A^2 . \qquad (8.47)$$

Here, the subscript alpha is dropped as being understood. The subscript L on the expectation indicates that the quantity averaged over a line is L-weighted in analogy with the A-weighted section (Chapter 4). The notation $L(\cdot)$ means "the length of". The subscripts 2 and 1 indicate the dimensionalities of the section and probe. The symbol E_A indicates the true areal mean. This should be compared with Eqn.(8.44) rather than Eqn.(8.45) because it is limited to a plane rather than an volume. As stated it seems no less solvable than Eqn.(8.40) since it requires detailed knowledge of $\varepsilon(x, y)$ (where x and y are the coordinates in the plane) to solve.

Hilliard and Cahn took a "back door" approach to solving what amounts to Eqn.(8.47). Let $\bar{l}_\alpha$ and $\bar{l}_\beta$ be the average intercept lengths in the two constituents of a binary material. The lineal fraction in the alpha constituent is then,

$$L_L = \frac{\bar{l}_\alpha}{\bar{l}_\alpha + \bar{l}_\beta} . \qquad (8.48)$$

If $\bar{l}_\alpha$ and $\bar{l}_\beta$ can be assumed independent, then it is possible to apply Eqn.(8.20). For a fixed length of total scan, $\bar{l}_\alpha$ and $\bar{l}_\beta$ are not independent but linearly related. However, if as in Eqn.(8.47) the length of the line is the entire linear intercept with the specimen, total length is independent. Since total length is the sum of the lengths in components, the necessary extra degree of freedom is present. The result is

$$\sigma^2(L_L) \approx \frac{\bar{l}_\beta^2 \sigma^2(\bar{l}_\alpha) + \bar{l}_\alpha^2 \sigma^2(\bar{l}_\beta)}{\left(\bar{l}_\alpha + \bar{l}_\beta\right)^4} . \qquad (8.49)$$

Noting that the product of the number of intercepts times the average length gives the total intercept length in a given constituent and that the number is the same for both constituents,

$$\sigma^2(L_L) \approx L_L{}^2(1-L_L)^2\left[\frac{\sigma^2(\bar{l}_\alpha)}{\bar{l}_\alpha^2}+\frac{\sigma^2(\bar{l}_\beta)}{\bar{l}_\beta^2}\right]. \tag{8.50}$$

Applying Eqn.(8.11) for sampling gives,

$$\sigma^2(\hat{L}_L) \approx \frac{L_L{}^2(1-L_L)^2}{N}\left[\frac{\sigma^2(l_\alpha)}{\bar{l}_\alpha^2}+\frac{\sigma^2(l_\beta)}{\bar{l}_\beta^2}\right], \tag{8.51}$$

where N is the number of alpha intercepts.

The relationship between Eqn.(8.51) and Eqn.(8.47) may seem tenuous. It is based on the presence of the indicator function reducing the integral to a discrete sum and the subsequent neglect of cross-covariance terms by arguing that the sampling is dilute. We see that we have reduced Eqn.(8.47) to a relationship involving merely the linear fraction of the alpha phase, the number of intercepts in the alpha phase and two properties of the structure which are measurable or knowable. We have not supposed anything about the dimensionality; Eqn.(8.51) could apply to properties defined in terms of a planar section or the bulk. For the case where the where the fraction of alpha is small, the beta intercepts will follow an exponential distribution and the second term in the brackets will go to unity. For the case of equal-sized spheres, from Eqn.(6.56) we find that the value of the first term in the brackets is 1/8. It follows that for this case $\sigma^2(\hat{L}_L) = 1.1(L_L)^2 / N$ where the variance is referenced V_V.

5.3.4 Variance of an areal analysis

The same approximation technique described for a lineal analysis can also be applied to an areal analysis. In order to estimate the variance of an areal analysis, it is necessary to make further assumptions about the form of the structure being analyzed. The simplest is that the alpha constituent appears as discrete profiles randomly distributed on the section. Because of impingement, this can only be satisfied in the limit $A_A \to 0$ for any specified shape or for the restriction to convex shapes. But these restrictions are not always needed. The process of generating a section can be described with two variables. Let $\bar{a}$ be the average area of an alpha profile and N be the number of alpha profiles in the subarea. The total area of the alpha profiles is simply,

$$A_\alpha = N\bar{a}. \tag{8.52}$$

Applying Eqn.(8.20) while assuming the number of α intercepts is independent of $\bar{a}$ (which it is not except for very large samples) gives

$$\sigma^2(A_A) \approx \bar{a}^2\sigma^2(N) + N^2\sigma^2(\bar{a}) . \tag{8.53}$$

The next assumption is again one of dilution. We presume that the distribution of N is Poisson. Experience has shown that Poisson distributions hold for many "random" processes such as photon scattering up to larger numbers than we care to think about here and in the arrival times of the most common queuing problems. We invoke Eqn.(8.11). Together these give us

$$\sigma^2(A_\alpha) \approx N\bar{a}^2 + N\sigma^2(a). \tag{8.54}$$

On dividing through by $A_\alpha^2 = N^2\bar{a}^2$ and noting that $A^2 = A_\alpha^2 / A_A^2$, we obtain the sampling variance,

$$\sigma^2(A_A) \approx \frac{A_A^2}{N}\left(1 + \frac{\sigma^2(a)}{\bar{a}^2}\right) . \tag{8.55}$$

We remind the reader the quantities in parentheses refer to properties of individual features, e.g. alpha phase profiles. We see here an expression very similar to Eqn.(8.51) containing in square brackets a term which is a property of the structure. As before, we can evaluate this expression for the case of spheres of constant diameter. The result is $\sigma^2(A_A) \approx 1.2(A_A)^2 / N$. This result is similar to that for the linear analysis; it contains a factor close to unity times what is essentially the volume fraction squared divided by N.

5.4 Model-based Estimation of Volume Fraction

In this section we are going to compare analytical approximations with Monte Carlo simulations. Monte Carlo simulations of complicated stochastic processes tend to reproduce experiment with accuracy superior to analytical approximations, since the simplifying assumptions needed are fewer. The major difficulty with this approach is that it can only evaluate the accuracy stereological measurements for known structures. If the structure is not known, they are of little use. Within these limitations, they can be used to test the accuracy of semi-empirical models, including the ones just given, for a family of known structures. If the model proves accurate, the only supposition becomes one of presuming that an unknown structure is similar to a member of the family for which the model is accurate.

5.4.1 Linear analysis

We will examine both a continuous Monte Carlo model of a line scan, and a somewhat abstract discrete Monte Carlo model of a line scan with several predictive indices. A couple of these predictive indices result from semi-empirical models to be developed later. The continuous Monte Carlo model is strictly geometrical and involves randomly distributed spheres. It concentrates its assumptions in the familiar. The discrete Monte Carlo model might be better introduced after examination of the predictive indices. But, the indices are best understood after understanding the model. However, presenting the indices first requires presenting them twice: once to say what they are and again to say how well they did. Since that wastes valuable paper, we will begin with the model.

In truth all digital models are discrete. In constructing the discrete model, we reexamine some assumptions that numerical modelers make every day without thinking much about them: First, we live in a zero-dimensional world and the illusion of dimensionality is the result of periodicity in correlations, that is, at least to a computer all arrays are (1-D) vectors; higher dimensions come from systems of assigning array indices. This vision describes the universe computer arrays (which are all one-dimensional in the language of computer science but are zero dimensional in the language of sets). Second is the hypothesis that for any probe, no matter how faithful, its intersection with some *reality* (a term we invent which is similar to "variety" in meaning except that it represents some real feature rather than a mathematical idealization) is always reducible to a discrete finite point set. For example, the intersection of the linear probe and the specimen generates a line segment along which values of one and zero are taken according to whether or not an alpha phase is present at each point. We divide up that line into small segments, cells or pixels, of equal length. The result is a sequence of ones and zeros, which may be regarded as a numerical message in which one and zero are its characters. Third, we posit that there exists an *ultimate resolution.* Further subdivision beyond this ultimate resolution gives no additional information in the message not only because the additional characters are totally correlated with their neighbors, but because there is an actual limit to the amount of information obtainable. That is, they are redundant. One may compare this case with the limiting spatial resolution, $\Delta x \geq h / \Delta p_x$, obtained from quantum mechanics to be reassured that there is such a thing as an ultimate limit to spatial resolution. Were we to attempt to divide a line into segments smaller than Δx, the additional segments would be restricted by quantum mechanical law from conveying additional information. This is a wave-related effect. Since all photographs are derived from a wave process, a similar but much larger limiting resolution applies to pictures. A sequence of cell values will be termed *complete* whenever it represents a scan taken at the ultimate resolution. We can apply this same notion to any dimensionality -- not just a line. This discrete hypothesis is convenient in that it greatly simplifies the task of calculating variances. It also has a considerable amount of verisimilitude.

5.4.1.1 Continuous Monte Carlo simulation

We want a model of a line scan that has minimal suppositions. For the sake of familiarity we will generate an actual three-dimensional space with embedded features and take random linear intersections through it. Because it is most easily done with spheres, this will be called the SphereModel. This model is actually a lot of fun to play with since it allows the user to visualize linear and areal probes, their intersections and the resulting statistics.

The model begins by defining a cubic space with an edge length of say 2000 units. Into this space is embedded a sphere (a ball actually) having a radus of 1000 units. We fill the sphere with say 10,000 points by using uniform distributions in x,y, and z directions to generate random points. Points falling inside the sphere are counted. Those outside the sphere are rejected. The result is a binomial point process since the total number of points is determined. But, this process is in all practicality indistinguishable from the Poisson point process discussed in Chapter 7. The coordinates of each point are recorded. Then a search is used to locate the nearest neighbor or the surface of the bounding sphere whichever is closer. Although this search involves measuring over 10^8 point pairs, it only requires 35 seconds on a home desktop computer at the date of writing. Were more points desired, they could be divided into families to reduce the search time. Within the search process, the features are inserted.

The table of points inherently lists the points in a random order. This is advantageous since it reduces the correlation between feature diameters. The features are themselves balls centered on the points. Thus to the record of each point's coordinates and its nearest neighbor distance, we add a radius. This radius is a fraction of the nearest neighbor distance or the distance to the surface of the bounding sphere, whichever is smaller. This fraction is a function of the local volume fraction and was found experimentally to be a second-degree polynomial function of the volume fraction. The local volume fraction is forced to be a beta-distributed variate with a mean equal to the global volume fraction. This adds a supervariance to the distribution of feature diameters above that that results from the process of inserting the features in random order. Since the radius of a feature must be less than the distance to the nearest neighbor point or surface, the program initially inserts large diameter balls and then fills the remaining points with progressively smaller ones. The maximum volume fraction obtainable is about 0.3. (See Problem 7.4). We should note that this process is not the same as one where new points are nucleated within the unoccupied interstitial space because the inserted spheres are always bounded by their nearest neighbors in a fixed configuration. In the case where interstitial points are added, the structure is correlated in such a way that the upper bound on volume fraction approaches unity.

Once the table is constructed, the volume fraction of the features is typically not exact. The volume fraction is adjusted iteratively using the derivative of the polynomial relating volume fraction to available radius fraction. In this adjustment, the previously set clearance may be consumed and a small amount of crushing (tiny overlap equivalent to elastic deformation) of the spheres facilitates obtaining volume fractions near 0.3.

Construction of the table is the most difficult part of the process. Once the table is constructed, the next task is to pick a random line. This is done implementing the geometry of Fig. 4.7. Here we first find a random plane (including the origin) and then select a line parallel to the plane's normal passing through a random point in the plane but within the bounding body. The choice of a sphere facilitates this since the bounding body is at all times a constant radius from the origin. Selection of the random plane is as follows. We begin with an equatorial plane. In that plane,

$$\begin{aligned} x' &= r\cos\theta' \\ y' &= r\sin\theta' \\ r &\in [0, D/2] \\ \theta' &\in [0, 2\pi) \end{aligned} , \qquad (8.56a\text{-}d)$$

where D is the diameter of the bounding sphere. The line, normal to the plane, is best defined by giving two points that belong to it. The point in the plane is $z' = 0$. A second point has the same coordinates except $z' = 1$. The plane is then rotated based on

$$\begin{aligned} \phi &= \arccos(1 - U[0,1]) + \frac{\pi}{2} U\{0,1\} \\ \theta &= U[0, 2\pi] \end{aligned} , \qquad (8.57a,b)$$

where $U[\cdot]$ denotes a uniform distribution on the stated interval; and, $U\{\cdot\}$ denotes a uniform distribution over the stated discrete set. But, since the plane is free to rotate about its axis, the rotation is still ambiguous. Hence we arbitrarily decide that the starting plane of Eqns.(8.56) was obtained from the true plane by a planned sequence of rotations, first by a rotation about the z-axis by θ and then by a rotation of ϕ about the x-axis. Undoing these rotations gives a transformation matrix, which looks like

$$T = \begin{bmatrix} \cos\theta & -\sin\theta\cos\phi & \sin\theta\sin\phi \\ \sin\theta & \cos\theta\cos\phi & -\cos\theta\sin\phi \\ 0 & \sin\phi & \cos\phi \end{bmatrix} . \tag{8.58}$$

Once we have defined the line, which is to be the linear probe, we next find the distances between this line and each point in the table. If the distance is less than the radius of the ball centered on that point, then an intercept results. The distance between a line given by two points (P_1, P_2) and a third point, P, is given in vector notation by

$$d = \left| \frac{\overrightarrow{P_2 P_1}}{\left\|\overrightarrow{P_2 P_1}\right\|} \times \overrightarrow{P P_1} \right| . \tag{8.59}$$

Since we have set the distance between P_2 and P_1 to be unity, the denominator of the first term in Eqn.(8.59) is unity. The length of the intercept is then found using the geometry of Fig. 6.1. The total length, number and moments of the intercept distribution are found be summation. Where only a small number of interceptions is desired, the distance of the probe line to the origin can be fixed by setting r in Eqns.(8.56) to a value to give the desired total intercept length.

5.4.1.2 Discrete Monte Carlo simulation

We may also consider a more abstract model of a line scan. In this model, we divide the line into discrete elements. We call this the LineModel. It is an abstract model in that it avoids any specifics about shapes or sizes of the three-dimensional features probed. As in the case of the SphereModel, we generate a population from which to sample. In the case of the SphereModel, the population represented a set of randomly placed and sized balls (alpha phase) inside a three-dimensional sphere, which were probed with one-dimensional probes. In the discrete line Monte Carlo model, the population is a long linear array consisting of ones and zeros. The ones represent the alpha phase. It is sampled by randomly selecting a portion of it. We set the size of this array to about 20,000. We construct it so that its fraction of ones equals the true volume fraction of alpha phase. Were the ones distributed binomially, there would be a defined scale relating the size of features to the resolution. Since this is unrealistic, we provide the model with a correlation function, which defines the distribution of the lengths of runs of sequential ones. In other words, since this is a discrete model, we need some way of specifying the scale of the resolution of the line scan relative to the scale of the structure being sampled. We want the scale of the structure independent of resolution of the scan.

If we think of the population as simply a long line (Borel set), the output of a scan is a sequence of ones and zeros according to the value of the indicator function at each resolved point along the line. Because the correlation parameter adjusts the distribution of the number of same-kind cells to be found in a row, it determines the number of intercepts in a scan line, which is directly related to the surface area of features per unit volume. We now have three parameters defined, the volume fraction, the surface per volume and the resolution. We wish to conduct our sampling in such a fashion that we can ignore this latter parameter. Given the first two parameters, we still need to specify a distribution of lengths for the intercepts. Rather than simply impose one, the distribution evolves in a two step process intended to model the behavior of real materials. (Imitating nature is convenient in an algorithm.) First, we position the desired number of intercepts by locating an internal point called the "germ", which is not necessarily the center, for each intercept. Second, we grow the intercept under the constraint that it meets the volume fraction requirement. We could proportion each interval according to the volume fraction. This structure has verisimilitude for many real cases, but for values of volume fraction away from 0.5, it displays too much regularity. Consequently, instead of simply proportioning off the interval, the growth takes place according to a beta distribution. This distribution has two parameters. Since we have specified the volume fraction, there is one remaining parameter. This is related to the standard deviation of the beta distribution and is called "roughness" in the program. A similar parameter appeared in the SphereModel. One reason why it is not simply variance is that if a beta distribution is to be well behaved, the range of allowed variance depends on the mean. This is a consequence of the distribution being confined to specific interval. Roughness amounts to the proportion of the maximum allowed variance that is applied. We now have sufficient parameters to characterize the distribution of intercepts. As in the SphereModel the generated population is termed "the table". Unlike the SphereModel, which has a defined bounding body, the LineModel uses a circular boundary condition so that samples having a defined length can originate at any point.

The LineModel can accommodate any volume fraction. Growth is a growth process for volume fractions of 0.5 and greater, but for volume fractions less than 0.5, the process used is actually one of erosion. In either case the program examines the space around each point. A beta-distributed variate having a mean equal to the volume fraction and a variance defined by the roughness parameter is then found to proportion off the amount of space available. If the volume fraction is less than 0.5, an alpha region is sized. It is centered on the randomly (uniformly) placed data point chosen in random sequence. If the alpha region would impinge on the previously set alpha region, the new region is shifted (to the right) to avoid impingement. In the event that the volume fraction is less than 0.5, a beta region is similarly grown around the data point. A running record is kept of the actual volume fraction. And, at the end of constructing the table, the volume fraction is adjusted to exactitude. The size of the table used was generally 10,000 elements. The lengths of scans drawn from the table are typically several hundred elements but varied widely to test the suppositions.

The choice of the beta distribution is arbitrary but convenient in that all variates fall in the interval [0,1]. The beta distribution is flexible and skewed to the left or right depending on the mean. Because of its similarity to the gamma and exponential distributions, it can be expected to approximate a wide variety of real-life intercept distributions. Although specific techniques are available to generate beta distributed variates [Atkinson, 1979; Cheng, 1978; Dagpunar, 1988] these were not used in favor of assembling an algorithm out of reusable code applicable to any density function. Since this algorithm gives insight and might be of interest to the experimentally minded reader, it will be described. First a uniformly distributed variate on the [0,1] interval is found. It is then matched to the cumulative distribution function and the variate having the desired density is found by inversion.

$$F(x)=U[0,1] \tag{8.60}$$

Equation (8.60) states the basic idea. The beta-distributed variate, x is given implicitly. The symbol $U[0,1]$ denotes a variate uniformly distributed on the [0,1] interval. Since the beta cumulative distribution function cannot be explicitly written, two iterations are required in the form of two nested loops. The outer loop inverts the distribution using Hamming's modified regula falsi root finder [Hamming, 1973] The inner loop finds the cumulative distribution by integrating the density function using Gaussian quadrature. Specific procedures are available for calculating the beta distribution function [Didonato and Morris, 1992], but the increased accuracy of the longer code did not offset its reduced speed of execution. The density function itself involves evaluation of the gamma function. This was done using a well-known standard routine called "Gautschi 221", which is a polynomial approximation; it is extended to a larger domain through the gamma recursion relation.

In this model, there are two parameters which may be used to check the results of various approximations to be discussed, the lineal fraction, which is set in the table exactly equal to the desired volume fraction and the standard deviation of lengths, which is verified to be that intended during a table-testing step prior to the table's use as a source of random scan data. (Actually all moments of the intercept length are available from the table but only two can be set independently.) Scans numbering from 10 to several hundred elements are then randomly drawn from the table. Their moments are computed and the variances of the lineal fraction of groups of these scans are compared with the variances estimated by way of the several approximations tested.

5.4.1.3 Entropies of a linear probe sample

We have written an expression for the variance of a linear analysis, Eqn.(8.51), in terms of structural parameters which although generally not known *a priori*, happen to reduce to a factor near unity times a function of known parameters: the length fraction and the number of intercepts. In deriving this and deriving other estimators of sampling variance, we have directly inserted Eqn.(8.11), which holds only for discrete sums. But, although we are working with a continuum, we allege we are evaluating Eqn.(8.40) or some analog of it with these sums. Since there is no immutable reason for choosing to reduce a continuum to sums in the particular way did, we need to examine the justification for doing it. There are several other estimators that can also be related to the variance of a linear scan, which are either not explicitly dependent on unknown parameters or not dependent on Eqn.(8.11). We need to examine these estimators; they are all based on discretization or, from an imaging standpoint, *pointillism*.

Before proceeding with estimators, we will review discretization through the relationship between the global variance of the material being sampled and the sampling variance. Consider a block of material four units on a side and having 64 cells in all. Following Miles's Eqn.(8) in [Miles 1980], we define the *global grade variance* as

$$\Sigma^2 = \frac{1}{V}\iiint(\lambda(\mathbf{x}))^2 dx_1 dx_2 dx_3 - \Lambda_V^2, \tag{8.61}$$

where $\lambda(\mathbf{x})$ is a function, the grade, of position vector $\mathbf{x}$, x_i, are its components and Λ_V is the average of λ over volume. To justify the word, "grade", we may imagine the block to be of some ore and λ to be the indicator function. We have two phases, the desirable metallic one and the "other" phase. Lambda equals one in cells having the metallic phase. Since each cell is one unit in volume, $\Lambda_V = V_V$ of the metallic phase. Now let us presume that $V_V = 0.5$. There are then 32 cells of each phase. Equation (8.61) reduces to a sum. The integral has a value of 32. Notice that the order in which the individual cells are arranged has no bearing on the value. For any arrangement of the cells,

$$\Sigma^2 = \frac{1}{2} - \left(\frac{1}{2}\right)^2 = \frac{1}{4}.$$

Let us now consider the section grade variances, $\sigma^2(X \cap T)$, given by Eqn.(8.41), where X is the set comprising the block of ore and T is the set comprising some plane passing through X. We first arrange the cells so that each has nearest neighbors of the opposite phase and then take sections parallel to a face of the block. Each section then has 16 cells. Eight of these are the metallic phase. The *sectional grade variance* is defined analogously to the grade variance but only over a section. Hence the sectional grade of each section is equal to the *mean grade* of the block in this case. The sum of the squares corresponding to the integral over T is 8, which also equals the sum over the same area. The area of the intersection of the plane T with the block is 16. The calculation repeats the one just given and the sectional grade variance equals the grade variance, 1/4. Miles showed that in general the sectional grade variance estimates the grade variance with a negative bias.

Let us now re-arrange the cells so that the 32 metallic ones comprise one face and a parallel plane below. If we take a section parallel to and through that face, the sum of squares is now 16 since all the blocks are metallic and the sectional grade variance is zero. In fact this will be the case for all plane probes parallel to the layers having the 32 metallic cells. We may observe by way of Miles's calculation that the grade variance must be at least the maximum of all possible sectional variances. We are now in the position to intuit something about the variance problem. The sectional grade variance of sections having the greatest diversity is closest to the grade variance itself. Sectional grade variance refers to the variance within the sample. In contrast, the sampling variance is the variance between samples.

We now look at indices or measures called "entropies"[2] in order to show a connection between continuum sampling and the assumptions which led to the use of Eqn.(8.11). We will do this by examining the information content of a continuum sample from a pointillist standpoint. We will not be able to establish that Eqn.(8.51) is the correct evaluation of the continuum case, since, given the assumptions, this may not even be possible. But, we will show that in terms of information content, choosing N in that expression to be the number of intercepts is consistent to within a near-unity multiplicative factor with a true continuum evaluation of the variance of a line scan.

The measure of a section's diversity, when that section is discretized as in the example just given, may be related to its information content or entropy provided we could distinguish between useful and irrelevant information, that is, if we can weight the information with respect to its importance. Information content as defined by Shannon was formulated as an entropy measure [Shannon, 1948]. We have already described the line scan as a sequence of ones and zeros. We expand this idea by dividing the line into linear elements; call them "lixels" in analogy with pixels. If we had four different kinds (phases) of lixels, the number of possible different scan lines N lixels long would be given by the multinomial coefficient

$\frac{N!}{n_1!n_2!n_3!n_4!}$, where N is the sum of the n_i, the number of times each type

of lixel appears in the sequence. This would also be the number of messages (in some code) N letters long given an alphabet of four letters. If we were to list the number of decimal places required to enumerate this number it would be the base ten logarithm of the multinomial coefficient. If we wanted to express our result as a binary number, the number of places would be the base 2 logarithm of the multinomial coefficient. The number of places is defined as the measure of the amount of information. If we convert this expression to one involving the probabilities of the various kinds of lixels and apply Stirling's approximation for the factorial, we obtain the Shannon entropy, which is quite similar in form to the entropy of a thermodynamic system,

[2] Information theoretic measures of this type are sometimes called Shannon-Wiener measures.

$$H = \log\left(\frac{N}{\prod_{(i)} n_i}\right) \approx -N\sum_{(i)} p_i \log p_i\,, \tag{8.62}$$

where $p_i = n_i / N$. This gives equal weight to every lixel. But, in the line scan, there is a very high probability that if a given lixel is of a particular phase, the next lixel will also be of that same phase. This correlation certainly will increase as we refine our subdivision of the line. We can examine the case where lixels are correlated through a Markov process representation. [Khinchin, 1957; Gnedenko, 1967] This process is illustrated in Fig. 8.2.

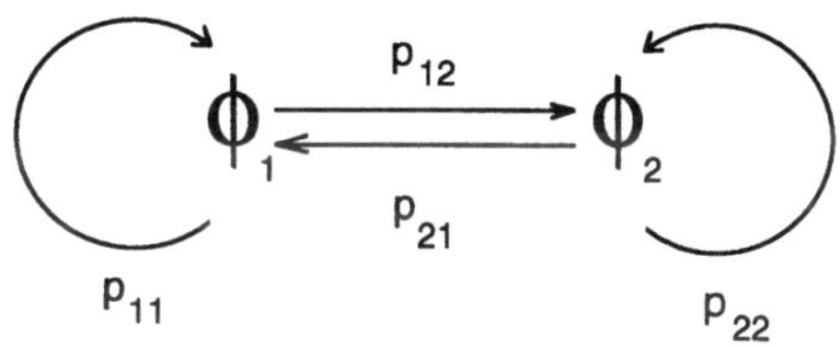

Figure 8.2: Two-state Markov process for two-phases

In this process the state of the system is the state of a cell. As the cell number is advanced in increments of one, the state of the system in the higher numbered cell is treated as a random function of the state of its lower numbered neighbor. The two phases, ϕ_1 and ϕ_2, constitute the two states of cells in the system. The parameters, p_{ij}, in Fig. 8.2 represent the transition probabilities following the paths indicated by the arrows. Thus, for example, p_{11} is the conditional probability of a return to state 1 in the next cell given that the system is presently in state 1. We assign such a system to each cell in our sequence and use Khinchin's [1957] conditional entropy formula,

$$H(AB) = H(A) + H(B\,|\,A)\ . \tag{8.63}$$

This formula gives the entropy for the product of two systems, A and B. Since, in our case one system is assigned to each cell, Eqn.(8.63) gives us the entropy of a pair of cells and can readily be extended to sequences of any length by making A the product of all systems preceding B. When all cells are iid, Eqn. (8.63) reduces to Eqn.(8.62). But this is not the case. From Eqn.(8.63) can be derived,

$$H = -\sum_{(i,j)} p_i \left(p_{ij} \log p_{ij} \right) \tag{8.64}$$

for the entropy of a single cell in a Markov process in which p_i is the total (or unconditional) probability of the state i. If we identify p with the phase fraction of ϕ_1, then we can use Eqn.(8.63) to write an expression for the entropy of the Markov process of Fig. 8.2.

$$H^{(1)} = -\left[p\left(p_{11} \log p_{11} + p_{12} \log p_{12}\right) + (1-p)\left(p_{22} \log p_{22} + p_{21} \log p_{21}\right)\right] \tag{8.65}$$

The superscript, one, indicate that this is for a single cell. Since the process is the same for all cells, the total entropy of a single line scan of n lixels becomes simply $nH^{(1)}$. If we specify that there are just ν groups of adjacent same-type cells for all n cells, we have,

$$p_{12} = \frac{\nu}{2np}; \quad p_{21} = \frac{\nu}{2n(1-p)}, \tag{8.66}$$

where these conditional probabilities are the reciprocals of the expected number of cells in the corresponding group. Since the sum of the conditional probabilities for leaving a group must equal one (including the recurrence probability),

$$p_{11} = 1 - \frac{\nu}{2np}; \quad p_{22} = 1 - \frac{\nu}{2n(1-p)}. \tag{8.67}$$

We examine what happens to $H^{(1)}$, Eqn.(8.65) as n increases in order to see what would happen as resolution increases. As expected, the entropy per cell goes to zero as n goes to infinity. However, the total entropy $H = nH^{(1)}$ does not come to a limit. Rather it slowly diverges. There are several causes. One of which is that the ν groups are fixed by Eqns.(8.66 and 8.67) only in the average.

A calculation which holds ν fixed may be formulated by setting np in the previous model to a fixed sum N which is reached when counting the cells of the n= $\nu/2$ groups of ϕ_1. Since we are counting integers, the number of ways of reaching this specific sum of n independent integers is

$$\begin{pmatrix} N-1 \\ n-1 \end{pmatrix}.$$

(We are indebted to Gabriel D. Carroll of the Berkeley Math Circle for pointing this out. Such a sum is referred to as a part of a *partition*. General reviews of combinatorial number theory are not numerous. One is that of Adhikari [2002].) Following the derivation of Eqn.(8.62) we have

$$H = \log\begin{pmatrix} Np-1 \\ n-1 \end{pmatrix} + \log\begin{pmatrix} N(1-p)-1 \\ n-1 \end{pmatrix}. \tag{8.68}$$

This too diverges as n goes to infinity by the sheer fact that the total number of lixels is increasing; information is contained in the act of specifying the amount of redundancy. We need to compensate for this. We will establish a weight for the increasing number N to represent the excess information of increased redundant resolution based on this first term.

The form of this weight is not given to us directly except through the asymptotics of Eqn.(8.68). We will need to guess a weight and see whether it both makes physical sense and meets the asymptotic requirements. To establish the plausibility of this weight from physical argument we consider the contents of a bucket. If we put one drop in the bucket, the integer 1 represents the content in drops. If we fill the bucket and add a drop, we increase the bucket's content by one added to N drops already present. Our knowledge about the contents of the bucket has increased by at most $1/N$. We want to reflect this diminishing knowledge in our weighted information. In terms of the number of places required to express the number of drops, this is $\log N$, which represents the un-weighted information content of the bucket. Analogously, if we keep the amount in the bucket constant, but refine our measurement to say units of 0.01 ml, N is increased by a factor of four; but, if all we can distinguish is drops, no new information is added. A weight which would keep the information about the bucket asymptotically constant would therefore be $1/\log(N)$. Consider the entropy of a set of systems in which each system is analogous to one bucket. In order to account for the excess information of redundant resolution, we might want to weight each system by the reciprocal of the logarithm the average number of units in it, N. This would give us a weighted entropy measure,

$$H_N = \sum \frac{H_i}{\log N_i} \approx \sum \frac{H_i}{\log\langle N_i \rangle} \xrightarrow[N>>\nu]{} \frac{1}{\log N}\sum H_i\ , \tag{8.69}$$

where $N = \sum N_i$.

Comparing Eqn.(8.64) with Eqn.(8.68) we may infer that each term in Eqn.(8.68) represents a sum of system entropies. So, we may define an N-weighted entropy for the α phase as,

$$H_{N,\alpha} = \frac{1}{\log N} \log \binom{Np-1}{n-1} . \tag{8.70}$$

We have posited that there exists an ultimate resolution at which further subdivision gives no additional information. We see that

$$H_{\infty,\alpha} = \lim_{N\to\infty} \frac{1}{\log N} \log \binom{Np-1}{n-1} = n-1 . \tag{8.71}$$

This result does not depend on the choice of logarithm. If we include both terms of Eqn.(8.68),

$$H_{\infty} = 2n-2 . \tag{8.72}$$

We can compare H_∞ to another index, which we will call the *phenomenological entropy* of a line scan, which has been correlated with the sampling variance of line scans directly and as seen through Monte-Carlo simulations.[3] It is an entropy measure grounded in the reasoning which led to Eqn.(8.51).

$$H = \frac{-1}{\log 2} \sum_{(i)(\varphi)} p_\varphi \log p_\varphi . \tag{8.73}$$

In the above expression, p_φ is the relative frequency of lixels representing the φ phase. The sum is taken with respect to each intercept i with phase φ. This is thus a double sum taken separately with respect to i and φ; the order of summation doesn't matter. We can thus take the sum over the values of φ first and then multiply by the number of phase intercepts. (Each time the scan line enters a new phase, that entering is counted as an intercept provided that the scan line also exits that phase rather than just stopping inside it.) The form of Eqn.(8.73) for the two-phase mixture is

[3]This entropy concept was part of a class project in John Hilliard's stereology course. (L.L.)

$$H_2 = \frac{-n}{\log 2}\left(p_\varphi \log p_\varphi + (1-p_\varphi)\log(1-p_\varphi)\right), \tag{8.74}$$

where n is now the number of phase intercepts with both phases and φ is the phase of interest, twice the value in Eqn.(8.72) . Only relative frequencies are used, so the actual number of points is not needed but must be sufficient to prevent information loss. Obviously, H_2 is dependent on the scale of the structure (or microstructure) relative to the length of the scan. Because of the factor of 1/log 2, the result is in bits. Bits are handy since a two-phase mixture is a binary mixture in many senses. H_2 is then the effective number of independent lixels sampled. On this assumption, we may estimate the variance of a line scan (in which volume fraction f replaces p_φ) using the variance of a binomial distribution given above and Eqn.(8.33). The result is

$$\mathrm{Var}(\hat{f}) \approx \frac{\hat{f}(1-\hat{f})}{H_2}, \tag{8.75}$$

where $\hat{f}$ is the phase fraction as measured in the line. Equation (8.75) has on the average given a better than order of magnitude estimate of the variance between scans as seen through Monte Carlo simulations. Note that in the case that $\hat{f}$ is highly uncertain, the variance estimated by Eqn.(8.75) is bounded from above by $1/(4H_2)$.

To examine it further, we begin by examining the relationship between entropy and the number of variates in a sequence of iid Bernoulli distributed variates. A Bernoulli distributed variable is one which takes only values of one and zero. Its distribution is given by

$$f(x,p) = p^x (1-p)^{(1-x)}, x \in \{0,1\}.$$

The maximum likelihood estimator (MLE) for p is

$$\tilde{f} = \bar{x} = \frac{1}{m}\sum_{(i)} x_i$$

for a sequence of variates of length m. For a well-mixed mixture, we would expect the values of ones and zeros in the cells of a sample to be iid. The number of ways of mixing the ones with the zeros in such a case is

$$\binom{m}{mp_{\varphi}}.$$

The negative of the logarithm of this binomial coefficient is the total entropy of the scan.

$$H = -m\left(p_{\varphi} \log(p_{\varphi}) + (1 - p_{\varphi})\log(1 - p_{\varphi})\right) = -\log\binom{m}{p_{\varphi}m} . \tag{8.76}$$

Substituting the MLE and dividing gives,

$$\overline{m} = \frac{H}{H_0} = \frac{H}{\overline{x}\log(\overline{x}) + (1 - \overline{x})\log(1 - \overline{x})} . \tag{8.77}$$

Equation (8.77) estimates the length of a sequence based on its information content. Here the entropy need not be in bits but must be calculated using the same base in both the numerator and denominator. The utility Eqn.(8.77) lies in the fact that in a real scan the cell values are not iid but correlated so that $\overline{m} < m$.

In conclusion, intercept lengths in different phases are usually not well mixed on the scale at which they are measured. Consequently, there is correlation between the lixels in any discretization of the line. Let ν now equal the observed number of groups of sequences of cells of the same phase type in a discretization. Let us now call N the number of cells in total. If we keep the same probe at the same place on the specimen and count the number of groups, the smoothed graph of ν vs. N, will look something like Fig. 8.3.

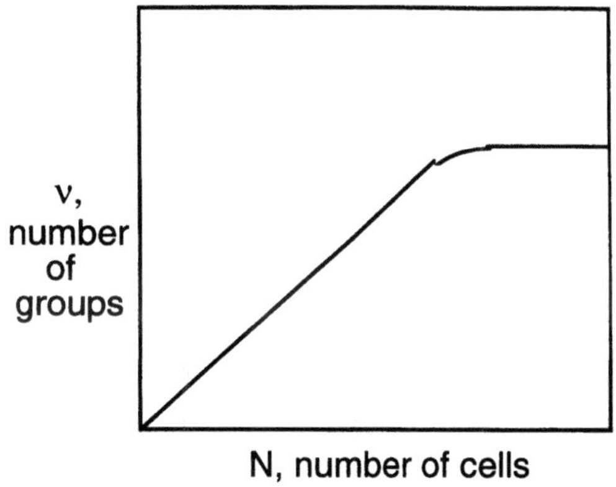

Figure 8.3: Number of groups vs. resolution

When the number of cells is small, the number of cells will be approximately the number of groups. As the number of cells increases, adding more cells will expose fewer and fewer new groups. When the number of cells exceeds that needed for minimal resolution, the number of groups will remain constant. It is the extension to infinity of this plateau on the right hand side of Fig. 8.3 that is approximated by a continuum model. Under pointillism there is an ultimate resolution beyond which no further information can be obtained from a system by additional subdivision. The very existence of such a condition reduces any continuum measurement to finite sums and justifies the use of discrete statistics. Furthermore, we see that the effective number of discrete variates is near to the number of intercepts.

5.4.1.4 Lumped covariance model

Our inquiry into the information content of a stereological sample did not give us any better or more useful estimator of its variance. Although simple, Eqn. (8.75) is as mediocre in performance as its broad assumptions would indicate. We would like to use what we have learned to forge a more useful estimator, preferably one that does not depend on unknown parameters. If we consider a line scan to be reducible to a cardinal number of linear elements, lixels, then the linear fraction is the sum of the number of these containing a particular (alpha) phase divided by the total number of lixels in the line. The variance of a sum is given by Eqn.(8.38) as the sum of the covariances. For the case of two phases, we use the indicator function to identify the alpha phase; the sum Λ is the count of the number of alpha lixels. Then, using subscripts to indicate the ordinal index of each lixel, the variance is

$$\begin{aligned}\operatorname{Var}(\Lambda) &= \sum \operatorname{Var}(\lambda_i) + 2E\left(\sum \operatorname{Cov}(\lambda_i, \lambda_j)\right)_{j \neq i} \\ \operatorname{Var}(\Lambda) &= NE\left(\operatorname{Cov}(\lambda_i, \lambda_j)\right)_{\text{allpermutations}}\end{aligned}, \tag{8.78a,b}$$

where λ is the value of the variate, either 1 or 0 , the summation is taken over all lixels, which number N in total.

We expect, as is common, that the variance of the sum grows linearly with the number of terms and that we can make use of the concept of an average variance per point. However, in any particular instance, we are restricted to the alpha phase since that is where the indicator function is non-zero. We can estimate the value of the expectation of the covariance of lixel pairs by lumping the correlation: pairs are either totally correlated or totally uncorrelated:

$$r(i,j) = \frac{\operatorname{Cov}(\lambda_i, \lambda_j)}{\sqrt{\operatorname{Var}(\lambda_i)\operatorname{Var}(\lambda_j)}} = \begin{array}{c|l} 1 & \lambda_i = \lambda_j \\ 0 & \lambda_i \neq \lambda_j \end{array}. \tag{8.79}$$

In this two-phase system, the variance of any random point drawn singly is the binomial variance,

$$p_{\phi\alpha} p_{\phi\beta} = p(1-p) \approx V_V (1 - V_V). \tag{8.80}$$

where p is the probability of drawing a point in the phase of interest, here the alpha phase. It follows that

$$\operatorname{Cov}(\lambda_i, \lambda_j) = \begin{array}{c|c} p(1-p) & (i=j) \cup (\lambda_i \in \phi_k \cap \lambda_j \in \phi_k) \\ 0 & \text{otherwise} \end{array}. \tag{8.81}$$

And consequently,

$$\operatorname{Var}(\Lambda_\alpha) = Np(1-p) + 2\sum_{\nu} \binom{n_i}{2} p(1-p), \tag{8.82}$$

where ν is the number and set of same-phase regions in the scan, n_i is the number of lixels in the i-th region, $1 \le i \le \nu$ and we recognize that we have summed only over the alpha phase. As seen in Section 8.5.3.3, contributions to variance arise from the variability of the beta phase. We approximate the total variance by that per element of the minority phase and therefore divide

Eqn.(8.82) by Np, to obtain the average variance per alpha lixel and then apply this value to all lixels of both phases so that $\mathrm{Var}(\Lambda) = \mathrm{Var}(\Lambda_\alpha)/p$.

Since $\sum_{\nu} n_i \sim O(N)$

and the sample average $\hat{p}$ is Λ/N, it follows that, via the variance scaling relationship Eqn.(8.6),

$$\mathrm{Var}(\hat{p}) \sim (1-p)E\left(\frac{\sum_{\nu} n_i^2}{N^2}\right), \tag{8.83}$$

where E denotes expectation and alpha is the minority phase. Since we are working with integers, a simplifying coarsening of the estimate can be applied to the variance given by Eqn.(8.83) using an inequality

$$\nu n_{\min}^2 \le \sum n_i^2 \le \nu n_{\max}^2$$

for the largest and smallest n_i, which lead to the relation

$$\left|\sum n_i^2 - \nu \bar{n}^2\right| \le \nu(n_{\max} - n_{\min})$$

and a bound on the corresponding minority phase fraction estimator $\hat{p}$

$$\mathrm{Var}(\hat{p}) \approx \frac{p^2(1-p)}{\nu}, \tag{8.84}$$

where, repeated for emphasis, ν is the number of α intercepts. This bound is wide, but Eqn.(8.84) appears to be a better estimator than Eqn.(8.83). If we substitute Eqn.(8.72) into Eqn.(8.75), we obtain a very similar looking estimator (for the case when V_V=1/2) using our current definition of ν,

$$\mathrm{Var}(\hat{p}) \approx \frac{p(1-p)}{2\nu - 2}. \tag{8.85}$$

5.4.1.5 Comparison of estimators with Monte Carlo simulations

The Monte Carlo simulations we derived in Sections 8.5.4.1.1 and 8.5.4.1.2 represent sampling of essentially stationary processes that are intuitively "typical". The SphereModel is slightly less stationary than the LineModel because the boundary condition in the SphereModel induces a thinning of the density as the boundary is approached because there are no balls nucleated outside the boundary and because the diameters of balls terminate at the boundary. However, the sampling is random throughout the simulated bounding body and that offsets any non-stationarity in the region being sampled. Note that were there no effect of translation on any scale, there would be no variance among samples since all samples would be the same. Through Eqns.(6.70 and 6.79) we see on one hand that the variances of intercept distributions generally depend on higher moments of the feature size distribution. Equations (8.51 and 8.55) appear to avoid dependence on the distribution function by incorporating the variance of the intercept distribution explicitly. Equations (8.75 and 8.84) make assumptions about the distribution that are tacitly supported by the central limit and Lindberg-Feller theorems, which connect most sampling processes to the normal or in the discrete case the binomial distributions. Connection to the normal distribution reduces the significance of higher moments in the calculation since they are predetermined. Because they contain superimposed beta distributions, the two Monte Carlo simulations we have chosen have complicated distribution functions, but nevertheless they cannot reproduce all possible sets of moments that could occur in stereological sampling. Random spheres inclusion models, such as SphereModel, are so common that they must be regarded as typical. In these, the ratio of the variance of the intercept size to the square of it's mean (the square of the coefficient of variation) is often near or less than 0.5. In the LineModel, this index frequently exceeds unity. Thus, we test our estimators with both high-variability and low-variability data even if we cannot test all possible combinations of higher moments.

For performance evaluation we chose Eqns.(8.51), (8.84) for graphing and Eqn.(8.85) to make comment on without graphing its data. To avoid confusion, we note that while N in Eqn.(8.51) is the number of alpha intercepts and is same as ν in Eqn.(8.84), the N in Eqn.(8.83) is the total number of cells. In testing Eqn.(8.51) with the SphereModel, we used values from the calculated global moments for $\sigma^2(l_i)$, $\bar{l}_\alpha$ etc. In testing Eqn.(8.51) with the LineModel, we used the moments of the samples since there is no explicit 3-D global model. (It is possible and even fun to construct 2-D data that match the LineModel 1-D simulated scans. The features so constructed are highly irregular.)

There are many possible ways to report performance data for an estimator. Since developing the reader's intuition is an important goal, we chose to compare the value of the estimator with the typical variance results of five runs of the simulations. Equation (8.51), Hilliard-Cahn for linear analyses, is plotted in Fig 8.4 predicting sample variance with respect to observed sample variances in five runs of SphereModel. Each bar represents a set of test conditions defined in Table 8.1:

Table 8.1: Test Conditions for Figs. 8.4 and 8.5

	Column 1	Column 2
Row 1	$V_V = 0.1,\ \nu=120$	$V_V = 0.3,\ \nu=210$
Row 2	$V_V = 0.1,\ \nu=12$	$V_V = 0.3,\ \nu=21$

The vertical axis is the negative of the base-10 logarithm of the sample variances and their estimators. The band at the top of each bar represents the range of the five typical test runs of Monte Carlo generated data. The inset represents the estimator. When the estimator depends on the simulated data, its median value is shown. However, estimators did not vary by more than a few percent between datasets. Equation (8.51) predicted a result that fell in the band in four out of four sets of conditions tested. The variables displayed, V_V and ν, were the ones most strongly affecting the results. Changing the variance of the beta distribution used in determining ball diameters appeared to have had negligible effect, so it not reported.

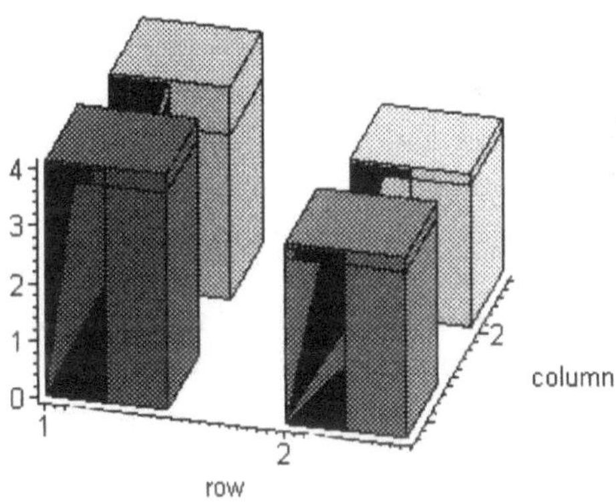

Figure 8.4: Equation (8.51) estimating the sample variances, SphereModel

Equation (8.84) has the utilitarian advantage of basing its guess of the variance on much less information than Eqn.(8.51). For the same test conditions its performance was expectedly somewhat poorer; it tended to underestimate the sample variance and fell in the band in only two of four tries, but it was quite close in the other two. It is plotted in Fig. 8.5.

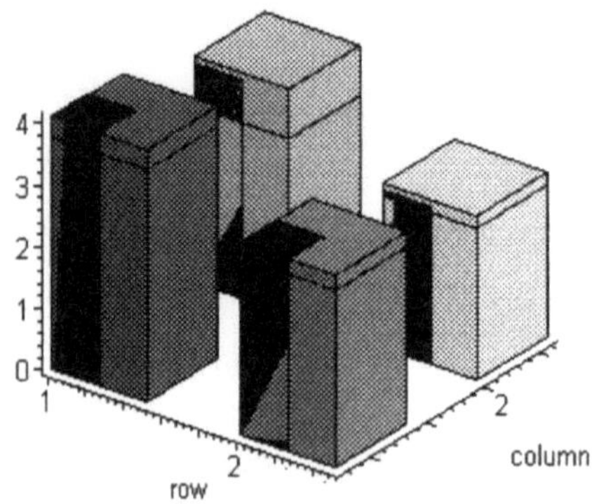

Figure 8.5: Equation (8.84) estimating the sample variances, SphereModel

Equation (8.85) overestimated the sample variance. It tracked the changes between the four sets of test conditions but was in the band in one in four tries. It therefore does not represent an upper bound; a meaningful one can not exist.

The SphereModel is limited to a range of volume fractions between 0.1 and 0.3. The upper limit is a packing limit while the lower limit is a practical one that results from the use of a numerically calculated beta distribution for sphere radii. The LineModel has the same limit on the lowest volume fraction but an upper limit of 0.9. Although an array of 20,000 lixels was used, problems with avoiding re-sampling the same elements limited the sample size to about 15 intercepts. The same estimators tried with this model give the results that follow. The vertical axis is again the negative base 10 logarithm of variance.

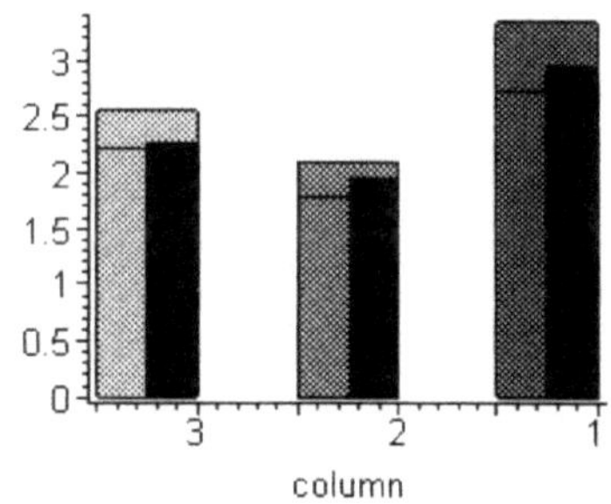

Figure 8.6: Equation (8.51) estimating sample variances, LineModel

Equation (8.51) again predicts results which are in the band for all three sets of test conditions, volume fractions of 0.1, column 1; 0.5, column 2; and 0.8, column 3. Equation (8.84) cannot be plotted directly for volume fractions of 0.5 and greater since it is derived only for a minority phase. Consequently the simplification of Eqn.(8.83) which led to Eqn.(8.85) has to be reformulated by exchanging a factor of (1-*p*) for one of the factors of *p*. The result is

$$\mathrm{Var}(\hat{p}) = \frac{p(1-p)^2}{v} : p > 0.5 . \tag{8.86}$$

The performance of Eqns.(8.84 and 8.86) in estimating variances using LineModel is plotted in Fig. 8.7.

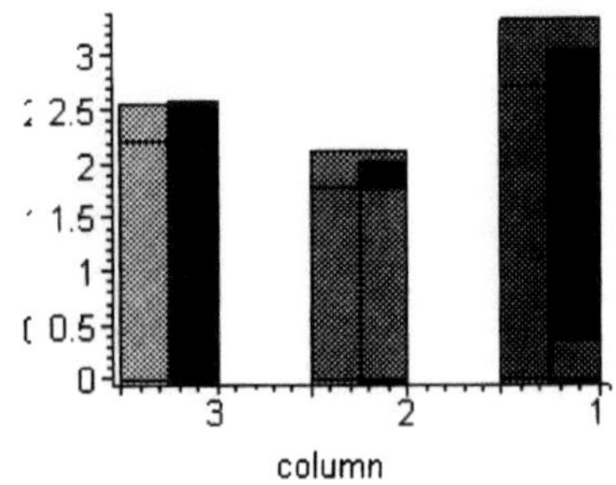

Figure 8.7: Equations (8.84 and 8.86) estimating sample variances, LineModel

The volume fractions and other parameters are again assigned just as in Fig. 8.6. Predictions fall in the band for two out of three sets of test conditions and are close in the third.

Equation (8.85) tends to overestimate the sample variance but its predictions fall in the band in one of three cases and are close in a second. As before, it replicates the order of ranking of the variances.

5.4.2 Areal Analysis

Equation (8.40) expresses the variance of an areal analysis but cannot be solved directly because it requires knowledge of unknowable covariances. The only expressions we have which are immediately applicable to areal analysis are Eqn.(8.55) and Eqn.(8.84). The former (Hilliard-Cahn) equation we derived for the purpose in Section 8.5.3.4. Its derivation is based on estimating the variances associated with the quantities used to calculate volume fraction from an areal analysis. The latter equation, Eqn.(8.84), was derived by considering the variances in an expression such as Eqn.(8.40), partitioning into discrete sums and estimating the variance of these sums by a lumping process while dividing out the size of the partition to allow it to be arbitrarily fine. Derived for linear analysis, it is applicable to areal analysis because it depends only on the number of groups of like elements and not on their dimensionality (or their size). Hence, it works as well with pixels as with lixels. Both equations were tested with the SphereModel previously discussed. The performance of Eqn.(8.55) is shown in Fig. 8.8 for the test conditions given in Table 8.2. The vertical axis is the negative of the base-10 logarithm of the sample variances and their estimators.

Table 8.2: Test Conditions for Figs. 8.8 and 8.9

	Column 1	Column 2
Row 1	$V_V = 0.1,\ \nu=500$	$V_V = 0.3,\ \nu=60$
Row 2	$V_V = 0.1,\ \nu=60$	$V_V = 0.3,\ \nu=21$

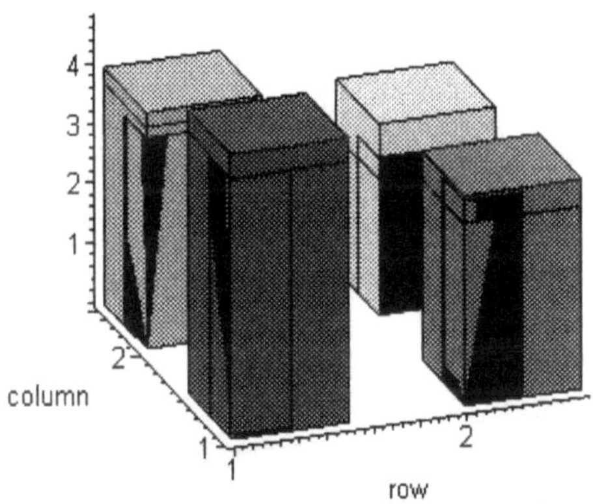

Figure 8.8: Equation (8.55) estimating sample variances, SphereModel

As in the case of linear analysis, the Monte Carlo simulation, SphereModel, generated a set of 50 scans for each set of test conditions and calculated their variance with respect to the true volume fraction. This variance was compared with that predicted by estimator, Eqn.(8.55). The process was repeated five times for each set of test conditions. For the four sets of test conditions, the median prediction (Eqn.(8.55) uses moments from the dataset.) was within the band of five runs twice and on the edge of the band once. In one case, where the volume fraction was large and the number of intersections small, the median prediction was significantly out of the band and overestimated the variance by 55% with respect to the nearest edge.

For the same set of test conditions, Eqn.(8.84) performed as shown in Fig. 8.9.

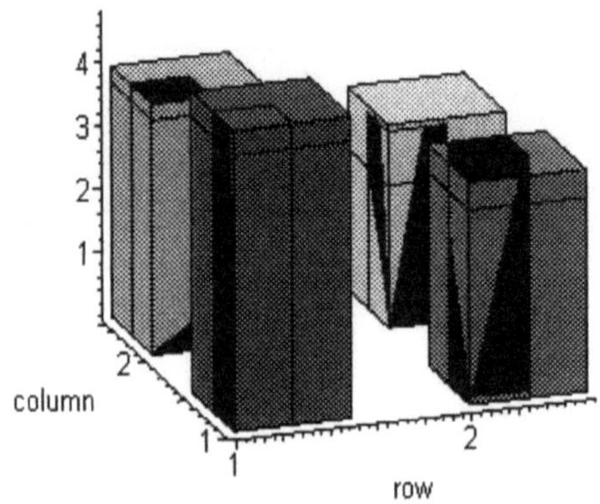

Figure 8.9: Equation (8.84) estimating sample variances, SphereModel

The overall performance of Eqn.(8.84) was similar to that of Eqn.(8.55). However the errors were differently distributed. For the four sets of test conditions, the prediction of Eqn.(8.84) was within the band twice and on the edge of the band once. In one case, where the volume fraction was small and the number of intercepts also small, the prediction was significantly out of the band and underestimated the variance by 46% with respect to the nearest band edge. When interpreting these performances we should be aware that the bands were up to an order of magnitude wide. The task of finding a meaningful confidence interval for estimators of the variance is frustrated by several factors including the need to know the variance of the quantity being measured within the material being examined. As mentioned earlier, specimen-to-specimen variation usually dominates the variances of actual stereological measurements.

If there is any generality to be concluded from these results for both linear and areal analysis, it is that, through Eqn.(8.84) albeit an approximation, the dominant factor determining variance -- apart from those extrinsic sources of variance just discussed -- is the number of independent intercepts. Hence, to answer the question raised at the beginning of Section 8.5, the efficiency of any test procedure may be best rated in terms of cost or effort per intercept evaluated.

6. ESTIMATION OF BOUNDARY AREA PER UNIT VOLUME

The boundary area per unit volume S_V can be determined either from the relationships: $S_V = 2\overline{P}_L$, Eqn.(5.59) or from: $S_V = (4/\pi)\overline{L}_A$, Eqn.(5.99). Since it is far easier to measure P_L than L_A, we will consider only the latter measurement. From the relationship for S_V and Eqn.(8.7) we have

$$\frac{\sigma^2(S_V)}{S_V^2} = \frac{\sigma^2(\overline{P}_L)}{\overline{P}_L^2} , \tag{8.87}$$

where $\overline{P}_L$ is the average number of intersections per unit length of a probe. Eqn.(8.87) expresses a relationship between the expectations for global values being equal to a similar relationship for sample values; $\overline{P}_L$ is a sample value. We consider the sampling process to work like this. A fixed-length probe line is applied to the section and its intersections with the lineal features in the section are counted as P. This process is repeated n times. The resulting total number of points counted is P_T. Dividing P_T by n gives $\overline{P}$. Dividing $\overline{P}$ by the length of the probe lines gives $\overline{P}_L$. From these definitions it follows that

$$\frac{\sigma^2(\overline{P}_L)}{\overline{P}_L^2} = \frac{\sigma^2(P_T)}{P_T^2} . \tag{8.88}$$

From the variance scaling relationship, Eqns.(8.10, 8.87 and 8.88) we obtain

$$\frac{\sigma^2(S_V)}{S_V^2} = \frac{\sigma^2(P)}{\overline{P}\,P_T} = \frac{k_S^2}{P_T} , \tag{8.89}$$

where k_S^2 is to be evaluated for the structure. We now consider the value of k_S^2 for certain types of probes and structures.

6.1 Circular and Straight Probes Applied to a Contiguous Grain Structure

Consider a structure composed of contiguous grains (or cells) whose boundary area per unit volume we wish to estimate. We will assume that three and only three grains meet along an edge. (This is usually so if surface tensions are a significant factor, since that is the smallest number that allows mechanical stability.) We first consider the case where a probe is in the form of a circle and will derive a theoretical estimate for k_S for the limit when the expected number of intersections $\overline{P}$ on the probe approaches zero.

The condition $\overline{P} \to 0$ is equivalent to the diameter of the circular probe going to zero. We therefore need only consider three possible outcomes: 0, 2 and 3 intersections as illustrated in Fig. 8.10.

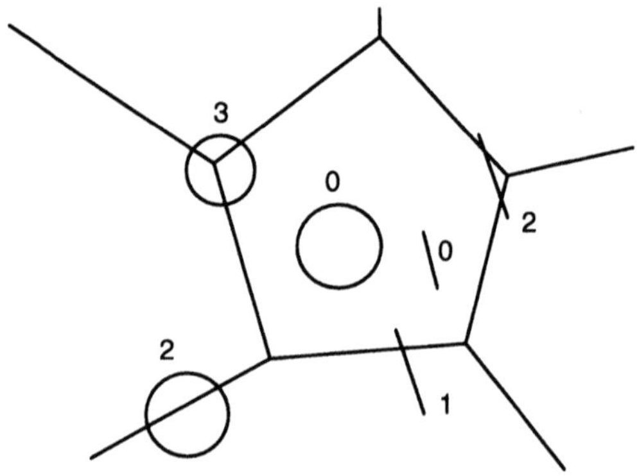

Figure 8.10: Small probes near grain boundaries

Let $p(0)$, $p(2)$ and $p(3)$ be the corresponding probabilities of these three outcomes, then

$$\overline{P} = 0p(0) + 2p(2) + 3p(3)\,. \tag{8.90}$$

Applying Eqn.(2.40) for the variance, we obtain

$$\sigma^2(P) = 0p(0) + 4p(2) + 9p(3) - \overline{P}^2\,. \tag{8.91}$$

Multiplying Eqn.(8.90) through by 2 and subtracting from Eqn.(8.91) we find, after rearrangement,

$$k_S^2 = 2 + \left\{ \frac{3p(3)}{\overline{P}^2} - 1 \right\} \overline{P} \tag{8.92}$$

under the condition that the number of intersections remains small. The next step is to estimate $p(3)$. This probability is equal to the area of the circular probe times the number P_A of triple junctions per unit area of the section.

Thus $p(3) = (\pi d^2 / 4)P_A$. Now, if L_V is the edge length per unit volume -- where we mean the actual joining together of the grain surfaces and not edges as seen on the plane of polish-- then from Eqn.(5.106) $P_A = L_V / 2$. And, since on the plane of polish, $P_L = S_V / 2$

$$k_S^2 = 2 + \left\{ \frac{3L_V}{2\pi S_V^2} - 1 \right\} \overline{P} : \overline{P} \to 0 . \tag{8.93}$$

The ratio L_V / S_V^2 is a measurable property of the structure. In order to obtain a rough estimate of k_S^2 we will assume that the specimen is composed of equal sized tetrakaidecahedral grains. From the data for their surface area and volume given in Table 6.3 and noting that each grain has 36 edges which are shared by two other grains we find

$$L_V = \tfrac{3}{2}(\sqrt{2}a^2)^{-1} \text{ and } S_V = (1 + 2\sqrt{3})(8\sqrt{2}a)^{-1},$$

where a is the edge length. Substituting these values into Eqn.(8.93) we obtain ,

$$k_S^2 = 2 - 0.639\overline{P} : \overline{P} \to 0 . \tag{8.94}$$

A plot showing this theoretical k_S is shown in Fig. 8.11. The points are from measurements made by Hilliard [1962] on a sample of silicon iron. It will be seen that both the theoretical and experimental curves display an initial decrease in k_S with increasing $\overline{P}$, or equivalently with increasing size of the probe. However, around $\overline{P}$ = 6 there is a leveling off, and thereafter k_S remains constant at a value near 0.63 with increasing probe size.

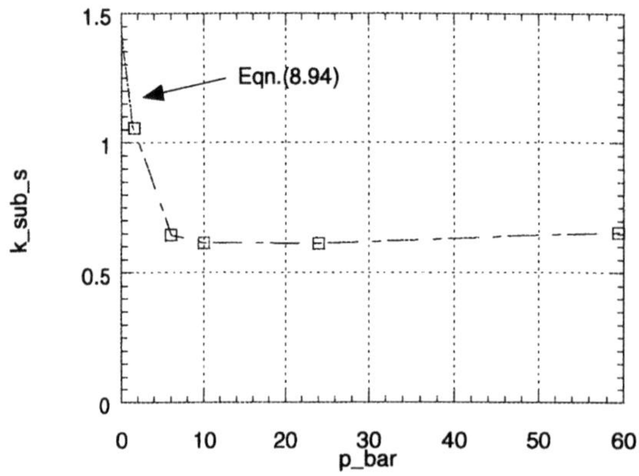

Figure 8.11: Circular Probes

We next consider the case of a straight test probe. At the limit $\overline{P} \to 0$ we can neglect any possibilities other than 0,1 or 2 intersections (Fig. 8.10). If $p(0)$, $p(1)$ and $p(2)$ are the corresponding possibilities, then

$$\overline{P} = 0p(0) + 1p(1) + 2p(2) \ , \tag{8.95}$$

and

$$\sigma^2(P) = 0p(0) + 1p(1) + 4p(2) - \overline{P}^2 \ . \tag{8.96}$$

Taking the difference between the two equations and dividing by $\overline{P}$ we obtain,

$$k_S^2 = 1 + \left(\frac{2p(2)}{\overline{P}^2} - 1 \right) \overline{P} : \overline{P} \to 0 \ . \tag{8.97}$$

In order to estimate $p(2)$ we will assume that the angle between boundary traces is 120°. Consider a probe of length l orientated at an angle θ to one of the boundary traces. Then as shown in Fig. 8.12, in order to obtain two intersections, the center of the probe has to fall within the triangle ABC where the construction is such that AP = PB = QC = CS = $l/2$, and AB is parallel to QS.

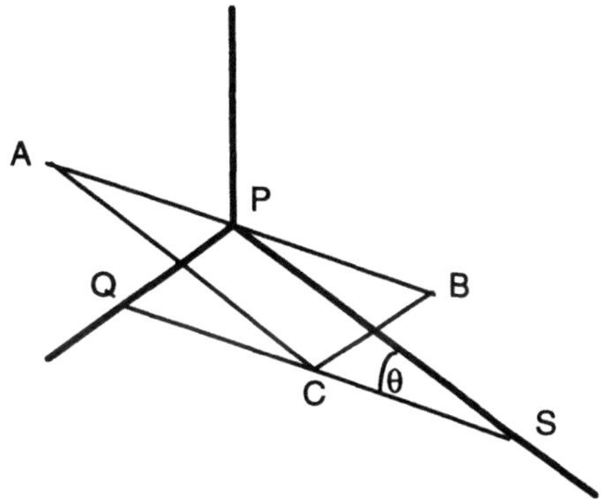

Figure 8.12: Construction for straight probe

Thus the probability that a probe randomly oriented over the range $\theta \in [0, \pi/8]$ will yield two intersections is

$$p(2) = \overline{A} P_A , \tag{8.98}$$

where $\overline{A}$ is the average area of the triangle and P_A is the number of triple points per unit area.

Noting that $PQ = l \sin\theta / \sin(2\pi/3) = (2l/\sqrt{3}) \sin\theta$

and that the height of the triangle is given by $h = PQ \sin(\pi/3 - \theta)$, we obtain for the area of the triangle ABC:

$$A(\theta) = \frac{l^2}{\sqrt{3}} \sin(\theta) \sin\left(\frac{\pi}{3} - \theta\right) . \tag{8.99}$$

We next have to determine the average area of the triangle for a randomly oriented probe. For $\theta \in [0, \pi/3]$, the density function $f(\theta)$ is 3/π. Hence,

$$\overline{A} = \frac{3}{\pi} \int_0^{\frac{\pi}{3}} A(\theta) d\theta .$$

Integrating gives

$$\bar{A} = \frac{l^2\left(3\sqrt{3}-\pi\right)}{4\sqrt{3}\pi} = 0.943951\, l^2 \ . \qquad (8.100)$$

Following the same path that gave Eqn.(8.93) we obtain

$$k_S^2 = 1 + \left(0.3776\frac{L_V}{S_V^2} - 1\right)\bar{P} : \ \bar{P} \to 0 \ . \qquad (8.101)$$

If we again assume that the grains are equal-sized tetrakaidecahedra then

$$k_S^2 = 1 - 0.714\bar{P} \ . \qquad (8.102)$$

A plot of this theoretical k_S is shown in Fig, 8.13. The experimental points were determined by Philofsky [1966] on the same specimen of silicon iron that was used for the circular probe measurements. The grains in this specimen were somewhat elongated and Philofsky found that for probes parallel to the direction of elongation k_S was about 15% lower than for those in the transverse direction. Consequently to provide the least biased estimate, the values plotted in Fig. 8.13 are the average for the two directions. It will be seen that for large $\bar{P}$, k_S for the straight probe is slightly less than for the circular probe. I (J.H.) believe this is an artifact due to the anisotropy of the grain shape since in general one would expect the properties of a straight probe to approach those of a circular probe as the diameter of the former is increased.

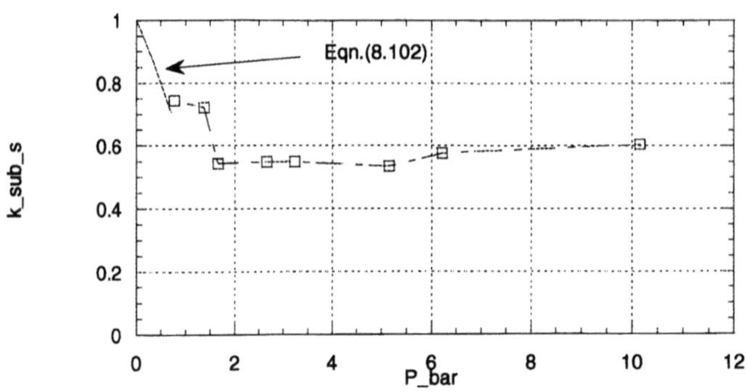

Figure 8.13: Straight probes

6.2 Straight Probe Applied to a Structure Containing Particles

In the calculation of k_S in the estimation of the surface area of discrete particles, we will restrict the treatment to linear probes that are long compared with the mean free path between particle profiles. With this condition it is permissible to neglect instances when the probe starts or terminates within a profile and assume that all intersections occur in pairs. Thus if $\overline{N}$ and $\sigma^2(N)$ are the average and variance of the number of intercepts on the probe, then with respect to the number of point intersections on profile boundaries:

$$\overline{P} = 2\overline{N} \ , \qquad (8.103)$$

and

$$\sigma^2(P) = 4\sigma^2(N) \ . \qquad (8.104)$$

We cannot proceed further without some assumption about the dispersion of the profiles on the section. A reasonable hypothesis is that the profiles are randomly dispersed. The assumption of random dispersion requires that the volume fraction of the particles be small (say, <15%) as otherwise there is an excluded area effect. For a random dispersion, the number N will follow a Poisson distribution. Hence from Eqns.(2.64, 2.65, 8.103, 8.104)

$$k_S^2 = \frac{\sigma^2(P)}{\overline{P}} = 2 \ . \qquad (8.105)$$

Thus for non-contiguous particles, k_S is about twice as large as it is for a long probe on contiguous grains.

7. ESTIMATION OF LINE LENGTH PER UNIT VOLUME

From Eqn.(5.106) we have $L_V = 2\overline{P}_A$, where $\overline{P}_A$ is the number of point intersections that lineal features make with a unit area of the section. The density of intersections $\overline{P}_A$ can be determined from a count of the number of intersections falling within a test figure of known area applied to a series of non-overlapping regions of the section. (For example, a reticule inscribed with a circle is suitable both for this purpose and the estimation of S_V.)

From Eqn.(8.7) and Eqn.(5.106) for any arbitrary area,

$$\frac{\sigma^2(L_V)}{L_V^2} = \frac{\sigma^2(P_T)}{P_T^2}\,, \tag{8.106}$$

where P_T is the total number of intersections counted in the arbitrary area. From Eqn.(8.89) we see that $k_S^2 \equiv \sigma^2(P)/\overline{P}$. Application of this definition yields by analogy,

$$\frac{\sigma^2(L_V)}{L_V^2} = \frac{k_I^2}{P_T}\,. \tag{8.107}$$

For the limit as $\overline{P} \to 0$, viz. the area of the test figure goes to zero, we need only consider the two outcomes of zero and one intersection falling in the test figure. Under these conditions it is easily shown that,

$$k_I^2 = \frac{\sigma^2(P)}{\overline{P}} = 1 : \overline{P} \to 0\ . \tag{8.108}$$

If the point intersections are Poisson dispersed, k_I^2 will be unity irrespective of the size of the test figure.

In an analysis of the edge length of grains in a polycrystalline metal sample Hilliard [1966] determined k_I for test circles of four different diameters with $\overline{P}$s ranging from 0.6 to 69. He found that the Ps were Poisson distributed for all four test circles and consequently, for this particular test specimen, the intersection of the grain edges with the section were randomly dispersed is the Poisson sense.

8. BIBLIOGRAPHY

S.C. Adhikari, *Aspects of Combinatorics and Combinatorial Number Theory* (CRC Press, Boca Raton, FL, 2002).

A.C. Atkinson, "A family of switching algorithms for the computer generation of beta variates", *Biometrika*, **66** (1979) 141-145.

E.O. Brigham, *The Fast Fourier Transform* (Prentice-Hall Inc. Englewood Cliffs, NJ, 1974).

F. Chayes, "A Simple Point Counter for Thin Section Analysis", *The American Minerologist*, **34** (1949) 1-11.

R.C.H. Cheng, "Generation of Beta Variates with Non-Integral Shape Parameters", *Communications ACM*, **21**(4) (1978) 317-322.

J. Dagpunar, *Principles of Random Variate Generation* (Oxford University Press, Oxford, 1988).

A. Delesse, "Procede mechanique pour determiner la composition des roches", *Annales des Mines* , Fourth Series, **13** (1848) 379-388.

A. Didonato and A.H. Morris, "Algorithm 708: Significant Digit Computation of the Incomplete Beta Function Ratios", *ACM Transactions on Mathematical Software*, **18**(3) (1992) 360-373.

A.T.J. Dollar, "An integrating micrometer for the geometrical analysis of rocks", *Mineralogical Magazine*, **24** (1937) 577-594.

A.A. Glagolev, "On the geometrical methods of quantitative mineralogic analysis of rocks", *Transactions Inst. Econ. Min.* (Moscow), **59** (1934) 1-47.

B.V. Gnedenko, *The Theory of Probability*, B.D. Seckler, trans. (Chelsea, NY, 1967).

H. J. Gundersen and R. Osterby, "Optimizing sampling efficiency of stereological studies 'Do more less well'", *Journal of Microscopy,* **121** (1981) 65-73.

R.W. Hamming, *Numerical Methods for Scientists and Engineers, Second Edition* (Dover, New York: reprint of 1973 edition issued 1986).

J. Hilliard and J. Cahn, "An Evaluation of Procedures in Quantitative Metallography for Volume Fraction Analysis", *Transactions of the Metallurgical Society of AIME,* **221** (1961) 344-351.

J. Hilliard, "Grain Size Estimation", General Electric Research Laboratory Report # 62-RL-3133M (1962).

J. Hilliard, "A user-oriented automatic scanning optical microscope", in *Fourth Intl. Congress for Stereology*, NBS Special Publication 431 (1976) 173-176.

R.T. Howard and M. Cohen, "Quantitative metallography by point-counting and lineal analysis", *Transactions AIME*, **172**, (1947) 413-426.

W.F. Hunt, "An improved Wentworth recording micrometer", *The American Minerologist*, **9** (1924) 190-193.

A.K. Huntington and C.H. Desch, "The planimetric analysis of alloys and the structure of phosphor-copper", *Transactions of the Faraday Society, 4* (1908) 51-59.

C. Hurlbut, "An electric counter for thin section analysis", *American Journal of Science*, **237** (1939) 253-261.

M.G. Kendall and A. Stuart, *The Advanced Theory of Statistics,* Volume 1, 4th Edition (Chas. Griffin, London; McMillan, New York, 1977).

A.I. Khinchin, *Mathematical Foundations oflnformation Theory,* R.A. Silverman and M.D. Friedman, trans. (Dover, New York, 1957).

A.I. Khinchin in *Mathematical Foundations of Information Theory* includes a separate section "The entropy concept in probability theory", reproduced from *Uspekhi Matematicheskikh Nauk*, **8** (1953) 3-20.

R.E. Miles, "On the Underlying Relationship of Plane Section Stereology to Statistical Sampling Theory", *Mikroskopie*, **37** (Suppl.) (1980) 13-18.

J.R. Newman, *The World of Mathematics,* Vol. 4, (Simon and Schuster, New York, 1956) p.2382.

E. Philofsky, *Specification and Measurement of Line and Boundary Anisotropy*, M.S. Dissertation, Northwestern University, Evanston, IL (1966).

L. Regeur and B. Pakkenberg, "Optimizing sampling designs for volume measurements of components of human brain using a stereological method", *Journal of Microscopy*, **155** (1989) 113-121.

A. Rosiwal, "Ueber Geometrische Gesteinanalysen: ein einfacher Weg zur ziffermaessigen Festellung des Quantitaetsverhaeltnisses der Mineralbestandteile gemengter Gesteine", *Verhaendlungen der Keiserlich-Koeniglichen Geologischen Reichanstalt, **5-6*** (1898) 143-175.

C.E. Shannon, "A Mathematical Theory of Communication", *Bell System Technical Journal*, **27**(1948) 379-423 .

J. Serra, *Image Analysis and Mathematical Morphology* (Academic Press, London, 1982).

E. Thompson, "Quantitative microscopic analysis", *Journal of Geology*, **38** (1930) 193-221.

H.B.M. Uylings, C. van Eden and M. Hofman, "Morphometry of size/volume variables and their bivariate relations in the nervous system under different conditions", *Journal of Neuroscience Methods*, **18** (1986) 19-37.

W. Yourgrau, A. van der Merwe and G. Raw, *Treatise on Irreversible and Statistical Thermophysics* (Dover, New York, 1982).

9. PROBLEMS

8.1) Consider the expansion of the square of the mean. The mean is formed by summing some number of elements and dividing by their number. If we consider the square of the associated sum we see that it is formed of all possible products between elements including permutations. If there are 100 terms in the sum, there are 10,000 elements in the sum forming the square of the mean. In contrast, the mean of the square uses the sum of a subset numbering 100 of these 10,000 elements. When we evaluate the variance we, apart from a negligible factor of $N/(N\text{-}1)$, subtract the square of the mean from mean of the square and obtain a non-zero answer. a) Explain how this happens through partitioning the sums involved. b) A researcher uses the just-mentioned formula for the variance to calculate the variance of a linear analysis for volume fraction but substitutes the known volume fraction for the sample mean in order to obtain the true variance with respect to the true volume fraction. The result is a negative variance. Explain. How should the true variance have been calculated?

8.2) A troubled electrical engineer is losing his hair. He consults a dermatologist who suggests a plug-graft hair transplant. She shows him a map of a scalp after surgery in which plugs are arranged in rows like a cornfield. The engineer is upset with the obvious periodicity and suggests placing the grafts by a random process. The dermatologist is skeptical. The engineer recalls that the power spectrum of a random process (the square of the magnitude of the Fourier transform of whatever function the process generates) is a constant for a Poisson process and therefore he reasons, devoid of periodicity. Hoping to find an optimal arrangement of grafts, he programs his computer to place grafts on a plane by a Poisson process, calculate the power spectrum of the result and determine the maximum deviation of the power spectrum from its average. He then programs the computer to save and display the result if the maximum deviation for a particular run is less than that for all previous runs. The program generates a few plots with a decidedly mangy appearance. Soon new plots are rarely generated. The program runs for days. When a plot is generated, it looks as irregular as the ones before with bald patches, sparse spots and hideous clumps. And, the power spectra for all plots are anything but constant with respect to frequency. A secretary suggested that his hairs were connected to his brain and the consequent loss was shared. Did the engineer have lapse of memory about Poisson processes? What does this have to do with sampling variances anyway?

8.3) Jim is a graduate student in Economics and lives on Foster Street in Evanston which is roughly midway between two stations for the elevated train, Noyes and Davis Streets. He does his grocery shopping either at D&D's on Noyes or at Dominicks on Davis. In bad weather, he takes the El train. Since he does not strongly prefer one store to the other, he boards the first train to arrive. If it is northbound, he shops at D&Ds. If it is southbound he shops at Dominicks.

One evening while mulling over a choice of breakfast cereal from a 100-yard display, it occurs to him that when he takes the El he shops far more frequently at Dominicks than at D&D's, even though the frequency of northbound and southbound trains is the same and his arrival time at the station is random. (The El originates from a yard at the Chicago terminal and travels through Evanston to the end of the line at Linder Avenue in Wilmette, a short distance north of Foster Street.)

a) How do you explain this? Hint: assume that the train does not go into the yard at Linder Avenue but simply reverses direction.

b) Quoting Hilliard's syllabus: "...as was pointed out by Larry Lawson during class discussion, if you take the realities of running a railroad into account, there are other possible factors." In fact the train always goes into the yard for cleaning when it reaches Linder Avenue and a new southbound train emerges from the yard as determined by the schedule. Trains in each direction are nominally 7 minutes apart. During normal traffic trains alternate at 3.5-minute intervals. Now explain why Jim is so often at Dominicks. Hint: In the evening the trains are mostly empty.

8.4) An interesting example is described by DeMorgan (cited in [Newman, 1956]) as follows:

"Suppose a planked floor of the usual kind, with thin visible seams between the planks. Let there be a thin straight rod or wire not so long as the breadth of the plank. This rod being tossed up at hazard will either fall quite clear of the seams or will lie across one seam... In 1855 Mr. Ambrose Smith of Aberdeen made 3,204 trials with a rod of 3/5ths of the distance between the planks: there were 1,213 clear intersections and 11 contacts on which it was difficult to decide. Divide these contacts equally and we have 1,218 divided by 3,204 for an estimate of 6/5 times pi, presuming that the greatness in the number of trials gives something near the final average, or result in the long run: This gives pi = 3.1553. If all the contacts had been treated as intersections, the result would have been pi = 3.1412, exceedingly near."

Discuss whether you would have bought a used buggy from Mr. Ambrose Smith.

Chapter 9

MULTIDIMENSIONAL GENERALIZATIONS

This chapter is intended to introduce the reader to some of the more general results that are part of the broad field of stochastic geometry. We will generalize many of the results of Chapter 5 into a capsule consisting of just a very few very general formulas in which the dimensionalities themselves are the major variables.

1. INTRODUCTION

In Chapter 4, we introduced some conventions and a "limited axiom" of uniformity to support the IUR and A-weighted section definitions for a random probe, which are expressed through the probability density functions defined there. At this point, we are in a position to improve on this definition. We will define a random probe in terms of a hitting probability proportional to a ratio of mean projections. We can use the hitting probability of a multidimensional probe simultaneously as an axiomatic condition defining an isotropic probe and as a way of extending Eqns.(5.12) and (5.87), which allow us to calculate hitting probabilities in terms of mean caliper diameters, into N-dimensional space. Following Miles [Miles, 1972] we say:

Hitting Lemma:

Given convex bodies $\mathbf{X}$, $\mathbf{Y}$ of dimensionality N such that $\mathbf{X} \supset \mathbf{Y}$ and also an isotropic P-subspace composed of P-dimensional flats, the probability that a P-flat through $\mathbf{X}$ hits $\mathbf{Y}$ is the ratio of measures,

$$P(P \uparrow \mathbf{Y} \mid P \uparrow \mathbf{X}) = \frac{v\left|\bar{S}'^{N}_{N-P}(\mathbf{Y})\right|}{v\left|\bar{S}'^{N}_{N-P}(\mathbf{X})\right|}, \tag{9.1}$$

where $\overline{S}'^{N}_{N-P}(\cdot)$ denotes the mean projection of the designated body onto flats of the orthogonal complement of the P-subspace. Equation (9.1) is a necessary condition that the P-subspace be isotropic. To avoid confusion in later formulas, the prime indicates an orthogonal subspace.

The Hitting Lemma concentrates on the isotropic aspect of the isotropic uniform random (IUR) probe just as, in retrospect, the "Limited Axiom" concentrated on capturing the attribute of uniformity. An example of an orthogonal complement is two lines in two dimensions meeting at the origin at right angles; $N = 2, P = 1$ $N\text{-}P = 1$. Another is a line and a plane in three dimensions meeting at the origin at right angles: $N = 3, P = 2$ $N\text{-}P = 1$. Equation (9.1) is of course the generalization of the idea of mean caliper diameter.

Most of the relationships developed in Chapter 5 can be grouped into five classes. First, there are those that relate the expected density of interceptions on a probe to the density of structural features. Examples of these are $\overline{L}_L = V_V$ and $\overline{P}_L = S_V/2$. The second class comprises those expressions relating an average projected property of a body to its surface content. For example, $d = c/\pi$ and $\overline{S}_P = S/4$. This class includes expressions for the mean caliper diameter of a body, since this is equivalent to projection onto a line. The inclusion of mean caliper diameter expressions in this class is obvious for a two-dimensional figure and is easily visualized for a three-dimensional body if one considers the projection as a two-step process. The first step is projection of the body onto a plane; the second is a projection of the previously projected area onto a line within that plane. The mean caliper diameter is then the average projected length for a random orientation of the body. In the third class are those relationships for the expected length or area of interception of a probe with a body. For example, $\overline{L} = 4S/V$ and $\overline{A} = V/D$. Fourth, we have the expressions such as $\overline{L}_V = (\pi/4)S'_V S''_V$ and $\overline{P}_V = (\pi/8)S'_V S''_V S'''_V$ for the density of interceptions generated by the superposition of two or more arrays. Finally, in a class by themselves are the results for the $(N+1)$ moment of the intercept length distribution.

In a remarkable paper Miles [1972] demonstrated that for each of the first four classes, a general expression for isotropic sectioning could be derived that is valid for any dimensionality of probe, body (or feature) and space. For the fifth class, he obtained a generalization of Crofton's second theorem for any N. At this point it is natural to inquire what the value is, other than that of the intellectual satisfaction, of considering relationships for other than one two or three dimensions? Or, as Miles put it with tongue in cheek, engaging in this "apparently frivolous exercise". The first justification is that the reduction of a group of apparently unconnected relationships to a single schema exposes the essence of the mathematical argument and immediately reveals any missing members. An example of missing members are the two results, Eqn.(5.102) and (5.103) for intersecting arrays (repeated in the examples above) which are of practical interest. They are immediately obvious from the multidimensional generalization but had not been previously discovered. By the same token, we can now conclude that it is unlikely that any more expressions of the types considered in Chapter 5 remain to be discovered.[1]

Secondly, the multidimensional treatment provides an economy in derivations. Admittedly, in the present case, that economy is achieved at the expense of an increase in complexity. However there are other instances where the N-dimensional derivation presents no more difficulty than one with a specific dimension. The determination of the properties of a random dispersion of points in N-dimensions is one that we examined in Chapter 8. Finally, the N-dimensional relationships are not entirely without practical applications. It is sometimes desirable to compare an observed structure with a mathematical model having well-defined properties. For example, one might model a cellular structure by a tessellation of random planes. The problem quickly runs out of ideas for models based in one, two- or three-dimensional space. However, a three-dimensional section through some regular structure in N-dimensional space can generate an infinite set of models. The availability of the expressions developed by Miles permits the properties of such models to be calculated. There is also the occasional need to determine the dimensionality of an embedding space. Multidimensional relationships provide direct solutions to this problem (in integer dimensions).

In considering the multidimensional extensions, we will make one concession to practicality by omitting the derivations. We will list only the results. The interested reader can consult the original paper, which should be read in conjunction with an earlier contribution [Miles, 1969]. The references given in Chapter 10 may also be helpful.

---

[1] Several new shape-independent relations have been discovered and two (star volume and the geometry of the Bitterlich sight) are discussed in this book. Both are related to "local stereology". Neither fit into Chapter 5 exactly, and both may be argued as being not really new.

2. EXPECTED DENSITY OF INTERCEPTIONS ON A RANDOM PROBE

Using the notations *I*, *P*, *F* and *N* introduced in Section 4.2 to denote the dimensionalities respectively of the interception, probe, feature and (embedding) space in which they reside, we can define a general symbol,

$$\rho_I^P ,$$

for the density of interceptions of dimensionality *I* per unit measure of a probe of dimensionality *P*. For example, in terms of our previous notation, in the new notation

$$\rho_0^1 = P_L \text{ and } \rho_1^2 = L_A \ .$$

With respect to the density of structural features, we define ρ_F^N as the density of structural features of dimensionality *F* per unit measure of space of dimensionality *N*. Examples:

$$\rho_3^3 = V_V \text{ and } \rho_2^3 = S_V \ .$$

The expression relating $\overline{\rho}_I^P$ and ρ_F^N obtained by Miles takes the form,

$$\overline{\rho}_I^P = K_N(P,F)\rho_F^N \ . \tag{9.2}$$

The constant in Eqn.(9.2) evaluates to

$$K_N(P,F) = \frac{\Gamma\left(\frac{P+1}{2}\right)\Gamma\left(\frac{F+1}{2}\right)}{\Gamma\left(\frac{I+1}{2}\right)\Gamma\left(\frac{N+1}{2}\right)} , \tag{9.3}$$

where $\Gamma(\cdot)$ is the gamma function defined in Section 2.1.3. With appropriate substitutions Eqns.(9.2) and (9.3) yield Eqns.(5.6, 5.7, 5.8, 5.28, 5.31, 5.57, 5.59, 5.93, 5.99 and 5.106).

3. MEAN PROJECTIONS OF A BODY

The average projected area $\bar{S}_{N-1}^{N}$ of an N-dimensional smooth convex body of surface area S_N onto an N-1-dimensional flat is,

$$\bar{S}_{N-1}^{N} = \frac{\Gamma(N/2)}{\Gamma\left(\frac{N+1}{2}\right)2\sqrt{\pi}} S_N \ . \tag{9.4}$$

This expression yields Eqn.(4.13) for the mean caliper diameter of a plane figure and Eqn.(4.24) for the mean area projected onto a plane. Miles showed that the expressions for a projection onto a flat of dimensionality less than N-1 involve a surface integral over the body. This explains why there is a simple expression for the mean caliper diameter of a plane figure but not for a three-dimensional body.

4. AVERAGE MEASURE OF AN INTERCEPTION IN A BODY

The average measure of an intercept inside a convex body is related to the body's average projection onto a suitable flat. Let V_N be the volume (or content) of an N-dimensional convex body and $\bar{S}'^{N}_{D}$ its average projection onto an orthogonal D-dimensional flat ($1 \le D < N$), then the size of the average interception $\bar{L}_{N-D}$ on a probe of dimensionality N-D is,

$$\bar{L}_D = K_N(D,F)V_F / \bar{S}'^{N}_{N-D} \ . \tag{9.5}$$

This formula yields Eqns.(5.17, 5.50 and 5.90). In all three of these cases the factor $K_N(D,F)$, which is evaluated using Eqn.(9.3), is unity. In fact, inspection makes clear that this factor is unity whenever the feature has the same dimensionality F as that N of the embedding space, since that makes the dimensionality I of the intercept equal to that P of the probe or in this case D of the D-flat. Note that the dimensionality of $\bar{L}_{N-D}$ is not always equal to unity. In fact, in the case of Eqn.(5.90) it is two: the expected area of interception in a aggregate equals the ratio of the average feature volume to the average mean caliper diameter of the features.

5. DENSITY OF INTERCEPTIONS PRODUCED BY THE SUPERPOSITION OF MULTIPLE ARRAYS

Consider the random superposition of in N-dimensional space of m arrays having dimensionalities $F_1, F_2, F_3, \ldots F_m$. The dimensionality I of the intersections generated is

$$I = (1-m)N + \sum_{i=1}^{m} F_i . \tag{9.6}$$

This expression is easily derived by considering the dimensionality of the intersection of the first two arrays and intersecting that with the third and so forth. For example, an areal array ($F_1 = 2$) and a linear array ($F_2 = 1$) intersect to form an array of points ($I = 0$). If we denote by $\overline{\rho}_I^N$ the expected density of interceptions per unit volume of space, then

$$\overline{\rho}_N^I = \left[\frac{\prod_{i=1}^{m} \Gamma\left(\frac{F_i + 1}{2}\right)}{\Gamma\left(\frac{N+1}{2}\right)^{m-1} \Gamma\left(\frac{I+1}{2}\right)} \right] \prod_{i=1}^{m} \rho_{F_i} , \tag{9.7}$$

where ρ_{F_i} is the density of the i-th array. This expression yields Eqns.(5.43, 5.85, 5.102 and 5.103).

6. CROFTON'S SECOND THEOREM

The generalization of Crofton's second theorem to an N-dimensional convex body of volume V and boundary content S is given by

$$\mu'_{N+1}(L) = 2N\Gamma\left(\frac{N+3}{2}\right)\pi^{\frac{1-N}{2}} \frac{V^2}{S} . \tag{9.8}$$

From Eqn.(9.8) can be derived Eqns.(5.26) and (5.55).

7. INTERCEPT DISTRIBUTIONS IN N-DIMENSIONAL SPHERES AND PLATES

In this section we will consider the generalization to N dimensions of some of the preceding results from Chapter 6. "The apologia for this section is similar to the one given at the beginning of this Chapter. Since, No new results of practical interest will be obtained, these generalizations can be omitted by those readers who do not have a taste for such matters." (J.H.)

7.1 The P-dimensional Probe with N-dimensional Spheres

It follows that so long as $I = P + F - N$, the dimensionality of an interception I of a P-dimensional probe in an N-dimensional space is $I = P$. We will assume, as in practical cases, that the probe is isotropic and uniformly distributed on an (N - P)-dimensional flat that passes through the center of the sphere. (For a linear probe the flat is a diameter for a circle and a diametrical plane for a sphere.) Accordingly, we will sketch a reduced-dimension reference drawing, one that brings out the essential elements common to cases in all dimensionalities. It is similar to and simply another version of Fig. 6.11.

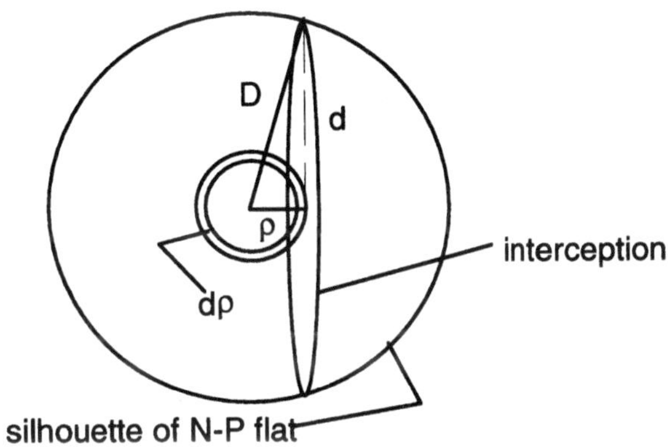

Figure 9.1: Reduced-dimensional geometry of N-sphere interception

Thus the probability $f(\rho)d\rho$ that the probe will be located at perpendicular distance $\rho \pm d\rho/2$ is equal to the ratio of the hyperspatial area of an annulus of radius ρ and thickness dρ on the diametrical flat (silhouette shown above) to the probe's total area available for interception with the sphere viz that of the "silhouette". Now this silhouette's "area" is in reality the content, expressed as a Lebesgue volume, of itself, an (N-P)-dimensional sphere of diameter D. The area of the annulus is its circumference times dρ. This hyperspatial "circumference" is equal to the surface area of an (N-P)-dimensional sphere of diameter 2ρ. It follows that this ratio can be written,

$$f(\rho)=S_{N-P}(\rho)/V_{N-P}(D/2)\ , \tag{9.9}$$

using Eqns.(2,155) and (2.156).

Let d be the diameter of the interception of the probe with the sphere. Then "as in the practical cases", because the space is Euclidean, the Pythagorean theorem applies,

$$d^2 = D^2 - 4\rho^2\ . \tag{9.10}$$

Recidivating to the curly-d notation used in Chapter 6 to avoid a bigger confusion at the expense of a lesser confusion,

$$\frac{\partial\rho}{\partial d} = -\frac{d}{4\rho}\ . \tag{9.11}$$

Substituting Eqns.(9.9) and (9.11) into the change-of-variable formula for probability density functions,

$$f_D(D) = f(\rho)\left|\frac{\partial\rho}{\partial d}\right|, \text{ yields}$$

$$f_D(d) = \frac{d\,S_{N-P}(\rho)}{4\rho\,V_{N-P}(D/2)}\ . \tag{9.12}$$

Finally, substituting from Eqns.(2.155) and (2.156), eliminating ρ with Eqn.(9.10) and using the recurrence relation for the gamma function Eqn.(2.5) we obtain,

$$f_D(d) = (N-P)\frac{d\left(D^2 - d^2\right)^{\frac{N-P-2}{2}}}{D^{N-P}}\ . \tag{9.13}$$

John Hilliard comments: "This is another example of rediscovery. I came across this result while working on an unrelated problem in 1974. Later that year I attended a meeting in Australia and had the opportunity of discussing it with R.E. Miles. He pointed out the simple derivation given above and also mentioned that the result was vaguely familiar. A few weeks later he wrote saying that he had located the source and enclosed a Xerox copy; it was of his Ph.D. thesis."

With appropriate substitutions, Eqn.(9.13) yields Eqns.(6.5) and (6.71).

Following the pattern of Chapter 6, we can also write Eqn.(9.13) in terms of a reduced intercept diameter $\lambda = d/D$.

$$f(\lambda) = (N-P)\lambda\left(1-\lambda^2\right)^{\frac{N-P-2}{2}}. \tag{9.14}$$

This expression can be readily integrated; the range of λ is from 0 to 1. Integrating gives,

$$F(\lambda) = 1-(1-\lambda^2)^{\frac{N-P}{2}}, \tag{9.15}$$

which is the cumulative distribution function for intercepts with hyperspheres. The distribution defined by Eqn.(9.14) has several interesting properties. The first of these is that it depends only on the difference N-P. This property allowed the reduced-dimensional scheme of Fig. 9.1. It also explains why the distribution of linear intercept lengths in a 2-D circle (N = 2, P =1) are the same as that of profile diameters for a planar probe on a 3-D sphere (N = 3, P = 2). Secondly, it can be shown that if N = P, the distribution becomes a delta function located at λ = 1. The distribution also becomes a delta function when $N-P \rightarrow \infty$. But, in this case the delta function is located at λ = 0. It is also somewhat surprising that the two distributions of practical interest, Eqns.(6.58) and (6.73), which at first sight seem quite dissimilar, do have a common origin.

7.2 Moments of the Hypersphere Intercept Distribution

The moments of the N-dimensional intercept distribution are defined by

$$\mu_n'(\lambda) = (N-P)\int_0^1 \lambda^{n+1}\left(1-\lambda^2\right)^{\frac{N-P-2}{2}} d\lambda. \tag{9.16}$$

On inspection on may recognize that this is a beta function which spelled out is

$$\mu_n'(\lambda) = \frac{N-P}{2} \frac{\Gamma\left(\frac{N-P}{2}\right)\Gamma\left(\frac{n+2}{2}\right)}{\Gamma\left(\frac{n+N-P+2}{2}\right)} . \tag{9.17}$$

The value of n must be greater than or equal to -1.

7.3 Intercept Distribution in an Aggregate of Hyperspheres

In order to calculate the results pertaining to an aggregate of hyperspheres, we must first derive some simple extension of the formulas previously derived. First we consider an assortment of spheres of dimensionality N designated $\mathbf{Y}_i$ and contained within a sphere also of dimensionality N designated $\mathbf{X}$. From Eqn.(9.1) we find the expected number of interceptions of the $\mathbf{Y}_i$ with a P-flat as

$$\overline{N} = \sum_{(i)} p_i = \frac{\sum_{(i)} \overline{S}'^N_{N-P}(\mathbf{Y}_i)}{\overline{S}'^N_{N-P}(\mathbf{X})} . \tag{9.18}$$

From Eqn.(9.5) when $N = F$,

$$\overline{L}_P = \frac{V_N}{\overline{S}'^N_{N-P}} . \tag{9.19}$$

And,

$$\frac{\overline{N}}{\overline{L}_P(\mathbf{X})} = \frac{\sum_{(i)} \overline{S}'^N_{N-P}(\mathbf{Y}_i)}{V_N(\mathbf{X})} . \tag{9.20}$$

Equation (5.91) follows immediately as a special case of Eqn.(9.20).

Continuing with the matter of the intercept distribution, let $N_N(D)\partial D$ be the number of spheres (or balls to be "convexedly" precise) of diameter $D \pm \partial D/2$ per unit measure of space of dimensionality N. This might be better written N_V, but then the dimensionality N would be omitted. Let

$N_P(d,D)\partial d\,\partial D$ be the number of interceptions of size $d \pm \partial d/2$ on spheres of diameter $D \pm \partial D/2$ per unit measure of a P-dimensional probe. Then : $N_P(d,D)\partial d\,\partial D$ equals *the expected number of interceptions per unit measure of the probe with spheres of diameter* $D \pm \partial D/2$ times $f_D(d)\partial d$.

Seeing that this distribution is separable, we make the first factor (following Eqn.(9.20)) $N_N(D)\partial D$ times the projection of the *N*-dimensional sphere onto an (*N-P*)-dimensional flat. This projected area is in the case of a sphere equal to the volume of an (*N-P*)-dimensional sphere of the same diameter. The second factor is the intercept diameter distribution as shown. Accordingly,

$$N_P(d,D)=V_{N-P}(D)N_N(D)f_D(d)\,.$$

Substituting the value for *V* from Eqn.(2.155) and that for f_D from Eqn.(9.13) we obtain,

$$N_P(d,D)=\frac{2\pi^{\frac{N-P}{2}}}{2^{N-P}\Gamma\left(\frac{N-P}{2}\right)}d\left(D^2-d^2\right)^{\frac{N-P-2}{2}}N_N(D)\,.$$

Integrating with respect to *D* gives,

$$N_P(d)=\frac{2\pi^{\frac{N-P}{2}}}{2^{N-P}\Gamma\left(\frac{N-P}{2}\right)}d\int_d^{\sup D}\left(D^2-d^2\right)^{\frac{N-P-2}{2}}N_N(D)\partial D\,. \qquad (9.21)$$

The solution for this equation for the cases *N-P*={2,1} has been given in association with Eqns.(6.61) and (6.75). "The general solution is left as an exercise to those readers who can find a use for it" (J.H.)

7.4 Relationship between Sphere Diameter and Intercept Size Moments

Multiplying Eqn.(9.21) through by d^{n-N+P} and integrating with respect to *d* we obtain (after setting $N-P \equiv k$ to save space),

$$N_P \mu'_{n-k}(d) = \int_0^{\sup D} d^{n-k} N_P(d) \partial d =$$

$$\frac{2\pi^{\frac{k}{2}}}{2^k \Gamma\left(\frac{k}{2}\right)} \int_0^{\sup D} d^{n-k+1} \int_d^{\sup D} \left(\xi^2 - d^2\right)^{\frac{k-2}{2}} N_N(\xi) \partial\xi \partial d ,$$

noting that $N_P(d) = N_P f_P(d)$ relating the total number of interceptions per content of probe to the probability density function. Referring back to Fig. 6.3 as a reminder of how the limits change, we reverse the order of integration in order to obtain an inner integral which can be evaluated in the manner shown repeatedly in Chapter 6. The intermediate result is

$$N_P \mu'_{n-k}(d) = \frac{2\pi^{\frac{k}{2}}}{2^k \Gamma\left(\frac{k}{2}\right)} \int_0^{\sup D} N_N(\xi) \int_0^{\xi} d^{n-k+1} \left(\xi^2 - d^2\right)^{\frac{k-2}{2}} \partial d \partial \xi .$$

The inner integral, with respect to d evaluates using the same trigonometric substitution as before to, $\dfrac{\xi^n \Gamma\left(\frac{k}{2}\right)\Gamma\left(\frac{n-k+2}{2}\right)}{2\Gamma\left(\frac{n+2}{2}\right)}$. This yields,

$$N_P \mu'_{n-k}(d) = \frac{\pi^{\frac{k}{2}} \Gamma\left(\frac{n-k+2}{2}\right)}{2^k \Gamma\left(\frac{n+2}{2}\right)} N_N \mu'_n(D) , \tag{9.22}$$

after noting that as before the probability density function is related to the total number of spheres per volume, $N_N(D) = N_N f(D)$. We can turn this relationship around and solve for the moments of the sphere distribution. Before doing this, from this from Eqn.(9.20) we obtain,

$$\frac{N_P}{N_N} = \left\langle \overline{S}'^N_k (\mathbf{Y}_i) \right\rangle \tag{9.23}$$

in analogy with Eqn.(5.91). The ratio N_P / N_N is equal to the "average projected area" of the spheres onto a k-dimensional flat. The angle brackets are used to denote averaging over $\{\mathbf{Y}_i\}$ of the enclosed value which is itself an average. From, Eqn.(5.91) we can see that this value is related to the average mean caliper diameter but to the power k which need not always be unity. It is the content or volume of a k-dimensional sphere (ball),

$$\left\langle \bar{S}_k^N(\mathbf{Y}_i) \right\rangle = \frac{\pi^{\frac{k}{2}}}{2^k \Gamma\left(\frac{k+2}{2}\right)} \mu_k'(D) . \tag{9.24}$$

This obviously equals the average mean caliper diameter when $k = 1$. Substituting Eqns.(9.23) and (9.24) into Eqn.(9.22) yields

$$\mu_n'(D) = \frac{\Gamma\left(\frac{n+2}{2}\right)\mu_k'(D)}{\Gamma\left(\frac{k+2}{2}\right)\Gamma\left(\frac{n-k+2}{2}\right)} \mu_{n-k}'(d) . \tag{9.25}$$

The range of applicability of this expression in given by $n-k \geq -1$. Because of this limitation, it is only for the case $k = 1$ that one can obtain the total density of spheres N_N by setting $n = 0$ in Eqn.(9.22):

$$N_N = \frac{2\mu_{-1}'(d)}{\pi} N_P . \tag{9.26}$$

This relation has been seen before as Eqn.(6.80) in which the harmonic mean of the intercept diameters replaced $1/\mu_{-1}'(d)$.

8. BIBLIOGRAPHY

R.E. Miles, "Poisson Flats in Euclidean Spaces Part 1: A Finite Number of Random Uniform Flats", *Advances in Applied Probability*, **1** (1969) 211-237.

R.E. Miles, "Multi-dimensional perspectives on stereology", *Journal of Microscopy*, **95** (part 2), (1972) 181-195.

H. Hadwiger, *Alttes und Neues ueber konvexe Koerper* (Birkhaeuser, Basel, 1955).

H. Hadwiger, *Vorlesungen ueber Inhalt, Oberflaeche und Isoperimetrie*, (Springer Verlag, Berlin, 1957).

9. PROBLEMS

9.1) Using the random number generator on your calculator, select any five relationships in the review table, Section 5.7, and verify that they are derivable from the multidimensional expressions.

9.2) Verify Eqn.(9.4) for an N-dimensional sphere using only the results given in Section 2.5.

9.3) From Eqn.(9.17) the n-th moment of the linear intercept distribution in an N-dimensional sphere of unit diameter is

$$\mu_n' = \frac{\Gamma\left(\frac{N+1}{2}\right)\Gamma\left(\frac{n+2}{2}\right)}{\Gamma\left(\frac{n+N+1}{2}\right)} . \tag{9.27P}$$

Using this result, verify Eqns.(9.5) and (9.8) for spheres.

9.4) As noted in Section 2.4.12, from comparing Eqn.(9.13) with Eqn.(2.121) we see that Eqn.(9.13) for the intercept distribution of an N-dimensional sphere becomes a delta function when $k \to 0$. Verify that this is so by showing that for k = 0, $\mu_n'(\lambda) = 1 \forall n \geq -1$. Discuss the physical significance of $k \to 0$.

9.5) Comparing linear intercept distributions (probability density functions) written in terms of a reduced variable for a single circle with a single stripe in two dimensions (Eqns.(6.9 with 6.37)) and for a single sphere with a single plate in three dimensions (Eqns.(6.58 with 6.108)) one may notice that they are the same for spheres and plates independent of dimensionality. It has been proposed that this "reciprocity" may hold for arbitrary dimensionalities. Thus, starting with Eqn.(9.14) a multidimensional distribution function for plates has been proposed,

$$f_W(L) = \frac{(N-1)W^2\left(L^2 - W^2\right)^{\frac{N-3}{2}}}{L^N} , \tag{9.28P}$$

where $W \leq L < \infty$. Investigate the validity of this claim.

9.6) If Eqn.(9.28P) is accepted then it follows that a relationship between moments can be derived in the same manner as previously. Such a relationship is proposed to be

$$\mu_n'(W) = \frac{\Gamma\left(\frac{3-n}{2}\right)}{\Gamma\left(\frac{N+1}{2}\right)\Gamma\left(\frac{4-n-N}{2}\right)}\mu_n'(L) \;, \qquad (9.29P)$$

where $n+N \leq 3$. Check that this relationship is consistent with Eqn.(9.28P).

9.7) We define an N-dimensional plate as a body in which the dimensions in one direction are small compared with those in the remaining N-1 dimensions. The distribution function for N-dimensional plates has been given as Eqn(9.28P). Verify the validity of this distribution without reference to the apparent similarities discussed in Problem 9.5.

Chapter 10

TOPICS IN STOCHASTIC GEOMETRY

This chapter discusses topics that relate stereology to the larger subject of stochastic geometry. It also provides a niche in which to display the topics of conversations between the authors and between John Hilliard and his students. The sections of this chapter are in no particular order since each topic is largely independent of the others.

1. GEOMETRIC INTEGRALS[1]

Here we will review some total integral measures (or just total integrals if you wish) of geometrical bodies. In Chapter 9 we looked at multidimensional generalizations of the shape-independent stereology formulas of Chapter 5. We referred to Miles's derivations but did not attempt to derive the generalized formulas ourselves other than to show that these generalizations were consistent with results we had derived in detail in two and three dimensions. These new formulas provided a way of: 1) summarizing these earlier results and 2) treating the number of dimensions as a variable. We now get to acquaint ourselves with a little bit of the theory underlying Miles's results and with the higher common sense of stereology as it is a part of the broader subject of stochastic geometry. You may wonder why I seem to be equating stochastic geometry with etiquette by using the phrase "higher common sense"? They have two things in common. One is that they both express a formalism. Two is that the role of the formalism is to bind together, to smooth, to eliminate possible inconsistencies and to insure that nothing essential is omitted.

At this point I want to inform the reader that we are not yet moving into a higher realm that picks up where we left off in the previous chapter. Rather, the concepts we will examine are more primitive than those to which we have become accustomed. Underlying the higher common sense are several

---

[1] This section was originally titled "Integral Geometry", but following the suggestion of Ivan Saxl [2003] (I.S.), I (L.L.) have chosen a less specific title. "Integral geometry" is properly associated with a more narrowly defined body of related work. See Santalo [1976].

notions one of which is orthogonal projection onto a subspace. This was introduced in Chapter 9. We now re-introduce another symbol, $\nu_d(\ )$ which we call the *d*-dimensional Lebesgue measure of whatever set is contained in the parentheses. This is simply shorthand for the Lebesgue integral. We presume that the nature of the set tells us what the limits of integration are and that *d* matches the dimensionality of the set. If we were to want to measure the content of a set in a different (lesser) dimensionality than its characteristic dimensionality, e.g. the two-dimensional content of a cube, we would have to first project the cube in some way in order to define how its two-dimensional content would be measured.

1.1 Minkowski Functionals

What are perhaps the most important functionals of integral geometry are the quermassintegrals or Minkowski functionals. The word quermassintegral literally translates "cross-measure" integral. We are in a sense sighting across the body in question, which is presumed to be smooth and convex. The Minkowski functionals can be defined in terms of integrals of projections onto orthogonal subspaces [Stoyan, Kendall and Mecke, 1995], integrals of projections without the need for defining orthogonal subspaces [Saxl, 1989: q.v. Eqns. 2.124 with 2.127] and at least formally through measures of intersections. Since the method of intersections has the most in common with classical stereology, we will pursue it. Gaining an overview of Chapter 4, we here define *S(d,k)* to be the family of all *k*-dimensional affine "planes" or flats ("affine" to include translations away from the origin) in $\mathbf{R}^d$. We note that for *S(d,k)* and its subsets there exists a good measure in the sense of Chapter 3 and the Appendix, (a Lebesgue measure) which is invariant under translations and rotations, and is unique to within a multiplicative constant. We apply this measure to parameterizations of *S*. (E.g. if *S* is the family of lines in 2-dimensions we can parameterize it with ρ and θ as in Chapter 4 and measure by integrating over these parameters. (How we define our parameterization would affect the choice of multiplicative constant.) Using this measure, which we call $\mu_{d,k}$, we define the Minkowski functionals formally as:

$$W_k^d(K) = \mu_{d,k}\{T \cap K \neq \varnothing : T \in S(d,k)\}, \tag{10.1}$$

where the *d*-dimensional body *K* is intersected with *k*-dimensional probes, *T*. The result is a *(d-k)*-dimensional measure. For details on the evaluation of this measure in the general case, the reader should see either the above work by Saxl or that by Stoyan et al. When we know what *d* is, we may omit it as below. The Minkowski functionals are properties of shape alone in the sense that, if they are determined for a "unit-sized" shape, they can readily be converted to appropriate values for different sizes of that shape. This is because they are homogeneous with respect to an isotropic multiplier, viz.

$$W_k^d(\alpha K)=\alpha^{d-k}W_k^d(K). \text{ (I.S.)}$$

This property has been has been tacitly applied in this book in several cases where dimensional scaling has been used.

In Eqn.(10.1), if k equals zero, W_0 is simply the volume of the body K in d dimensions. If k equals one, we have for W_1, the surface measure δK divided by d. For example, consider a three dimensional ball. We obtain $4\pi R^2/3$ for W_1^3.

If we set k equal to d, we obtain the volume of a unit sphere in d dimensions since K drops out of the equation and that is how the Minkowski functionals are defined in terms of a multiplicative constant. Note that this unit sphere diameter equals two when d equals one. Compare this with Eqn.(2.155).

If we set k equal to d-1, we can obtain the mean caliper diameter as

$$D=2\frac{W_{d-1}(K)}{W_d}\,. \tag{10.2}$$

Whether we write using intersections or projections, we can view Eqn. (10.1) as an integral over a (d-k)-dimensional subspace orthogonal to S. How would we assign this orthogonal subspace if say d = 3 and k = 2? There is only one-dimension left available. The answer lies in how the class of linear subspaces is defined. If we consider only the method of projections, we do not need translations; consider a space of flats passing through the origin, what they all have in common is a complementary (and hence perpendicular) linear subspace, which in Eqn.(10.1) is of dimensionality k. For this class of two-dimensional linear subspaces in a three-dimensional space, the complementary subspace is a line threading of all the normals. This is illustrated in Fig. 10.1.

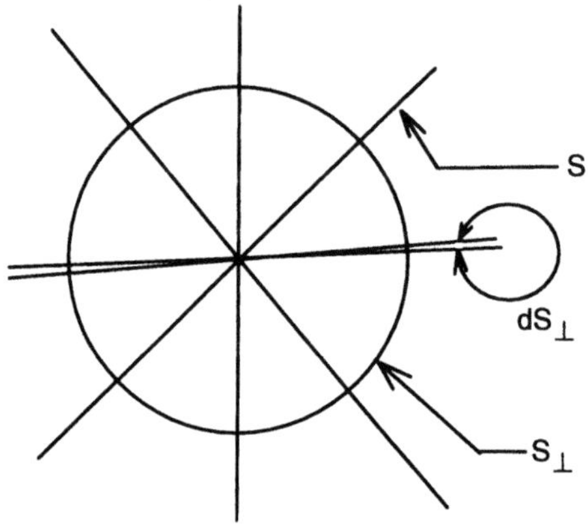

Figure 10.1: A class of linear subspaces viewed in section.

Considering all orientations of these planes, gives a bundle of such lines. The space needed to contain the bundle is d-dimensional but the bundle is of k-dimensional elements. Figure 10.1 shows a section through some example planes and through the differential of such an orthogonal subspace $dS^{\perp}$. The planes appear all perpendicular to the axis, which is normal to the page; they are not. The differential $dS^{\perp}$ appears to be an arc; it is in reality a cone. Some things are difficult to draw.

Minkowski functionals were used by Miles to derive the results given in Chapter 9 [Miles, 1972]. They frequently occur in generalized expressions for volume, area, boundary length, which are essential stereological quantities. These and similar quantities when expressed through functions of measures of convex sets are often termed *convex body functionals*. An important result called the *Hadwiger characterization theorem* states that almost[2] any of the convex body functionals can be written as a weighted sum of the Minkowski functionals:

$$g(K) = \sum_{k=0}^{d} a_k W_k^d(K) \ , \qquad (10.3)$$

[2] The reader should be alerted to the fact that the Hadwiger theorem applies to any convex body functional meeting the assumptions of the theorem. (I.S.)

where g is a convex body functional meeting some requirements to be enumerated and a_k is a real constant. The requirements to be met by g are that 1) the functional is monotone

$$(g(K_1) \geq g(K_2) \mid K_2 \subset K_1);$$

2) the functional is C-additive, cf. Chapter 3,

$$(g(K_1 + K_2) = g(K_1) + g(K_2) - g(K_1 \cap K_2);$$

additionally the functional is non-negative; and, the functional is isometry-invariant, that is, it does not depend on the orientation or position of the set K but only the relative location of its points. Since these requirements are rather broad the reader should feel some temptation to invent some new functional and see how it would be defined using Eqn.(10.3). The index k ranges from 0 to d but since W_d^d is a constant, the g form a d-dimensional function space to within an additive constant.

1.2 The Steiner Formula

In Chapters 1 and 3 we discussed the Minkowski operations on compact sets. These included the Minkowski sum and difference. The Steiner formula [Stoyan et al., 1995] (actually there are two of them plus a generalized form) relates the Lebesgue measure of a Minkowski sum to a sum of the Minkowski functionals:

$$\nu_d\left(K \oplus b(o,r)\right) = \sum_{k=0}^{d} \binom{d}{k} W_k(K) r^k \quad . \tag{10.4}$$

The Steiner formula gives us the measure of the Minkowski sum of a convex set K and a ball of radius r centered at the origin. So we are basically constructing an r-thick layer around the set K and finding the measure of the set and its layer. Since this new set is everywhere parallel to the set K, it is called a *parallel set* of K^3. If $d = 2$, the measure $v_d(K \oplus b(o,r))$ is an area. From Eqn. 10.4 we can deduce that

$$v_2(K \oplus b(o,r)) = A(K) + L(\partial K)r + \pi r^2 . \tag{10.5}$$

Figure 10.2 illustrates how this works for the case of a polygon.

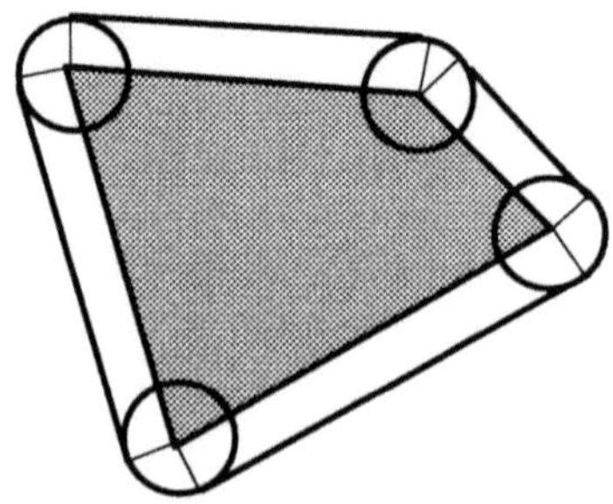

Figure 10.2: Parallel set

The set K is shown shaded. It is contained inside the parallel set. The spacing of the parallel set is r from K. The flat portions of the sides of the parallel set together with the dashed lines form a set of rectangles. Since the lengths of the sides of these are either r or the length of the corresponding side of the polygon, the total area of the rectangles is r times the perimeter of the boundary of K $L(\delta K)$. Examining the circular arcs at the vertices, we see that the total arc for a closed figure must be a complete circle. Its area is πr^2. For a smooth convex set, we can imagine the area of the limit of a set of approximating polygons.

Similarly for a three-dimensional convex set the Steiner formula gives,

$$v_3(K \oplus b(o,r)) = V(K) + rS(K) + 2\pi Dr^2 + \frac{4}{3}\pi r^3 , \tag{10.6}$$

where S denotes the surface of K and D is its mean caliper diameter. We can easily understand the first second and fourth terms of Eqn.(10.6) as

[3] Parallel sets were introduced in Section 3.3.3.4. Of course parallel sets are not necessarily parallel in the Euclidean sense; "only lines and planes etc. are strictly parallel." (I.S.)

equivalent respectively to the three terms of Eqn.(10.5): the set K itself, its surface and its vertices. We want to know where the term containing D comes from since it distinguishes two-dimensional objects from three-dimensional ones. It comes from the edges. To see this consider a cube. It has a mean caliper diameter of $(3/2)a$ where a is the length of an edge. There are twelve edges to a cube. The sides of the cube meet at right angles; so, the cross-section of each edge represents a quarter of a circle. Taking the product of the area of this circle divided by four, times the total length of the edges, we obtain $3\pi r^2 a$, which is exactly what we would obtain by substituting the mean caliper diameter of a cube into the third term of Eqn.(10.6). From Eqn.(10.4) we see that for d dimensions there are going to be d+1 terms. Hence, apart from the set K, for a four-dimensional object we would have besides vertices, area and edges, an additional source of content to envision.

1.3 Kinematic Measure

In Chapter 5 Section 5.4.2.3 we investigated the probability of intersection of two linear arrays. In Fig. 5.10 we considered the intersection geometry of two line segments. We took a fixed line segment BB and a movable line segment AA. The latter was identified by its midpoint. We constructed the area in which the midpoint would have to fall in order for intersection to take place. Based on the discussion of Minkowski algebra in Chapter 3, we see now that this area was the content of $\{AA\} \oplus b_1(r,o)$, the Minkowski sum of the line AA and a one dimensional ball having a radius equal to half the length of BB and the angular orientation specified in the figure. Or in equivalent words, the set comprising the line AA was dilated by the probe. The measure of the dilated set, its area, was proportional to the probability of intersection. This example illustrates what is called *kinematic measure.* Kinematic measure is "kinematic" because it involves Euclidean motions: translation and rotation. (Ivan Saxl commenting on this section said, “It was introduced to describe the motion, namely to measure $A \uparrow gB$ (where g is a Euclidean motion) or equivalently $A \cap gB \neq \emptyset$ for a fixed A and moving B”.) Translations and rotations of the probe comprise sets called *motions* and these motions can be shown to have the properties of a mathematical group.

In Chapter 8 Section 8.6.1 we wanted to calculate the probabilities of intersection between circular or straight probes and grain boundary triple points. We again found that the probability of intersection was the ratio of the area of the triple points dilated by the probe to the area of the surface being probed. This was discussed in connection with the geometry given in Fig. 8.10. In the case of dilation by circular probes, the matter of orientation was not important; the dilation of a point by a filled circle is a disk. In a plane, a disk is isotropic. This was not the case for the straight probe. In that case, see Fig 8.12, we had to consider what amounts to dilation by a straight probe of each specific orientation and then average over distribution of possible orientations. These examples illustrate how kinematic measure works. We have already used it. Now we give it a name and formalize it.

We have from time to time used the notation $A \uparrow B$ to express the notion that "set A hits set B", namely that the sets A and B have a non-null intersection. If we make one of these sets the probe and the other the feature, we can express the notion of the probe intersecting the feature in this way. Associating a probability of occurrence with this intersection, this "hitting probability" can be expressed in terms of a dilation of the feature by the probe. For a probe of fixed orientation we can write an expression for the measure of hitting as,

$$H_t(tB \uparrow K) = v_d(K \oplus \check{B}_t) \ . \tag{10.7}$$

The set $\check{B}$ is the reflection of the probe, t is the translation and the set, K is the feature; we imagine the probe as having a fixed orientation and centered at the origin. We see at once that if the probe were a ball, the right hand side of Eqn.(10.7) would be the left hand side of Eqn.(10.4), the Steiner formula. In fact if K is convex, the Steiner formula could be used to evaluate Eqn.(10.7). In the left hand side of Eqn.(10.7), the H_t is the kinematic measure **with respect to translation only** of the probe hitting the feature. Kinematic measures are not probability measures by themselves since they are not normalized. They may be associated with "Haar measures". Haar measures are often assumed to be normalized making them probability measures when integrated over all possible isometries. (See Mattila [1995].)

If the probe set B is capable of being oriented, then it is necessary to integrate with respect to the distribution of all possible orientations in order to obtain the full kinematic measure that includes orientation as well as translation. We see that whether we consider only translations or translations and rotations together, the measure H_t is a measure of d-dimensional volume where the d is the dimensionality of the space containing the probe and the feature. In order to be interpreted as a probability, H_t must be divided by an appropriate reference volume and each probe must be associatable with a single point, as the midpoint in the example just given from Chapter 5. In this way, the hitting measure becomes the probability of a point in the set enclosed by the bounding figure -- as we have used this term-- falling in the subset which is the dilation of K. If K itself is a set of features we see that we can write either a single Eqn.(10.7) for the entirety of K, or sum a set of such equations, one for each feature. Although the features do not overlap, their dilations are very likely to overlap. Thus we see that this sum is of C-additive measures. It is possible to elaborate on this subject at great length, since it summarizes all that we have done and allows extensions into dimensionalities that are difficult to imagine. Before leaving this topic, we should ask, "What is an appropriate volume"? It is the hitting measure of what we have called the bounding figure.

2. TRANSLATIONAL AND ROTATIONAL ORIENTATION: ANISOTROPY REVISITED

In the previous section we mentioned the subject of motions. The group of motions can be partitioned into two subgroups, translations and rotations. The kinematic measure given in Eqn.(10.7) pertains to translations. If we want to make a kinematic measure which includes rotations we have to rotate Eqn.(10.7). In retrospect, we have done this many times in Chapters 5 and 6 without abstracting on it. Integrating over the rotations of the probe is a natural consequence of the definitions of probes given in Chapter 4. It possible to generalize Eqn.(10.7) further to include rotations, but the result is more formal than interesting. Instead, in this section we will look at cases where we translate a probe but keep its orientation fixed. Under these circumstances a rotation of a probe is to be regarded as a new probe.

2.1 Orthogonal Sketches

In this section we look, for its aesthetic merit, at the idea of constructing a probe for a purely psychological property. In Chapter 5 we looked at how to determine orientation. We saw that the property of orthogonality can provide us with an assortment of ways of constructing series representations that are usually dependent on orthogonality. We explored Fourier series, Legendre polynomials and spherical harmonics. Orthogonality is not that difficult to come by, so to speak. It is that property of a set of functions that the integral of some binary operator of any two different members of the set is zero while the corresponding integral of any one of them associated with itself through the same operand takes a non-zero value. This property allows us to easily construct series representations of functions provided that what is being represented can be faithfully reproduced as a weighted sum of the member functions of the set. These member functions are called *basis functions*. Of course we also must suppose some linearity -- enough to allow us to use weighted sums. A good bet is that the integral of any member function of a set of orthogonal functions by itself must be zero or tend to zero in some way. Why? One reason is that because usually the binary operator is the product and unity is likely to be a member of the set by implication. How so? The answer is that the set and the operator are likely to form a group and groups contain the identity operator. Do basis functions have to be orthogonal to form series representations? No. But, they need to be in some way limited or partitioned so they "don't all run together". But, for example, non-orthogonal wavelets are popular basis functions for representing some types of images. However, wavelets are spatially limited. Do they have to form a group? (The author Hubbard [1996] provides a relaxed introduction to wavelets, if not specifically to the non-orthogonal ones.)

Fourier series use sines and cosines as basis functions and therefore has all the desired properties mentioned above. What is sometimes overlooked is that there is no magic in sines and cosines. For instance, square waves also form a set of orthogonal basis functions, which has been used in optics for years to test lenses. It is much easier for the person setting up the test to have a lens to look at alternate bright and dark stripes than at a sine wave of brightness. Simple linear transformations exist to convert between square wave and sine wave transforms. Thus the frequency response of a lens can be measured with square waves and converted to its sine wave equivalent for the better appreciation of engineers accustomed to audio-style frequency responses. With a little adjustment it is possible to construct a "Fourier series" out of many different types of basis function besides sines and cosines. A natural desire would be to find the set of basis functions that would yield the shortest series for any given task or representation. Making the basis functions look like the thing we want to represent is one approach to this problem.

Equation (5.43) hints at a relationship that could be used to form an orthogonal basis for analyzing line drawings. It says that that the number of intersections per unit area is proportional to the product of the length per unit area of two intersecting arrays. Taking random arrays in pairs of equal length and making one positive and the other negative, we can infer that the expected net number of intersections between the pair and any third array will tend be zero unless the third array is not random with respect to the pair. This is just what we did in Figure 5.17.b. We can extend this idea by devising pairs of linear arrays each of which is tuned to some feature in the drawing we want to analyze. Lines drawn as segments parallel to a feature for which we would like to construct a basis function will intersect it minimally. Line segments drawn perpendicular to the contour of a feature will intersect it maximally. To begin, if we made two arrays, one of maximally-intersecting segments and the other of minimally-intersecting segments, and defined the maximally-intersecting one as positive and the other as negative, we would have made a pair in some sense tuned to the contour of some feature. This captures the essence of what in some form or other is the foundation of most optical character recognition schemes. See Mori, Nishida and Yamada [1999].

To explore this idea further, let us indulge in a pastime that we will call *orthogonal sketching*. The idea is to find a two-color line drawing (a compact subcover-basis function) that will most economically pick a feature out of a picture. Depending on what we call "economical", there is an art as well as a science to it. One color we define as negative the other as positive. We make an overlay of our two-color drawing and drag it over the surface of some drawing we want to explore counting the intersections as we go.

Figure 10.3 shows a line drawing partitioned into twenty blocks. Figure 10.4 shows (in black and white) a two-color overlay, somewhat larger than the individual blocks. The top and left hand edges of the overlay are aligned with the corresponding edges of the blocks. Since the overlay is larger than a block, there is a certain amount of overlap. In this oversimplified example, the amount of overlap is minimal -- about 30% on in terms of area. (A better result would be had with at least 90% overlap. But, then the example would not be so simple.)

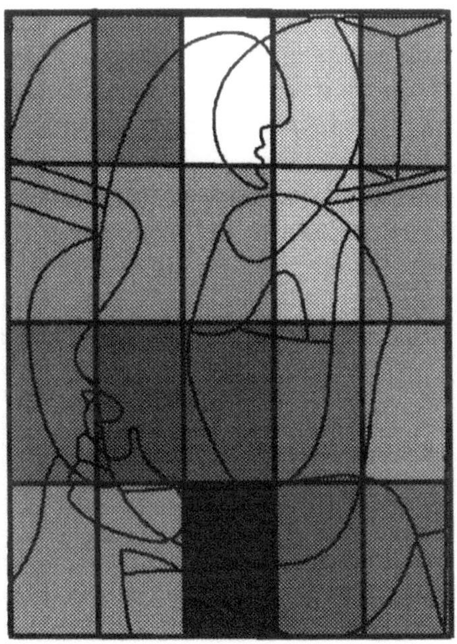

Figure, 10.3: Drawing divided into blocks
Shading is the negative log of the amount of "face" found in each cell according to Table 10.1.

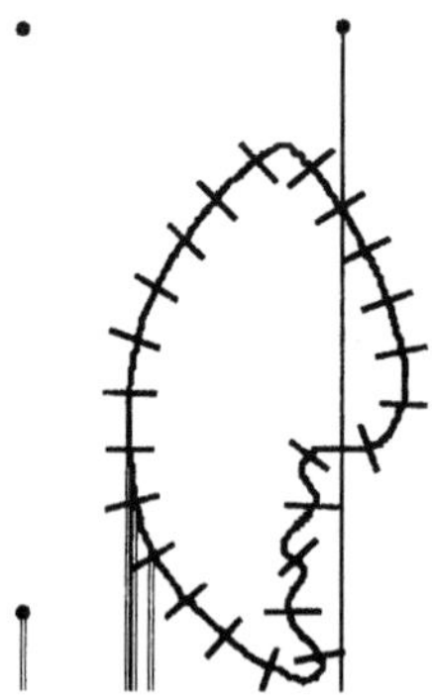

Figure 10.4: Orthogonal sketch to the face in 10.3

Figure 10.4 can be understood without color. What we want to find is the face. We draw the outline of the face. This is drawn in the negative color. We then draw filaments at 90 degrees to the outline in the positive color. We make the total length of filaments the same as that of the outline to the best of our abilities. The choice of the length and spacing of the individual filaments is a matter of discretion. In Fig. 10.4 the filaments are evenly spaced. The length of the filaments determines in part the range of the orthogonal sketch. It cannot effectively detect the feature if its filaments do not intersect it. For the purposes of example, we arranged the blocks in Fig. 10.3 so that at its closest, the orthogonal sketch probe element is displaced about 1/4 of a filament length from that position at which the contour in the orthogonal drawing and the face in the picture would be exactly aligned. The three dots in Fig. 10.4 are used to align the probe with the upper left corner of each box in Fig. 10.3. This deliberate misalignment facilitates intercept counting with lines of finite width. Table 10.1 shows the results of the tally.

Table 10.1: Net intercept counts

block 1 net: 2 total+:4 total-:2	block 2 net: 0 total+:4 total-:2	block 3 net: 21 total+:24 total-:3	block 4 net: 3 total+:8 total-:5	block5 net: 1 total+:7 total-:6
block 6 net: 1 total+:9 total-:8	block 7 net: 1 total+:3 total-:1	block 8 net: 2 total+:12 total-:10	block 9 net: 6 total+:12 total-:6	block 10 net: 2 total+:6 total-:4
block 11 net: 0 total+:12 total-:12	block 12 net: -2 total+:4 total-:6	block 13 net: -1 total+:4 total-:3	block 14 net: 0 total+:5 total-:5	block 15 net: 1 total+:3 total-:2
block 16 net: 1 total+:4 total-:3	block 17 net: 1 total+:12 total-:11	block 18 net: -3 total+:3 total-:6	block 19 net: -1 total+:3 total-:4	block 20 net: 0 total+:8 total-:8

The blocks are arranged as those in Fig 10.3. The face is located in block 3. The net count in this block is almost three times that for the next highest, block 9. The third highest count is three, in block 4. The average count is 1.75. Although not really apparent to this author (L.L.), a second face in blocks 4 and 9 an artist told me she recognized. It is associated with the arm, which ends in a hand in blocks 12 and 17. Whether or not you can see the second face, there are curves in blocks 4 and 9 that are substantially parallel to those in block 3. The average positive count per block is 7.25 while the average negative count is 5.35. We may presume that in the absence of a face, both counts are Poisson distributed. The variance of the difference is thus 12.6 and the standard deviation is 3.54. The face in block 3 is over six standard deviations from the mean. The face in block 9 is less than two standard deviations from the mean. We have too small a sample to assure us that the difference is normally distributed. It is not Poisson-distributed. Were the difference normally distributed, the face in block 9 would be barely significant at the 5% level. However we have 20 blocks and 1/20 is 5%. The alleged face in block 9 cannot be regarded as significant. But, the face in block 3 is significant at any reasonable significance level.

The purpose of this exercise is to acquaint the reader with an example that both illustrates results given in Chapter 5 and may stimulate some speculation as to how the mind might recognize faces and other features in a scene. It is evident that the recognition process depends on setting a threshold. If in this example, we set a threshold as low as 7 or as high as 20, we would "see" one face. Above 21, we would see no faces at all and at a threshold of 5 we would see two faces. If something were to lower the threshold further, quite a number of faces could be "seen".

2.2 Orientation of Features

We now build on the intuition of the previous section to quantify the orientation of particular features. In Chapter 5 we looked at the orientation of arrays. Our answer was in the form of a distribution of tangents. Now we look at a slightly more concrete problem. In 1998 the Donvers Company of Salamanca, NY created a large lumber pile approximately 25 feet tall and nearly 100 feet long. Figure 10.5 shows a picture of a portion of this lumber pile. We are of course looking at it in projection. Because the boards are of random widths, it is difficult to identify the scale. In fact, Fig. 10.5 looks very similar to a micrograph of glass reinforcement fibers in a polymer matrix. The orientation of reinforcing fibers has a strong effect on the shrinkage of reinforced polymer composites that in turn can have a significant impact on such economically important problems as the spontaneous fracture of overmolded ceramic assemblies. But, for our purposes, it may be just as interesting and a great deal more convenient to study a woodpile since no microscope is needed.

Figure 10.5: Woodpile, Salamanca, NY 1998
X0.025 magnification

The woodpile in Fig. 10.5 was assembled by throwing pieces of wood at it randomly. Random processes however, do not always lead to completely disordered results. Ordered molecular structures do form thermally in spite of the fact that the ordered structure would appear to have a lower entropy of mixing and therefore a higher free energy than an alternative disordered structure. A chemist might describe the seeming contradiction by noting that a box of pencils shaken up will tend to settle into relative order to minimize mutual mechanical interference. (There is a similar phenomenon with the vibrational modes in a plate.) They are not really independent. Take a piece of paper. It is easily bent both transversely and longitudinally. But once bent transversely, it is difficult or impossible to bend longitudinally.) Climbing on the woodpile makes clear that when boards lie parallel the structure is colloquially "compact". To form an isotropic structure, boards must become embedded normal to the surface of the pile to form pegs on which added boards can then be supported loosely and at unrelated angles. Inspection of Fig. 10.5 shows orientation domains. A particularly obvious one begins near the center, extends to the lower right hand corner of the photograph and is roughly lenticular in shape.

In the previous section we constructed a probe that was spatially and rotationally limited. The rotational orientation of the face was specified. Our question, only partially answered, was, "where is the face?" Here the question is: in terms of relative board feet, what is the anisotropy of the orientation of the boards. Following the method of the previous section, we could sketch out a probe to locate a board.

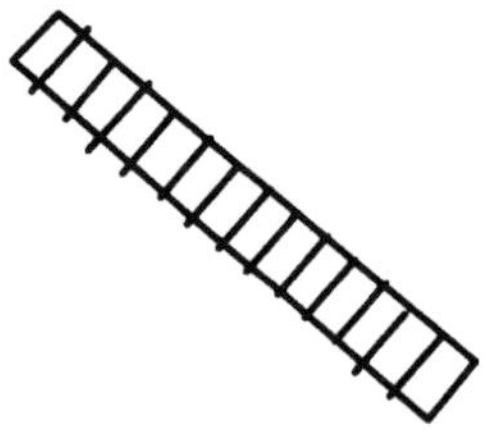

Figure 10.6: A board probe

The board probe in figure 10.6 has positive crossbars and negative sides. We could apply it at various orientations to Fig. 10.5 to identify boards having orientations similar to that of the probe. But, our question does not involve identifying specific boards. Thus without loss of generality, we could dissemble the probe into its line segments and re-assemble them into a linear array composed of a single positive and negative line at right angles. It would be redundant to apply such an array at a range of angles greater than that from 0 to $\pi/2$. In fact, we could simplify the problem further into one of applying a test single line at a range of angles from 0 to π. But, that is precisely the problem previously solved in Chapter 5. We apply our test line at various angles and record the number of intersections with boards rather than lines. This weights the result in terms of the circumferences of the boards (Eqn. 5.11), but this is not out of keeping with our purpose since boards are long in comparison with their transverse dimensions and these do not vary widely; most are two-by-fours.

We can take Eqn.(5.40) as a starting point for the analysis of anisotropy in the woodpile. Equation (5.134P) provides us with a more specific formula for the case of a straight-line probe. But, it uses the traditional sine and cosine Fourier transform while many pocket calculators now implement the complex Fourier transform as the "Fast Fourier Transform" (FFT). The FFT does not give the same result as the continuous transform but it is invertible and is a close enough approximation so long as the number of the frequency of the harmonic in question is sufficiently below the Nyquist frequency. We need to recall the value of the coefficients for the linear probe. We may remember from Problem (5.18) that if we choose our reference direction as the x-axis so that the probe is parallel to the x-axis, all the coefficients are real and equal $1/\pi$. Now, from Eqn.(5.40) we can immediately write the complex equivalent of Eqn.(5.143P) -- well, almost.

There are two practical concerns that must be handled before we can carry out the analysis. They are both closely related. The first and least important is that of the finite nature of the number of angular steps in which we may apply the probe. This limits our sample. Worse, this limit is abrupt. Its abruptness leads to aliasing. If we apply an equation of the form of Eqn.(5.40) to a finite data set for the case where the unknown array is a single line, our resulting angular $L_A(\theta)$ would have a large number of spurious lobes, part of a phenomenon called *leakage*. To eliminate these, we need to convolve our data with a "virtual aperture" or *window* at least as wide as the spacing between data points. This is analogous to the dithering of a low-resolution image on a computer screen to eliminate its block-like appearance. The edges of the blocks are artifacts. Window functions called *Hanning* and *Hamming* are often used to reduce leakage.

The second and more serious concern is noise. The collection of data in the form of numbers of intercepts as a function of angle is a form of linear analysis having a variance as discussed in Chapter 8. However the analysis leading to Eqn.(5.40) is based on differentiating this inherently noisy data set. Differentiation is a form of high-pass filtering. Differentiation of a noisy data set -- even in the Fourier domain -- enhances the noise. Direct implementation of Eqn.(5.40) to the analysis of the woodpile results is a $L_A(\theta)$ rosette which is all noise!

Perhaps the most traditional remedy for noise is data averaging through convolution with a gaussian window. This has similar but much stronger effect than Hamming or Hanning. Like the aperture in a pinhole camera, it blurs (or "averages") the image increasingly as the aperture gets larger. Gaussian windows are commonly used for smoothing time series in finance calculations. Convolution with a gaussian in the angular domain is equivalent to multiplying by a gaussian in the frequency domain. Consequently, the formula for determining angular distribution using straight-line probes becomes,

$$\alpha_n^\theta = \frac{1-4n^2}{2} e^{-\frac{n^2}{k^2}} \left[\alpha_n^\mu\right]^* \quad , \tag{10.8}$$

where the alphas are the complex Fourier coefficients as defined in Chapter 5, the star denotes complex conjugation, *n* is the number of the partial, and k^2 is twice the variance of the gaussian window.

Let us spend a moment to look at the gaussian window. Noting as in Section 5.4.2.3, that $n = \pi f$, where f is frequency, the inverse transform of the gaussian factor in Eqn.(10.8) becomes another gaussian,

$$e^{-\frac{n^2}{k^2}} \Leftrightarrow \frac{k}{\sqrt{\pi}} e^{-k^2\theta^2} . \tag{10.9}$$

The width of a window or aperture is typically defined in terms of "Full width at half maximum" (FWHM) of either the aperture function itself or its square depending on the context. Here we take the function itself envisioned as a "transmission peak". This is a signal-processing concept borrowed from spectroscopy. To find the half width at half maximum, we set the exponential equal to one half ignoring the prefactor because it is a constant: we want the find the point at which the transmission drops to one half from its peak value. We obtain $\log 2 = k^2\theta^2$. If we now specify that we are going to use Eqn.(10.8) with the Fast Fourier Transform (FFT) [Brigham, 1974] we know that the number of partials will equal the number of data points. If N is the number of data points, then n will range from -(N/2-1) through zero to N/2. We now identify the data points as being numbered n'. This makes $\theta = n'\pi / N$, and

$$\frac{n'_{FWHM}}{2} = \frac{N\sqrt{\log 2}}{\pi k} . \tag{10.10}$$

We may think of n'_{FWHM} as being the "effective" number of data points over which the weighted average of the number of intersections is taken. Integrating the right had side of the transform relation Eqn.(10.9) from negative to positive infinity, gives an integral that is unity. This tells us that the window is normalized and the inverse Fourier transform of the coefficients in Eqn.(10.8) is close to $L_A(\theta)$.

Examining the woodpile of Fig. 10.5, we may see what appears to be an isotropic zone on the left side in the middle. This zone is separated out and displayed in Fig.10.7.

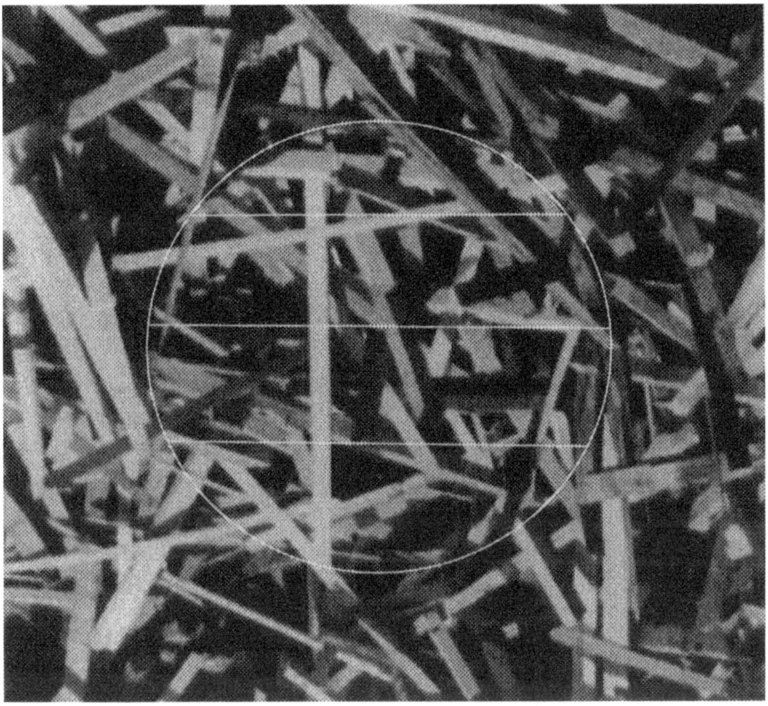

Figure 10.7: Region 1 of Salamanca Woodpile

The array shown superimposed in Fig. 10.7 is used to determine $L_A(\theta)$. It consists of three parallel but not exactly symmetrically (to reduce overlap) placed lines bounded by a circle. The intersections of boards with the lines are counted as the test pattern is rotated relative to the image. The circle fixes the location of the pattern. (A convenient way of doing the rotation is to use a program such as Corel Draw™ that allows the user to perform the rotation simply by entering the angle into the program.) In order to facilitate the use of the FFT, the number of data points should be a power of two. For this case 16 data points corresponding to 11.25° increments are used. Several values of k were tried in Eqn.(10.8). The FFT of the number of intersections vs. angle is taken (Although this calculation can be performed and displayed using a better-grade hand calculator, the package MAPLE™ is convenient since it "runs a tape" giving a printout of all the steps); Eqn.(10.8) is applied and the inverse transform taken. Best results were had with a window having an FWHM of six data points. The result is shown in Fig. 10.8 as "region 1".

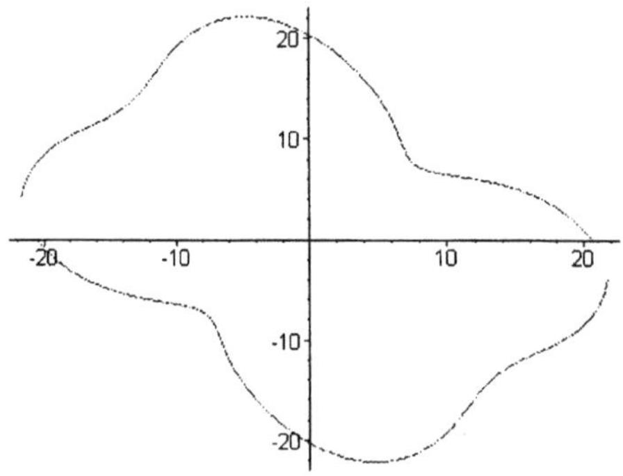

Figure 10.8: Plot of $L_A(\theta)$ for region 1 of woodpile

Because the scale of the photograph of the woodpile was not accurately recorded, the numerical values for the plot, Fig. 10.8, are only relatively accurate. What is shown is the degree of isotropy. For angles between 90 and 180 degrees, the degree of isotropy is good. Near-vertical and near-horizontal boards are present in about equal numbers. However there appear to be fewer boards at angles near 45 degrees than would be expected for an isotropic array.

For comparison a second region, one within the "pencil box domain" at the center, is taken. Its rosette is shown in Fig. 10.9.

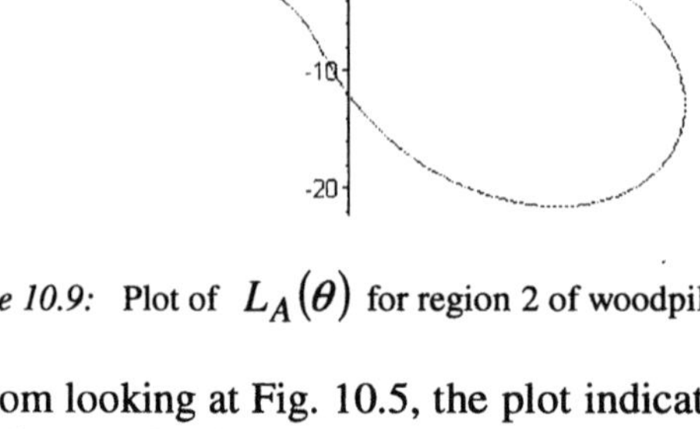

Figure 10.9: Plot of $L_A(\theta)$ for region 2 of woodpile

As expected from looking at Fig. 10.5, the plot indicates that most of the boards are oriented near 135 degrees. There is a distinct shortage of boards oriented near 45 degrees. Since region 1 and region 2 are contiguous, one may suspect that interference from the large number of boards at 135 degrees in region 2 may have reduced the number at 45 degrees in region 1.

Seeing this, one can imagine how orientation domains in short fiber composites can strongly affect their mechanical properties and durability.

3. TOPOLOGICAL RELATIONSHIPS FOR SOME TESSELLATED STRUCTURES

Contrary to that one might have expected, applications of topology to stereology have been limited to a few elementary results. The probable reason for this is that in stereology we are mainly concerned with metric properties and these are outside the domain of topology. In this section we will be considering the application of topological principles to tessellated cellular structures having nearly straight edges. Such structures arise frequently under conditions where surface energy is significant such as in polycrystals. The treatment is based on a review by Smith [1952] and retains, by John Hilliard's choice, some of Smith's terminology. For a more advanced and up-to-date treatment see Okabe et al. [1992].

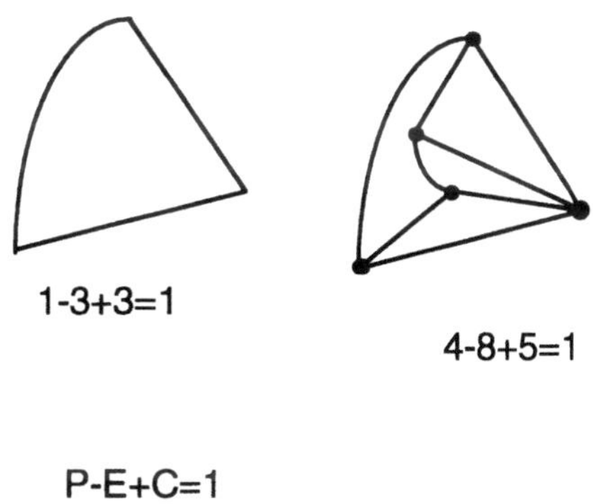

Figure 10.10: Polygons Edges and Vertices

In a two-dimensional lineal array (Fig. 10.10) we can distinguish between three different types of features. (1) Cells (P) (or polygons) which are totally enclosed regions; (2) Edges (E) which are lineal segments; and (3) vertices (C) which are the termination of one or more edges. The basis of Euler's relationship is that the number of these three types of features cannot be varied independently, but must satisfy:

$$P - E + C = 1 \quad . \tag{10.11}$$

The development of this expression is illustrated in Fig. 10.11.

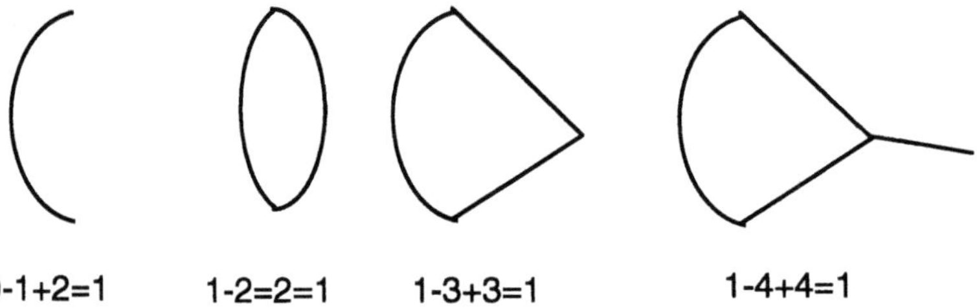

Figure 10.11: Development of Euler's Relation, Eqn.(10.11)

Equation (10.11) holds rigorously for any network; a more complex example (based on a Picasso painting and adapted from Smith), shown in Fig. 10.3, has (without the superimposed grid) 35 cells, 86 edges, and 52 vertices totaling 1 according to Eqn.(10.11). For an isolated polygon (i.e., P=1), Eqn. (10.11) reduces to

$$C - E = 0 \ . \tag{10.12}$$

In a section through structures that are influenced by surface tensions it is rare to find any vertices other than those at which three edges meet. In fact such an arrangement is termed a *normal tessellation*. (As was first shown by Gibbs [1948] a four junction intersection is unstable with respect to decomposition into two triple junctions unless there is a large difference in the surface tension of the four boundaries.) If three edges meet at every vertex then, since each edge is a member of two vertices,

$$3E = 2C \ . \tag{10.13}$$

In addition each cell with k sides contributes $k/2$edges except for cells on the boundary of the network. Neglecting the boundary cells, we have:

$$E = \frac{1}{2}\sum kP_k \ , \tag{10.14}$$

in which P_k is the number of cells with k edges. Substituting from the last two equations into Eqn. (10.11) we find after some re-arrangement: $6P - \sum kP_k = 6$ or

$$\frac{1}{P}\sum kP_k = 6 - \frac{6}{P}. \tag{10.15}$$

The left-hand side of this equation is just the average number, $\bar{S}$, of edges per cell. Also, since we are neglecting cells at the boundaries of the array we can also omit the term $6/P$ since it is of the same order. Thus:

$$\bar{S} = 6 \tag{10.16}$$

or, in other words, the average cell[4] is hexagonal in an infinite array. Smith [1952] has described a method for correction for boundary effects that yield $\bar{S} = 6$ exactly even for a finite array.

Euler's relationship for three dimensions is:

$$C - E + P - B = 1, \tag{10.17}$$

in which B is the number of three-dimensional cells (or bodies) and C, E and P have the same significance as they do in the two-dimensional case except that the term "facets" is better applied to P than polygons. It will be noted that if B is set equal to 0, Eqn. (10.17) reduces to the two-dimensional expression Eqn. (10.11). Also, for an isolated polyhedron, B is unity and therefore:

$$C - E + P = 2 \tag{10.18}$$

For example, for a cube: $C = 8$, $E = 12$ and $P = 6$.

When junctions are determined by surface tensions, all edges are shared by three facets, P, and by three bodies. A polygonal facet with k edges will therefore contribute $k/3$ edges to the array. Thus, if P_k is the number of facets with k sides we find:

$$E = \frac{1}{3}\sum kP_k \;. \tag{10.19}$$

Assuming again that the surface tensions are in equilibrium, there are four edges at each vertex and since each edge joins two vertices:

[4] Cells cannot strictly be averaged; John Hilliard here used the term "average cell" in the sense of the "average man on the street". Here, Ivan Saxl also points out that for a normal tessellation, half the number of vertices per unit area is a convenient unbiased estimator of the number of grains per area and hence the reciprocal of an index of the grain size in metals and ceramics.

$$C = \frac{E}{2} \ . \tag{10.20}$$

Substitution of Eqns. (10.19) and 10.20) into Eqn. (10.17) yields:

$$6P - \sum kP_k = 6(B+1) \tag{10.21}$$

or, since the average number $\overline{S}$ of edges per facet is $\frac{\sum kP_k}{P}$, we have:

$$\overline{S} = 6 - \frac{6(B+1)}{P} \ , \tag{10.22}$$

from which it follows that $\overline{S} < 6$. If terms of order $1/P$ are ignored (i.e., for an infinite array)

the last equation reduces to:

$$\overline{S} = 6\left[1 - \left(\frac{B}{P}\right)\right] \ , \tag{10.23}$$

which can be rewritten as

$$\frac{P}{B} = \frac{6}{\left(6 - \overline{S}\right)} \ . \tag{10.24}$$

Combining Eqns. (10.20) and (10.17) we find:

$$P - C = B + 1 \ . \tag{10.25}$$

which, neglecting terms of order $1/B$, can be written as:

$$\frac{P}{B} - \frac{C}{B} = 1 \ . \tag{10.26}$$

and hence attaching Eqn. (10.24) we have for the average number of vertices per body:

$$\frac{C}{B}=\frac{\bar{S}}{(6-\bar{S})}\ . \tag{10.27}$$

From an argument based on the requirement that at equilibrium the total boundary energy should be a minimum, Smith [1952] has deduced that C/B should, ideally, be 6. This yields $\bar{S}=36/7$ and P/B = 7. Smith compiled data on the frequency of occurrence of S for vegetable cells, grains of β-brass and soap bubbles; it will be seen that $\bar{S}$ is remarkably close to 36/7.

4. PROJECTIONS OF CURVED SURFACES, A VOID MORPHOLOGY EXAMPLE

This is a continuation of Section 7.4, star volume. The star volume methods given in Section 7.4 including the popular Eqn.(7.63) provide, as discussed, an unbiased estimate of average volume. This works best under the condition that the image of the void when sectioned represents its orthogonal projection. Otherwise the star volume has in some cases a negative bias. Another limitation of the star volume approach as practiced is that while it can provide an unbiased estimate of the average volume, it is a very noisy estimator for the volume of individual voids. Consequently, it is not suited to finding second moments or distributions of voids. When voids occur on surfaces, however, improvements in estimation are possible since the observed profiles are almost always true orthogonal projections because the voids usually split in the middle. Additionally, measurements in the third dimension can be incorporated. Illustrating the use of these is the purpose of this section.

I (L.L.) once had a conversation with John Hilliard about approximate symmetry groups. I recalled it when in 1998, I was climbing a local mountain and discovered a rock formation containing what I will call "weathering voids", which resembled scanning electron micrographs I had taken previously of voids formed in ductile fracture. The rock formation voids had an advantage over the ductile fracture voids in that they could be closely examined and easily measured in three dimensions. Recognizing that it would be a good test bed for void morphology, I took a stereo pair of photographs the rock formation. The results proved interesting. So, several friends and I trekked back up there in 2003 to measure the formation again with large micrometer calipers to provide bench marks to calibrate and check the photogrammetric analysis. Photogrammetry, discussed in Chapter 1, is one of several ways of obtaining measurement data in the third dimension.

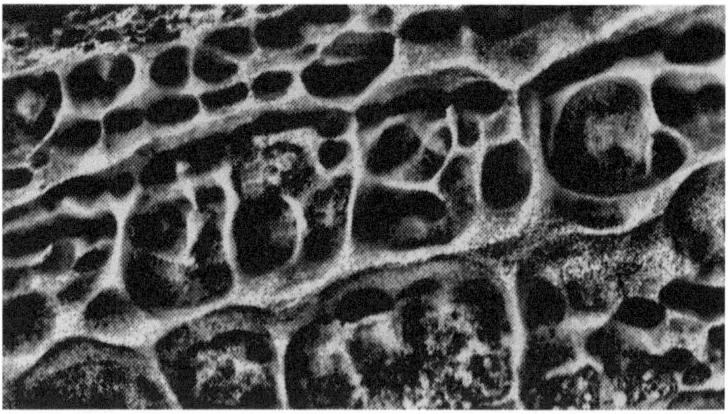

Figure 10.12: Rock formation at 2300 feet altitude, west of Bradford, PA X0.14 magnification (processed to simulate an SEM photo).

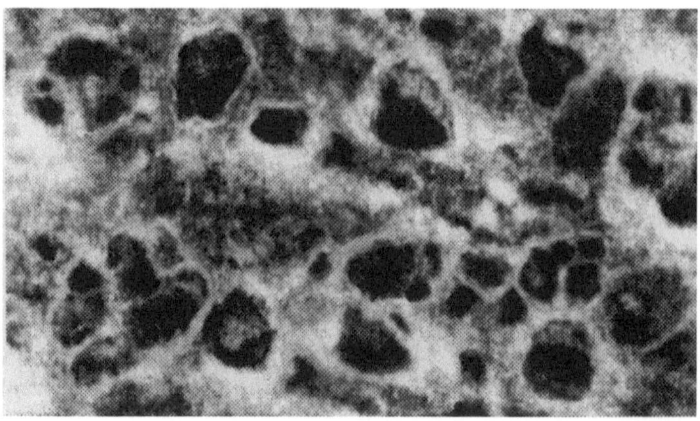

Figure 10.13: Voids formed on fracture surface of a thermomechanically fatigued lead alloy. X5000 magnification

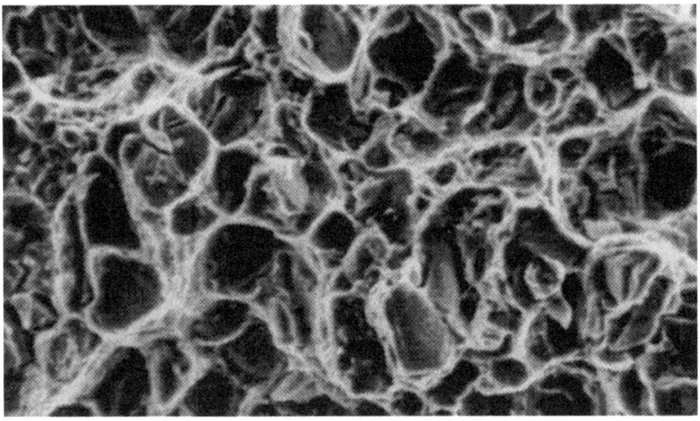

Figure 10.14: Voids formed in ductile fracture of an aluminum matrix composite. X600 magnification

Figure 10.12 shows the stereo pair of the rock formation with its weathering voids. For comparison Fig.10.13 shows the voids formed in lead by thermomechanical fatigue. Both illustrate hierarchical void networks, in which large voids are connected to progressively smaller ones. In both these figures the voids appear spheroidal. Figure 10.14 shows the void structure on the fracture surface of a particulate-reinforced aluminum matrix composite. Although similar, void hierarchy is less noticeable and most of the smaller voids connected to larger ones seem to originate at disbonds between the matrix and the reinforcement. The statistical characteristics of void hierarchies may be useful in elucidating the mechanisms that produced them.

Void networks arise in a variety of contexts. Figure 10.13 illustrating the voids formed in lead during thermomechanical fatigue (TMF) show them having a thick coat of oxide, which appears like snow on the surface. However, it is still possible to see that voids occur within voids as in Fig. 10.12. A likely explanation of the TMF void hierarchy is that the inner voids grew from vacancy coalescence before fracture took place but were surrounded by a local hardening (possibly caused by lattice vacancies pinning dislocations.) limiting their capability for further growth. At fracture a different set of voids grew and in their growth subsumed the smaller voids as features of their internal surfaces. Under these exact circumstances, there could be only two levels in the hierarchy.

In Section 7.4 we were introduced to a technique, star volume, commonly used to characterize void networks in solids. The technique is based on taking an IUR section through the solid, selecting random points and drawing probe lines through the selected points. The average cubed lengths of the interceptions are used to find a characteristic volume. We wish to apply this same idea to surface voids. We could apply it directly to voids on fracture surfaces because in the case of fracture there is a mating half to the surface being examined. The fracture acts to section the void -- not in a random way -- but in a way at least somewhat consistent with the integration formula that underlies the star volume concept. However, in the case of surface voids there is no mating half as well as no random sectioning. The absence of random sectioning means that we cannot assume the void profiles seen in the two-dimensional view represent random sections through the three-dimensional void structure. The absence of a mating half means that the formulas of Section 7.4 would give us values about double the actual void volume.

We would like to use a measure like star volume to evaluate the sizes of the voids. But, because of the absence of the isotropy of IUR sectioning, we need to obtain intercept lengths in three dimensions. So, we cannot use a flat star. Beyond that restriction, we are free to choose a wide variety of estimators. It is conceivable that one could locate a very large number of points on the surface of each void and numerically integrate to find the volume. In practice, while measurement of the void's silhouette is easy, determining depth is not. Shadows and ambiguities are likely to limit the number of depth measurements to about a single point per void -- most likely near the centroid of the void profile. If we were to locate the center of a half-star directly above such a point, so that the depth from the surface would conveniently be the length of ray from the star's center point, how many additional rays would we need? How should they be arranged? These questions are not easy to answer, but clearly the answers depend on the symmetry of the void. If the void were spherical, the one ray would be enough. If the void were ellipsoidal, two additional rays oriented to give the maximal and minimal breadths to the profile would be entirely sufficient. But, for greatest accuracy in the ellipsoidal case, we would have to take the cube of the geometric mean to find the appropriate volume rather than the arithmetic mean of the cube as is done with the star. If the rays are each given equal weight, each ray should be associated with an equal division of the 2π steradians available. Hence, the single ray in the z direction should be balanced by two orthogonal rays. Since the star may not be located exactly at the centroid and because measurements of the void silhouette are comparatively easy to make, two orthogonal lines could be used yielding four rays, each weighted one half. Figure 10.15 illustrates this. The thick portions of the lines indicate the intercepts. Their lengths are L_1, L_2 and L_3.

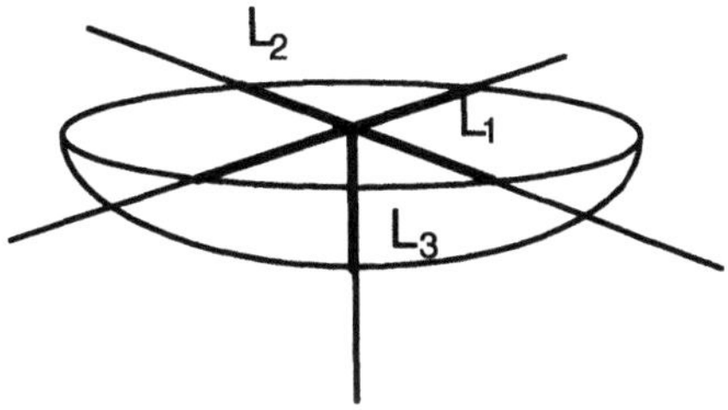

Figure 10.15: Half-star of orthogonal rays

We could use more rays in the plane of the surface of the void, but we would have to give each of them still less weight if we wanted to keep the weights in proportion to solid angle.

Orientation of the rays could be random but are best oriented to match the symmetry of the void. Since the star volume formula can be interpreted as superposition of spherical surfaces, let us consider for example a void having a square profile $2a$ on a side and depth of a. This choice should increase the estimating error over that of a spheroid. We orient the half star at the center such that all the radii are a. Since we are looking at a surface void, which does not have a matching other half, the half-star volume[5] formula becomes from Eqn.(7.60),

$$v\left(\frac{*}{2}\right)=\frac{2\pi}{3}E(r^3) \ , \tag{10.28}$$

where the appropriate weights are applied as discussed. In this first case all the r are the same so the weights are immaterial. Simple arithmetic shows that this orientation gives a value only about 50% of the correct one for a cube. Rotating the half-star by 45° lengthens the r perpendicular to the depth by the square root of two. This gives the maximum value of r. The result is an improvement in volume estimation, an overestimate by 16%. Choosing one of these r to be the minimum and the other the maximum improves the volume estimate further. The error is now an underestimate of less than 16%. However now our axes are skewed since one radius is to the side and the other goes into the corner. How can this give a correct result?

The answer to the skew axis question lies in the symmetry of the void. From the standpoint of the original integration formula, Eqn.(7.59) choosing two radii 45° apart accounts for only about $\pi/2$ steradians (including part of the radius to the depth). However, we have four-fold rotational symmetry for

[5] The ½ indicates "half-star". Originally it appeared as a superscript. Ivan Saxl referred to this as an "unhappy notation". I (L.L.) think this phrase is so à propos that it deserved a footnote.

the square profile. This, in effect, increases the number of radii counted from two in the surface plane to eight. It follows that choosing a location for the star center that has high symmetry is an advantage. On the other hand, choosing the radii to capture the minimum and maximum distances from center at the expense of making their axes skew needs support of symmetry to be justified. In cases such as the one we are studying, choosing voids at random may be expected to reduce bias. But, placement of the star center randomly within the void profile, while not introducing bias, is likely to introduce noise. On the other hand, choosing an orientation of the star that minimizes the measurement error in each individual case reduces noise but is likely to introduce a bias as our example with the square void just illustrated.

Real voids are usually only suggestively symmetric. Many of the voids in Fig. 10.12 seem rectangular. We have seen that exploiting symmetries can improve accuracy. What happens when we employ a symmetry that is not there? This is the question about approximate symmetry groups mentioned earlier. The question can be at least partially answered by asking what is the appropriate measure of the distance between a set and its rotated self? The Hausdorff metric for the distance between a set K and its rotation, $\mathbf{r}K$ is formally[6]

$$h(\mathbf{r}) = \inf\{r : K \subset \mathbf{r}K \oplus b(o,r) \text{ and } \mathbf{r}K \subset K \oplus b(o,r)\}, \tag{10.29}$$

where $\mathbf{r}$ denotes the rotation operator and r denotes the radius of the ball $b(o,r)$. That is, the distance h is the smallest r which when applied to dilate either the set or its rotation will allow the one to engulf the other. This naturally depends on what is chosen for the center of rotation. If we choose the centroid, we expect the error to be minimal. If we choose some eccentric point, we expect a greater error. Only when there is a center of symmetry and $\mathbf{r}$ belongs to the symmetry group of the set is h equal to zero. We observe that the following inequality must be true for the error of volume estimation, v_e, when a set is arbitrarily rotated about a point,

$$v_e < h(\mathbf{r})\delta K \ . \tag{10.30}$$

This terse expression is obviously capable of further refinement especially if the rotation is about an axis passing through a point rather than the point itself.

---

[6] This measure is that of Section 3.5.3 and is not here related to fractal measure. Sorry about the two "r"s being unrelated.

We continue this example by applying what we have discussed to one group of voids in Fig 10.12 shown in Fig. 10.16. It is convenient to diagram the group of voids as a tree as shown in Fig. 10.17.

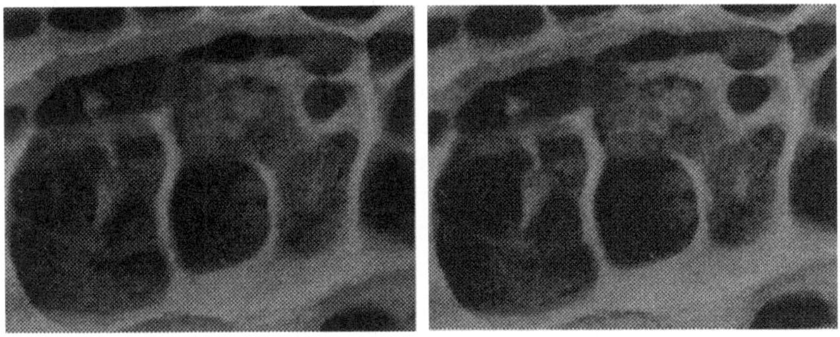

Figure *10.16:* Weathering void network as a stereo pair X0.20 magnification

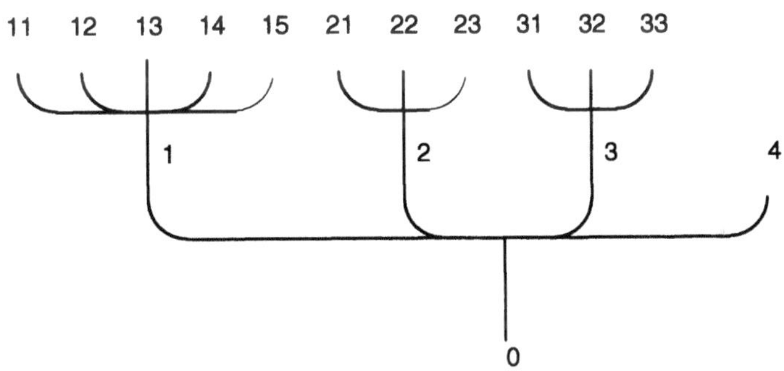

Figure 10.17: Void network tree

Each level of structure is distinguished by being inset into the stone by some deeper distance. If we were to lay the stone so that it was facing upward and sprinkle water on it, the voids designated by two digits would fill first along with void 4. Eventually the two digit voids would spill into the single digit voids connected to them. Finally when all the single digit voids were filled, the water would cover all of them and fill the root void indicated as 0. It is the fact that we can order the voids which gives rise to the tree structure. An ordering could also be performed on the TMF voids of Fig. 10.13, but that ordering could not be defined by the order in which the voids filled. So-called "rainflow" methods are but one way of ordering. Note that the voids with dashed branches, 15 and 23 appear to exist in the photographs but on close direct inspection had bottom contours without any changes in elevation. So, in the treatment that follows, void 15 was merged with void 14 and void 23 was considered an integral part of void 2. Figure 10.18 shows the void cluster indexed.

The study of branched structures and their geometries has a long and interesting past worth mentioning to the reader. Leonardo da Vinci in his notebook #394 stated that for actual botanical trees, the trunk diameter was equal to the sum of the branch diameters. Since we know that trunks and branches provide transport for water and nutrients, we might infer from Leonardo's rule that this transport takes place along the surfaces of the trunks and branches rather than in their interiors. This is in fact true. Transport is confined to the cambium, an outer layer below the bark of the tree. Because of the self-similarity of branching, trees have been extensively studied in connection with fractal geometries [Mandelbrot, 1977]. Fractal exponents and related constants are not the only way to summarize branched structures. They may be described by rule systems related to Chomsky grammars. The term L-systems has been applied to this algorithmic approach after Aristid Lindenmayer. An excellent and entertaining discussion of this approach to geometry is given in *The Algorithmic Beauty of Plants* [Prusinkiewicz and Lindenmayer, 1990]. A mathematical treatment of clustering which might be applicable to this example were the analysis carried out on a larger scale is given in *Fractals and Chaos in Geology and Geophysics* [Turcotte, 1997]. It considers the interrelationship between fractal geometries and the chaotic sets of differential equations. The latter may be used as causative models for the former.

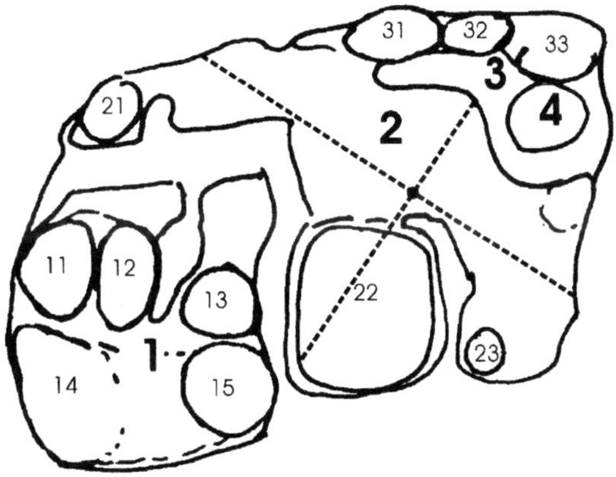

Figure 10.18: Indexing of voids in Fig. 10.16. Dotted lines indicate half-star for measuring void 2

The voids in Fig. 10.16 are flat-bottomed, as seen by direct inspection. For this reason, practical analysis allows the depth measurements to be taken at offset positions with respect to the half-star centers. Circles indicate the depth measurement locations. In this example, we apply star measures in a systematic rather than a random way. Following the discussion in Section 8.5.3.2, we hope by doing this that we may increase the accuracy of our estimates since we are working with a very small sample and because we would like to calculate the dispersion of the void volumes. Measures of dispersion are insensitive to bias of the mean. We shall measure each void using this rule: A point shall be located in each void near to the centroid of its profile under the condition that the distance from the surface plane to the bottom of the void is measurable there. This point shall be the center of a half-star. The surface plane shall be defined as one containing a contour line of constant elevation and shall be aligned in the average with the edges of the root void. A probe line shall be drawn through the point such that its intercept, L_1 in Fig. 10.15, with the boundary of the profile of the void is maximal. A second probe line shall be drawn at right angles to the first and its intercept, L_2, measured. The average cube of the radius shall be found as

$$E(r^3) \to \frac{L_1^3}{24} + \frac{L_2^3}{24} + \frac{L_3^3}{3}, \tag{10.31}$$

which the reader may verify.

By the arrow rather than an equals sign it is meant that the sum in Eqn.(10.31) shall be substituted as an estimator into Eqn.(10.28) to find the half-star volume. This assignment of weights gives an equal solid angle to each radius. We could also use L_1 and L_2 to find a star area using Eqn.(7.61) if we wished. How can we prove that our choice of orientation is optimal or unbiased? In general we cannot. This is an important lesson: In this particular case, access can be had to the structure we are measuring and by other means we know that the maximum error in our volume estimations is on the order of 45%. Were we to randomize our choice of origin for the half stars, and randomize the orientation of our linear probes, we could comfort ourselves with the thought that in principle our measurement would give an unbiased average were we to have one hundred or more samples. But, for most voids individually, it would be grossly in error and since the sample is small, the lack of bias could never be appreciated. The results of the measurement are tabulated below.

Table 10.2: Weathering Void Volumes

Void	Mean cubed radius (in.)	Volume (cubic inches)
0	67.53	141.4
1	11.07	23.2
11	4.13	8.6
12	4.10	8.6
13	7.06	14.8
14	5.89	12.3
2	26.73	56.0
21	1.03	2.2
22	9.33	19.5
3	4.90	10.3
31	1.86	3.9
32	1.78	3.7
33	1.93	4.0
4	2.52	5.3

From Table 2 we can calculate the mean void volume for the first level above the root as 23.7 cubic inches with a standard deviation of 22.8. For the second level above the root (two-digit void numbers) the mean in 8.6 cubic inches with a standard deviation of 5.9. Since the void volumes are C-additive rather than σ-additive, it is not possible to determine the total void volume without calculating the volumes of the various intersections. The reader may ponder what these statistics mean. The purpose of the exercise is to show that they can be gathered.

Photogrammetry was introduced into electron microscopy in the 1970s but did not gather much interest [Howell and Boyd, 1972; Boyd and Ross, 1975]. Terrestrial photogrammetry (as opposed to aerial photogrammetry) is gaining in popularity and specialized cameras are produced for it. Several proprietary computing packages also are available for digital photogrammetric processing which were not available in the 1970s. Photogrammetry is thus applicable to both microscopic and mesoscopic as well as cartological length scales. In optical microscopy, depth is readily measured by focusing. Thus the application of 3-D star measures to surfaces is not difficult.

5. SECOND ORDER STEREOLOGY

Contributions to second order stereology come from two different directions. One is precise and comes from stochastic geometry. It will be discussed second. The other is more intuitive and appears under the heading of "geostatistics", a subject created largely by Matheron who was one of the founders of mathematical morphology. Geostatistics emphasizes approximation methods used for predicting. These methods can be computationally intensive and generally have to be fitted to the available data. But, because they are useful, they comprise an extremely important part of the current literature on geometric sampling methods. Consequently, some introduction is needed. Many would put geostatistics outside the pale of stereology. We think this is unfortunate because stereologists may have occasion to map stereological quantities on large specimens. Geostatistical methods are the best available for this. Furthermore, as we will see, they make a good introduction to the basic concepts of the stochastic geometric theory.

5.1 Some Geostatistical Methods

Geostatistics grew out of the need for mining engineers in the South African gold fields to predict the grade of ore that mining a particular site would produce. The subject began with the practical approximations of Krige and DeWijs later codified by Matheron [1963]. The concept of regionalized variable introduced in Chapter 1 is a geoststistical concept. Because very little seems to be able to be said about shape independent second order stereology, geostatistical methods are founded in many assumptions some of which are subject to criticism [Henley, 1981]. The literature is quite extensive. (As an aside, the reader who chooses to read Matheron's paper just cited might notice a similarity between Matheron's Eqn.(12) and our Eqn.(8.84). Matheron's approximation deals with a continuous variable while Eqn.(8.84) deals with a discrete variable. Matheron's "extension variance" is replaced by the numerator of Eqn.(8.84) while Eqn.(8.84)'s number of same-phase regions ν corresponds to Matheron's number of "influence zones".) In what follows we will restrict our discussion to regional variables whose distributions are fully characterized by a mean and a variance. This restriction constitutes part of the "normal theory" assumptions that have been criticized. A non-parametric approach is also available [Henley, 1981]. In order to justify the normal theory technique, these variables are often transforms (for example they are frequently transformed by taking the logarithm) of directly measured counterparts.

5.1.1 Variograms

We begin our glimpse at geostatistics with a statement of orientation. The subject deals with spatial estimations under circumstances where the sampling is insufficient. In practice this limitation often arises when the specimen is very large. Censoring of the data may also result in a similar problem. The methods presented so far in this book postulate the existence of a constant mean value for some parameter we wish to find and that this mean can be found through sampling. This is certainly the case for small specimens. In common with classical stereology the geostatistical approaches we are going to discuss postulate that there is a mean (for some parameter such as volume fraction) and this that mean is a constant. However, from the geostatistical viewpoint, this mean may be unknown and unknowable. Instead of trying to find the mean, the geostatistical approach concentrates on characterizing the geometry of sample-to-sample variations. Stereologists may also be interested in characterizing spatial variability, but may also have access to the mean as well. The following shows that the two types of problem can be separated from one another.

First, we consider the covariance of two samples taken at locations $\mathbf{x}_1$ and $\mathbf{x}_2$, where these are vectors. Let the parameter we are measuring be $y(\mathbf{x})$. Formally, the covariance is given by

$$\mu_{12}' = E\{[y(\mathbf{x}_1) - m][y(\mathbf{x}_2) - m]\} , \tag{10.32}$$

since the mean m is a constant. Let $\mathbf{x}_1$ equal $\mathbf{x}_2 + \mathbf{h}$ and let $\mathbf{x}_2$ range through every point in volume V. Then,

$$\mu_{12}'(\mathbf{h}) = E\{[y(\mathbf{x}+\mathbf{h})-m][y(\mathbf{x})-m]\} \ . \tag{10.33}$$

It is apparent that

$$\mu_{12}'(0) = \mu_{11} = \mathrm{Var}(y(\mathbf{x})) \ . \tag{10.34}$$

By hypothesis, we do not know what *m* is. So, a different but related statistic is used to relate separated points, $y(\mathbf{x}+\mathbf{h}) - y(\mathbf{x})$. This is a function of both **x** and **h**. Both **x** and **x** + **h** range over *V*. Since the mean is constant, $E\{y(\mathbf{x}+\mathbf{h}) - y(\mathbf{x})\} = 0$. However, the variance of this statistic has a named definition,

$$E\{[y(\mathbf{x}+\mathbf{h}) - y(\mathbf{x})]^2\} = 2\gamma(\mathbf{h}) . \tag{10.35}$$

When meaningful, $\gamma(\mathbf{h})$ is termed the *semivariogram* (because of the factor of 2) or loosely "the variogram". In order to be meaningful, the form of $\gamma(\cdot)$ should remain the same as that on *V* for any sufficiently large subset $v \subset V$ where we use *V* to denote both the set and its measure. The term "sufficiently large" should embrace any contemplated inter-sample spacing. At the outset, this does not seem a realistic assumption. However, the technique that results has certainly passed the test of time. In addition to the restriction just given it is often required that $\gamma(\cdot)$ be invariant under translation in space and time, viz. stationary, and under highly desirable conditions, invariant under rotation so that it is isotropic. With both these restrictions, γ is termed *homogeneous* and **h** can be reduced to a scalar variable *h*. We can relate Eqn.(10.35) to Eqn.(10.33) for the covariance using the identity for the variance of a sum,

$$2\gamma(\mathbf{h}) = 2\mathrm{Var}(y(\mathbf{x})) - 2\mathrm{Cov}(y(\mathbf{x}+\mathbf{h}), y(\mathbf{x})), \tag{10.36}$$

where the factors of two have been retained so that the reader might recognize both the identity and the reason for the presence of the factor of two in the variogram. This equation strictly applies to punctual support where there is no "nugget", a term to be explained. A plot of γ vs. *h* is also termed a "variogram" or semivariogram.

Semivariograms have frequently occurring features that are illustrated in Fig.10.19.

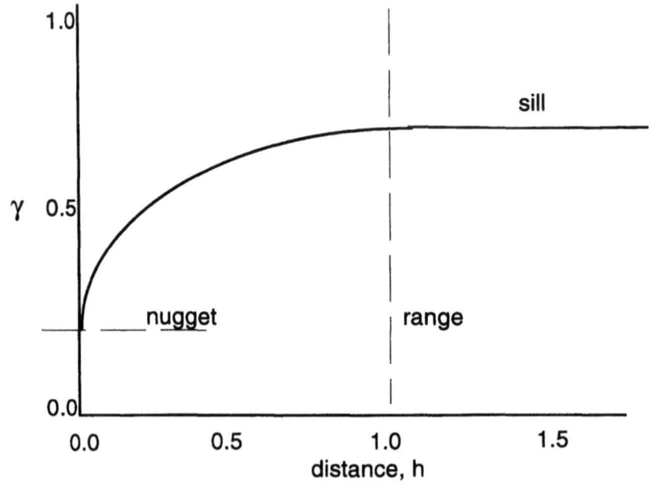

Figure 10.19: Features of semivariograms

Inspecting Eqn.(10.36) we see that we would expect the semivariogram to go to zero when the separation went to zero. In practice this does not always appear to be so. Practicality limits the smallest possible value of separation h. Samples have finite size; so, going to zero exactly may be impossible. A discontinuity near the origin of γ can result when the variable y is discontinuous. Discontinuities can make the correlation between nearby values of y poor even for small values of h. For instance, consider a case where volume fraction is the variable y being estimated and the phase of interest is distributed in nuggets. Recall for a general comparison the sin ϕ weighting that appears in spherical coordinates. It appears because near the poles integrating θ over 2π corresponds to integrating over a much smaller area than near the equator. Similarly when h is small the integrated area is also going to be relatively small even though the actual details depend on the definition of sampling used in the particular case. So, in the case of nuggets the oscillation of y, the grade, between one and zero averages less for small h. The resulting correlation between samples falls short of a smooth approach to unity as h vanishes. When it is non-zero the apparent limit of γ as $h \to 0$ is called the *nugget*.

We generally expect the difference between samples to increase with increasing distance but only up to a limit. In most real scenarios only so much difference is possible. For instance, volume fraction cannot be less than zero nor greater than unity. This suggests that the semivariogram should level off eventually. In some cases that leveling off takes place near a specific value of separation. When that happens, that amount of separation is termed the *range*. The plateau beyond the range is called the *sill*. From Eqn. (10.35) we can deduce that the sill is equal to the variance of samples since as **h** goes to infinity we expect both the correlation and hence the covariance between samples to go to zero.

Many standard functions have been used to approximate experimentally derived semivariograms and many of these are tabulated in the references given. In general, a certain form of semivariogram is found to apply to a specific application. At this point few have been tabulated for phenomena on the small dimensional scale on which stereology is usually conducted. This gives future stereologists something to do. Semivariograms are often used to characterize spatial phenomena especially in two dimensions. Photographic film grain and crop density are two examples. In the former case the surface is conveniently flat. In the latter the surface is three-dimensional.

5.1.2 Kriging

The application of semivariograms that is probably of greatest interest to stereologists is kriging, the estimation of parameters in unsampled regions of a large specimen. Let us take an admittedly artificial example but one that is easily discussed. Say we want to characterize the grain size in a large casting. We might do this through the stereological parameter S_V of grain boundary as a function of position. Perhaps we have dissected some large castings and found it justifiable to define a semivariogram function for castings of this type. We may be concerned about the casting's minimum ductility, which we may believe related to its maximum grain size. (We could conduct a similar argument with inclusions in a casting. Practical metallurgists might like it better but the example would be too complicated.) Given a specific large casting, can we estimate the deviation of grain size from the mean at some inaccessible or inconvenient areas in the casting? In this example we will allow that we can measure grain size at several places by x-ray diffraction (or direct observation) and even obtain a good estimate of the mean grain size. This is the type of problem that can be solved by kriging: We would like to map confidence limits for the grain size throughout the casting. To do this we need both the estimate of the grain size at unsampled points in the casting and the standard error of such estimates. Both can be obtained by kriging.

Kriging is named after the South African mining engineer D.G. Krige who originated the idea and developed it on an intuitive and experimental basis in the 1950s. Since the time when it first proved its worth in gold mining, kriging has undergone mathematical refinement and has found its way into numerous applications. As of the date of this writing (2002) only a few stereological studies have used it. It is one of those areas waiting for development. One reason stereologists may have not used it more widely is that upon close inspection the theory of kriging is filled with inconsistencies which although smoothed over by the concept of a regional variable are not really eliminated. Some of these are apparent at the outset. Kriging tacitly presupposes that the variable we call *y*, the semivariogram, is continuous "in the average" enough that it can be treated by the methods used for the numerical solution of partial differential equations. This assumption itself seems to raise a red flag. But we proceed anyway. There are many types of kriging already and many more are possible. We will discuss two linear types: *ordinary kriging* and *simple kriging*, which should be an adequate beginning for stereologists wanting to try it out. Both ordinary and simple kriging are in essence based on a first order difference equation approach which when further simplified yields a very simple statement:

$$\hat{y}(\mathbf{x}) = \sum_{(i)} w_i \, y(\mathbf{x}_i) \ , \tag{10.37}$$

where $\hat{y}(\mathbf{x})$ is an estimate of the value of y at location $\mathbf{x}$, w_i are weights and $y(\mathbf{x}_i)$ are the measured values of y at the known locations $\mathbf{x}_i$. The first order difference idea is not the only way to view Eqn.(10.37). The estimate is a sum of random variables and therefore by the central limit and related theorems tends toward a normal distribution. Since this may be quite unrealistic, it is common to transform the real variable as previously mentioned. Thus y is often the logarithm of something that is lognormally distributed. Ordinary and simple kriging are thus parametric and so, for this reason too, intimately connected with "normal theory". In ordinary kriging,

$$\sum_{(i)} w_i = 1 \ . \tag{10.38}$$

This is termed *unbiasedness* since it results in a predictor that is unbiased with respect to the samples already taken. Simple kriging drops this assumption. That is the only difference between the two methods. Anecdotal evidence suggests that simple kriging performs better than ordinary kriging under conditions where the mean is known reliably from the samples taken. When used for mapping concentrations, it gives a result which is "more pleasing to the eye" than does ordinary kriging. But, under conditions where the mean is not reliably known, ordinary kriging is the better estimator since it is unbiased.

5.1.3 Derivation of Kriging Equations

Kriging is performed using the stock methods of linear algebra and as such is ideal for rapid machine calculations. The task is simply solving a set of linear equations to obtain a set of weights for each point to be kriged. The equations are simple yet most texts make their derivation seem needlessly obscure. For this reason, we choose to present a derivation of the ordinary kriging equations that we hope is easy to follow and will clarify the assumptions made.

The idea behind ordinary kriging is to choose the weights, w_i, so as to minimize a functional called the *prediction variance*,

$$\tilde{\sigma}_R^2 = E\left\{ \left(y(\mathbf{x}_P) - \hat{y}(\mathbf{x}_P) \right)^2 \right\}, \tag{10.39}$$

where $\mathbf{x}_P$ is the location where y is to be predicted. Equation (10.39) can also be interpreted as giving the square of the standard error of estimate from which, in principle, a confidence interval for the estimate (or if preferred, the deviation of y at $\mathbf{x}_p$ from the mean) can be constructed. We can expand $\tilde{\sigma}_R^2$ and rewrite it in terms of γ using Eqn. (10.36). In doing so we will simplify our notation such that $y_k \equiv y(\mathbf{x}_k)$ for any k.

$$\tilde{\sigma}_R^2 = E\left\{ y_P^2 - 2y_p\hat{y} + \hat{y}^2 \right\}. \tag{10.40}$$

Taking advantage of the common constant mean, we rewrite Eqn.(10.40) as,

$$\tilde{\sigma}_R^2 = \mathrm{Var}(y_p) - 2\mathrm{Cov}(y_P, \hat{y}) + \mathrm{Var}(\hat{y}) \ . \tag{10.41}$$

The first term equals the common variance $\mathrm{var}(y)$ for y at any arbitrarily selected point. We expand the second term substituting the sum, Eqn.(10.37) apply Eqn.(10.36). This gives us,

$$\mathrm{Cov}(y_P, \hat{y}) = \mathrm{Var}(y) - \sum_{(i)} w_i \gamma(\mathbf{h}_{pi}), \tag{10.42}$$

where $\gamma(\mathbf{h}_{pi})$ is the semivariogram between the location of the unknown and that of the known y_i . The third term can be expanded in just the same way to give,

$$\text{Var}(\hat{y}) = \text{Var}(y) - \sum_{(i)}\sum_{(j)} w_i w_j \gamma(\mathbf{h}_{ij}) \ . \tag{10.43}$$

Substituting these results into Eqn.(10.41) causes the variance of y terms to cancel leaving,

$$\tilde{\sigma}_R^2 = 2\sum_{(i)} w_i \gamma(\mathbf{h}_{pi}) - \sum_{(i)}\sum_{(j)} w_i w_j \gamma(\mathbf{h}_{ij}) \ . \tag{10.44}$$

This expression allows us to calculate the standard error of estimate. At a minimum, the derivatives of Eqn.(10.44) with respect to the w_i will be zero.

$$\frac{\partial \tilde{\sigma}_R^2}{\partial w_i} = 2\gamma(\mathbf{h}_{pi}) - 2\sum_{(j)} w_j \gamma(\mathbf{h}_{ij}) \tag{10.45}$$

It follows immediately that for simple kriging given n known values of y we have n equations containing n unknowns. These equations take the form,

$$\sum_{(j)} w_j \gamma(\mathbf{h}_{ij}) = \gamma(\mathbf{h}_{pi}) \ , \tag{10.46}$$

that can be solved by Gauss-Jordan elimination or some similar method.

For ordinary kriging we need to apply the constraint given by Eqn.(10.38). Using the method of Lagrange multipliers, we construct a functional,

$$\varphi = \sum_{(i)} w_i - 1 \ , \tag{10.47}$$

that equals zero. We take the derivatives of the functional, which also must equal zero and add a multiple of them to Eqns.(10.46) giving:

$$\sum_{(j)} w_j \gamma(\mathbf{h}_{ij}) + \lambda \frac{\partial \varphi}{\partial w_i} = \gamma(\mathbf{h}_{pi}) \tag{10.48}$$

The derivatives are all unity and this reduces simply to

$$\sum_{(j)} w_j \gamma(\mathbf{h}_{ij}) + \lambda = \gamma(\mathbf{h}_{pi}) \ . \tag{10.49}$$

We now have a set of n equations with n+1 unknowns since we need solve for λ. In order to make this system solvable, we append Eqn.(10.38) to the set. Solving this set of equations gives us the weights for ordinary kriging.

5.2 Second Order Point Processes

We have just examined an application of second order statistics through the variogram. We might envision a mining situation in which we had only information from boreholes and these were spread about haphazardly. We would know the grade at each borehole. Each borehole has a center, so each borehole has a point on the map that locates it. The grade at a particular borehole belongs to it and in the language of stochastic geometry would be called its *mark*. Looking at the map, we might conclude that the locations of the boreholes were random and the result of a random point process such as the Poisson point process discussed in Chapter 7. A point process where each point has a mark is termed a *marked point process*. Not all point processes are Poisson. Notable exceptions are point processes where there is some minimum distance between points. Such a minimal distance is termed a *hard core* distance. Clustering is another property of non-Poisson processes. Anyone who has examined the spacing of birds perched on a telephone wire will have noticed that birds will get only so close to each other and often seem to perch in clusters in which the distance between the birds' centers is fixed at the hard core distance, Fig. 10.20.

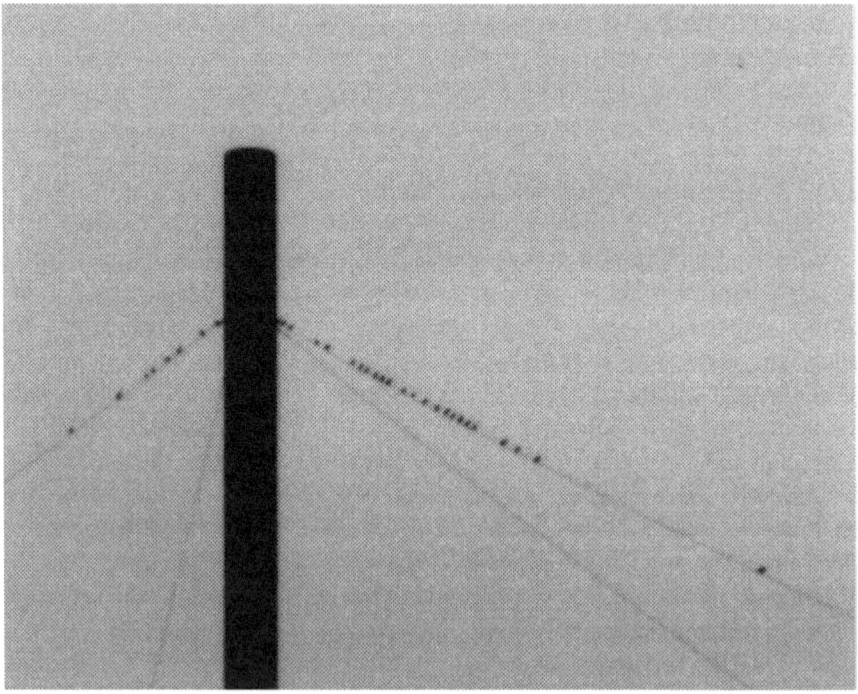

Figure 10.20: Birds on a wire.

The analytical description of second order properties of point processes begins with the so-called *K-function*, [Stoyan and Stoyan, 1994] where $K(r)$ is the expected number of points that have a distance less than or equal to r from a typical point divided by the intensity of a the point process. Thus for a Poisson process in the plane,

$$K_P(r) = \pi r^2 \ .$$

The function K is referred to as a *reduced* measure because the point at the origin is not counted (or K would not go to zero as r goes to zero). The mathematical apparatus implied in the phrase "from a typical point" is extensive and treated under the theory of *Palm distributions.* For the case of birds on a wire, K for a Poisson process is linear. The function K can be generalized to any number of dimensions. Were we to plot the average number of bird centers vs. r and normalize by dividing by the total number of birds per length, we would see that the result would not be linear as it would have been for a Poisson process. Rather there would be a value of r below which K would drop essentially to zero. This would be the hard-core distance. The Poisson process is often used as a null reference for analyzing point processes.

It is often customary to define point processes in terms of densities where the number of points is large. These densities can be integrated over Borel sets to obtain a measure of the number of points or of some property of the set of points. For some readers it may be helpful to note that in materials science the "dislocation density function" plays a similar role. Dislocations are usually thought of as discrete crystal defects but many important results depend on being able to smear them out and then count their probable number or some property such as strain by integrating that density as if it were a continuous function. If we can treat K as being defined in terms of a point density, then it becomes differentiable for an arbitrary point process. In that case additional functionals can be defined [Hanisch, 1983].

One such functional is the pair distribution function also called the *pair correlation function.* In two dimensions it is defined as

$$g_A = \frac{1}{2\pi r}\frac{dK_A(r)}{dr}, \tag{10.50}$$

where K_A is the K-function for some arbitrary areal point process. One decided advantage of working with pair distribution functions as opposed to K-functions is that they are bounded at infinity and in fact tend to unity as r goes to infinity. A second useful property is a rule of thumb [Stoyan et al., 1995],

$$g_A \sim g_V ,$$

that is, the pair distribution function of a projection through a slice of a three dimensional point process is similar to the pair distribution function for the volume process itself. Obviously some additional considerations are required in the case of anisotropic distributions, lattices for instance. The pair correlation function can be thought of as,

$$g(r) = \frac{\rho^{(2)}(r)}{\lambda^2}, \tag{10.51}$$

where, λ is the intensity of the point process and $\rho^{(2)}(r)$ is the second - order product density. It is a density as described above. Intuitively it is related to the probability that two points of the same distribution are found simultaneously at two locations separated by r where r may be considered to be a vector when the point process is not isotropic. The product

$$\rho^{(2)}(r)dF_1dF_2 ,$$

where dF_1 and dF_2 are differential areas surrounding the two points in question, is equal to the differential of probability for the two points being in the two locations. Clearly then, g is a second-order statistic. Figure 10.21 illustrates a typical plot of g vs. r.

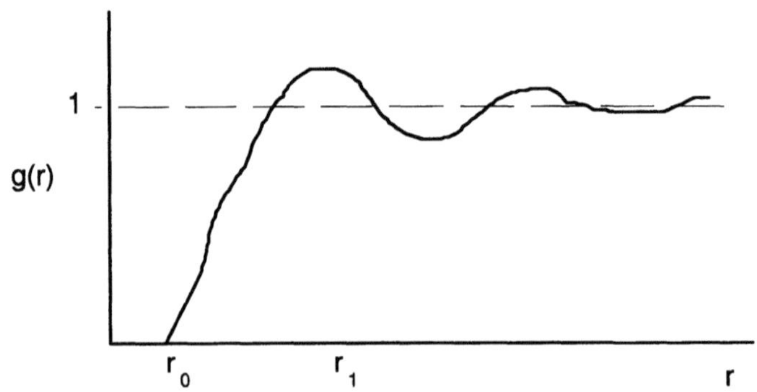

Figure 10.21: Typical pair correlation

In Fig.10.21, r_0 is the hard-core distance and r_1 indicates the location of the typical first nearest neighbor. Plots of this type may be obtained from both point and parallel projections as well as from sections. For glassy structures, they may be obtained on an atomic size range through special x-ray diffraction techniques.

6. FRACTALS IN PRACTICAL STEREOLOGY

There are some measurements (such as surface to volume ratio) in which the result will be strongly dependent on instrumental resolution. This is the equivalent to the problem in which the perimeter of an irregularly shaped figure increases as the scale of measurement is decreased. Such problems stimulated the development of the theory of fractals. At the present many more tomes have been written on the subject of fractals than on the subject of stereology. Mandelbrot [1977] gives a fully illustrated view of the fractal concept. For our purposes the importance of the theory of fractals is that it provides both a means and a justification for a power-law fit to stereological parameters as functions of instrumental resolution -- when such a fitting seems appropriate.

Fractals as a subject arose from the discovery that real-world phenomena were related to concepts of set theory once though to be "monsters". One of the first of these monsters was Peano's curve, which, although a line, passed through every point of a plane. What dimensionality should be assigned to it? Is it one-dimensional or two-dimensional? The importance of such questions took on new significance when studies of the length of coastlines revealed that their lengths did not seem to be rectifiable but instead increased indefinitely with the fineness of the resolution with which they were measured. Extrapolating these lengths to "absolute precision" caused them to go to infinity. Mathematical inquiries into the nature of coastlines assigned them a fractional dimensionality between one and two. From this arose the word *fractal* for a set having a non-integer dimensionality. Since real coastlines actually change minute by minute with each wave of the ocean, such an extrapolation may seem justifiable because it reflects the uncertainty of their location. For stereologists looking at time-independent features, singular length functions are only one way of representing certain types of “fuzzy” boundaries—those that appear to be measurable except that the measurement changes with the resolving power of the optics used to view them. One way to describe them is to give them non-integer dimensionalities. In any case, by the arguments given in Chapter 8, time-independent lengths, areas and volumes would not be expected to become infinite no matter what the resolving power in representations of "real-world" features. Consequently, we are aware at the outset that non-integer dimensionalities are approximations. They apply over some range of dimensional scaling.

In order to assign a dimensionality capable of taking non-integer values we must first assign a carefully defined measure. Many choices of measure are possible; two are significant in Euclidean spaces, and many other measures reduce to them [Zaehle, 1984]. The first historically is the Minkowski measure. This measure is obtained by dilating the set to be measured, taking the Lebesgue measure of the dilated set and finding the limit as the radius of dilation goes to zero. This limit may not always be reached in a unique way. So, upper and lower Minkowski measures are defined depending on whether we take the supremum or the infimum of the possible limits. The second type of measure is the Hausdorff measure. It is related the covering concept utilized in the Heine-Borel theorem. (see Appendix) We choose a system of convex compact sets as a covering. We then set the upper bound on their diameters to a value and determine the infimum of the sum of the Lebesgue measures of the covering sets when all possible systems of coverings are tried. The Hausdorff measure is the limit of such infima as the upper bound on diameter approaches zero. The dilation approach although older is suggestive of image algebra and has been used as an automated computational means of measuring sets. The Hausdorff approach is an example of what we have simply called "tiling" taken to a limit of infinitesimal tiles. These two measures yield two corresponding ways of assigning dimensionality.

The Minkowski dimension for a set (of non-integer) dimension, defined as either an upper or lower one depending on whether it is derived from an infimum or supremum, is defined in Eqn.(10.52). Since both are extrema which we may abbreviate, ext:

$$\mathsf{M}\text{-}\dim(A) = \lim_{r \to 0^+} \operatorname{ext} \frac{\log\left(\nu_N(A_r)/r^N\right)}{\log(1/r)}, \tag{10.52}$$

where A_r is the set A dilated by a ball or radius r, ν denotes the Lebesgue measure in this case and N is the dimension of the embedding space.

The Hausdorff dimension for a set of fractional dimension is defined [Mattila, 1995] in a different way since the Hausdorff measure is only finite for the Hausdorff dimension:

$$\mathsf{H}\text{-}\dim(A) = \sup\{s : \mathsf{H}^s(A) > 0\} = \inf\{s : \mathsf{H}^s(A) = \infty\}. \tag{10.53}$$

For reasons such as that the Hausdorff dimension is an outer measure and the Minkowski dimension is not, the Hausdorff dimension is more often identified as the *fractal dimension, D,* of a fractal set. For the case of a lineal set in 2-D, we have [Mandelbrot, 1977]

$$L = Cr^{1-D}, \tag{10.54}$$

where L is the measured length of a curve, C is some constant and r is the radius of the covering. From this we can see that

$$L = C'\varepsilon^{1-D}, \tag{10.55}$$

where ε is the smallest resolvable distance. Since we are anatomically restricted to viewing only two dimensional images, Eqn.(10.55) has significance in that it applies to the only fractal objects we can actually observe. Two questions arise immediately.

The first question is: does Eqn.(10.55) apply to the Minkowski dimension as well? Flook[1982] showed that both the Hausdorff and the Minkowski dimensions are likely to be nearly equivalent -- a least for any entity we would be likely to photograph. The second question is that, since we are restricted in our viewing to fractals having a dimensionality between one and two, can we apply Eqn.(5.1) to fractal intersections? The answer is both yes and no. Chapter 13 of Mattila [1995] treats this problem for general sets in which both the feature and the probe have non-integer dimensionalities. In general we may expect from Mattila's Theorem 13.11 that loosely,

$$E(I) \geq P + F - N \ , \tag{10.56}$$

where N is the integer embedding dimension, P is the not-necessarily integer dimension of a Borel set we call the "probe" and F is a fractal dimension of a feature which is also a Borel set in $\mathbf{R}^n$. The symbol for expectation is used to indicate that the probe is in general rotated and translated as defined in Chapter 4 with respect to the feature set being intersected. We see that the intersection of two fractals may be "fuzzier" than conventionally expected.

In the case that the probe is an m-plane (examples: a plane in $N=3$, a line in $N=2$) passing through a point in the feature set, the result is more certain since a lower bound on the dimensionality can be found. Paraphrasing part of Theorem 10.10 in the reference by Mattila[7],

$$E(I) = P + F - N \ , \tag{10.57}$$

where P is the integer dimension of the probe P-plane so long as the set F is sufficiently regular. The reader should be aware that Eqn.(10.56) and especially Eqn.(10.57) specify a dimension for the intersection through an indirect process based on the property of the Hausdorff measure that if the measuring dimension is too high, the measure will be zero. These intersection dimensions only assert the possibility of a finite Hausdorff measure. They may not have all the properties the reader might like to associate with a dimension.

As an example Lawson et al. [1991], using several different microscopes on the same specimens, found that the relationship between the length of a thermomechanical fatigue microcrack network per unit area and instrumental resolution took the form

---

[7] Quoting a letter from Prof. Mattila [2003] referring to 10.11in [Mattila, 1995]: "If A is flat (the probe) Theorem 10.11 says that (Eqn.) 13.1 (loc.cit.: equivalent to Eqn. 10.57 here with the understanding that the probe is rotated and translated as defined in Chapter 4 subject to the restriction that it intersects the feature) holds typically if B (the feature) has finite Hausdorff measure in its dimension. If not, it still holds up to a small error in many senses..." Words in parentheses were not in the letter.

$L \propto \varepsilon^{-0.988}$. From this relation we would find D to be equal to 1.988 for the cracks on a two dimensional surface, Eqn.(10.55), and we may expect that from Eqn.(10.57) the dimension of the entire three-dimensional crack network would be approximately 2.98. This should not be taken as an expression of the relative amount of the material that is cracked. Rather, it is an expression of the growth of apparent cracking with increasing fineness of resolution. At very high magnifications, cracks often appear to merge into such features as subgrain boundaries, intrusions and extrusions. At what scale cracks cease to be cracks is a question materials scientists have yet to fully resolve. The dimension 2.98 is certainly confined to a small region (having finite support- not punctual) near where the observations were made. In this region, possible cracks appear almost everywhere under fine examination. This is because the proximity to a free surface has a great deal of effect on the amount and nature of cracking. Because the protocols for examining fractures are standardized, the growth exponent itself has value in allowing comparison to be made between typical observations taken at different magnifications or with different optical systems even by different laboratories.

7. BIBLIOGRAPHY

A. Boyde and H.F. Ross, "Photogrammetry and the Scanning Electron Microscope", *Photogrammetric Record*, **8** (1975) 408-457.

A.G. Flook, "Fractal Dimensions: Their Evaluation and Their Significance in Stereological Measurements", *Acta Stereologica*, **1** (1982) 79-87.

J.W. Gibbs, *J.W. Gibbs, The Collected Works*, Vol. I, (Yale University Press, New Haven, CT, 1948) 289.

K.H. Hanisch, "On Stereological Estimation of Second-Order Characteristics and the Hard-Core Distance of Systems of Sphere Centers", *Biometrical Journal*, **25** (1983) 731-743.

S. Henley, *Nonparametric Geostatistics* (Applied Science Publishers, London, 1981).

P.G.T. Howell and A. Boyde, "Comparison of Various Methods for Reducing Measurements from Stereo-Pair Scanning Electron Micrographs to Real 3-D Data" in *Scanning Electron Microscopy*, Proceedings of the Seventh Annual SEM Symposium, IITRI, Chicago, April (1972) 233-240.

L.Lawson, M.E. Fine and D. Jeanotte, "Thermomenchanical Fatigue of a Lead Alloy", *Metallurgical Ttransactions A,* **22a** (1991) 1059-1070.

B.B. Hubbard, *The World According to Wavelets* (A.K. Peters, Wellesley, MA, 1996).

G. Matheron, "Principles of Geostatistics", *Economic Geology,* **58** (1963) 1246-1266.

B. Mandelbrot, *The Fractal Geometry of Nature* (W.H. Freeman, NY, 1977).

P. Mattila, *Geometry of Sets and Measures in Euclidean Spaces,* (Cambridge University Press, Cambridge UK, 1995).

P. Mattila, Personal Communication, comments on chapter 10 (2003).

S. Mori, H. Nishida and H. Yamada, *Optical Character Recognition* (John Wiley & Sons, Inc. NY, 1999).

A. Okabe, B. Boots and K. Sugihara, *Spatial Tessellations: Concepts and Applications of Voronoi Diagrams* (John Wiley and Sons, Chichester, UK, 1992).

P. Prusinkiewicz and A. Lindenmayer, *The Algorithmic Beauty of Plants* (Springer-Verlag, NY, 1990).

L. Santal6, *Integral Geometry and Geometric Probability* (Addison-Wesley, Reading, MA, 1976).

I. Saxl, *Stereology of Objects with Internal Structure* (Elsevier, Amsterdam, 1989).

I. Saxl, Personal Communication, comments on chapters 5 and 10 (2003).

C.S. Smith, "Grain Shapes and Other Metallurgical Applications of Topology" in *Metal Interfaces* (American Society for Metals, Cleveland, OH, 1952).

R.L. Smith, *Environmental Statistics,* Version 4.0 (Internet-published manuscript, Department of Statistics, University of North Carolina, Chapel Hill, 2001) also as presented at National Science Foundation-Conference Board for Mathematical Sciences Regional Conference: Environmental Statistics, June 25-29, University of Washington, Seattle, WA (2001).

D. Stoyan, W.S. Kendall and J. Mecke, *Stochastic Geometry and Its Applications, 2nd Edition* (John Wiley & Sons, NY 1995).

D. Stoyan and S. Stoyan, *Fractals, Random Shapes and Point Fields* (John Wiley & Sons, NY 1994).

D. L. Turcotte, *Fractals and Chaos in Geology and Geophysics, 2nd Edition* (Cambridge University Press, Cambridge, UK, 1997).

U. Zaehle, "Sets and Measures of Fractional Dimension", *Elektronische Informationsverarbeitung und Kybernetike,* **20** (1984) 261-269.

8. NON-PROBLEMS

All but one of these were garnered from handouts prepared by John Hilliard who titled them "non-assignments".

1) Smith [1952] has shown that for a finite array of cells the average number of sides per cell is given exactly by:

$$\overline{S} = \frac{6 - \left(6 + C_1' - C_2'\right)}{P} \tag{10.58P}$$

in which C_1' is the number of vertices on the boundary of the array that belong to only one cell and C_2' the number of vertices that are shared by two cells. You may verify this by yourself.

2) "The next time you visit the toilet in the Materials Science wing of Tech, contemplate the floor. You will note that it is made up of a mosaic of small tiles of several different colors. An attempt has been made to make the distribution of colored tiles apparently random. What test might you use to determine whether the distribution of tiles *is* random? Note that there is no single answer to this problem." - J.H.

3) Suppose we add a point at random to an existing distribution of points with an average spacing of $\overline{L}$. Define x as the distance in a specified direction between the added point and an existing point. As we have discussed, $\overline{x} = \overline{L}$ for a Poisson random dispersion and $\overline{x} = \overline{L}/2$ if the points are equally spaced. Since a Poisson random and a uniform distribution represent two extreme cases, would we be right in guessing that $\overline{L}/2$ and $\overline{L}$ are the lower and upper limits for $\overline{x}$ for *any* dispersion of points?

In order to answer this question it is necessary to determine the distribution $f(x)$ in terms of the distribution $f(L)$. {You are probably becoming somewhat weary of this topic. The reason I (J.H.) am setting the problem is that it provides an excellent example of the application of the techniques we considered in Chapter 6.} The first step is to determine the joint probability $f(x, L)dxdL$ that the added point will fall between an existing pair of points with a spacing of $L \pm dL/2$ and a distance $x \pm dx/2$ from the point on the left hand side. There is always a left hand side. Integrating with respect to L (Be careful about the limits!) yields the distribution $f(x)$. Following a procedure similar to that in Section 6.2.3 you can determine a relationship between the moments of the (marginal) x and L distributions. From this relationship it is easy to verify the earlier results and also whether the guess about the limiting value of $\overline{x}$ is correct.

4) Evaluate the following strategy for winning a jellybean counting contest. A cylindrical glass jar is filled with jellybeans. A contestant taking a straight edge to the jar counts the number of contiguous jellybeans along paths intersecting and parallel in the average to the straight edge. For each measurement, he records the count for the shortest of such paths He applies the straight edge to the diameter on the bottom of the jar and to the height on the side of the jar. He then estimates the number of jellybeans in the jar from averaged counts as

$$N = \frac{\pi}{4} E\left(n_d^2\right) E(n_h) ,$$

where n_d^2 is the square of the number of jelly beans in paths intersecting a diametral probe line and n_h is the number of jellybeans in paths intersecting a probe line in the height direction. (This strategy won a jellybean counting contest in McKean County, PA.)

Appendix

ADDITIONAL TOPICS IN ANALYSIS

This appendix contains material that might be helpful to the reader in understanding the real analysis aspects of Chapter 3. It is also a digest of the terminology of real analysis, which is basic to modern geometry.

1. SYSTEMS OF NUMBERS

The system of numbers used in stereology and stochastic geometry is primarily the positive real number system although complex numbers may be encountered in transformations and similar calculations. We also specifically use integers. Many of the distributions we have already encountered in Chapter 2 have been discrete, they take only integer values. Often probabilities represent the ratio of one integer to another and frequently points on the real number line represent these probabilities. But such probabilities are rational numbers and therefore not dense in the number line. That is, each rational number does not have a neighborhood comprised entirely of rational numbers. When only rational numbers are integrated on the real line, it turns out that the value of the integral is zero. Nevertheless, this is commonly overlooked when integrating distribution functions. The Stieltjes integral, to be discussed, provides a means for circumnavigating this obstacle. Much of geometrical probability deals with continuous number spaces. These include the real line and its n-dimensional space, $\mathbf{R}^n$. Hence we will be working with three number systems, the integers, **Z**; the rational (or aliquot) numbers, **A**; and the real numbers, **R**. Overlooking the distinction between **A** and **R** is common because Stieltjes integration is assumed in the context of probability theory. A number system includes at least by implication, an arithmetic to relate numbers to each other. With the advent of set theory the weight codification of these arithmetic relations has been transferred from the concept of number system itself to its related set theoretic concepts such as group and field.

The relationship between the rational and real number systems can be sketched through Dedekind cuts. This is a good introduction to the concept of density since the rational numbers are *everywhere dense* among real numbers. This property is required for the formation of a separable metric space. If a space were not separable, the terms “open” and “closed” set would not conform to our usual understanding. Sets of measure zero that are non-null are important to the development of measure theory. They are easily realized in separable metric spaces and allow an arbitrary refinement in any measured value that conforms to our everyday experience with measuring things. Because sets in a separable space can be measured with arbitrary precision, they can be similarly divided up with precision. We see that this creates the capability for them to be partitioned into a countable number of measurable disjoint subsets. In such cases the measure of the whole is the sum of the measures of the parts. This useful property makes sets measurable. We explore it here as in Chapter 3 from several angles.

1.1 Dedekind Cuts

Advances in set theory have pushed the theory of cuts into obsolescence, but it is useful precisely because it does not require much understanding of set theory. (See Rudin [1953].) There is a "chicken and egg" problem associated with the number systems since they existed prior to set theory but are now viewed as being the consequence of it. This situation perpetuates a subtle confusion that is resolved when number systems are viewed as existing separately from set theory. The only numbers we use in actual computations are rational.

A set, A, of “rational numbers” is said to be a cut if:

(I) A contains at least one rational but not every rational.

(II) if $p \in A$ and $q<p$, then $q \in A$.

(III) A contains no largest rational.

Given this definition, it follows for rationals p and q that if $p \in A$ and $q \notin A$, then $q>p$. For this reason the members of A are called the lower numbers of A while the rationals not in A are called the upper numbers of A. A (rational) smallest upper number need not always exist but may exist for certain cuts: we may define a cut as containing those numbers less than a specific rational number, r, which then becomes the smallest upper number of A. The cut so defined is designated r^*.

The relative size or measure of cuts is determined such that $A>B$ if there is a number contained in A not contained in B. Cuts can be ordered because if $A<B$ and $B<C$, $A<C$. Equality between cuts takes place when neither contains a number not contained in the other. Equality is thus identity for cuts.

1.2 Arithmetic of Cuts

The addition of cuts is defined as follows. The sum, C, of two cuts, A and B, is the set of all rationals, $p+q$ where $p \in A$ and $q \in B$. This set is also a cut. The null cut has a smallest upper number, zero, and is designated 0*. An interesting consequence of the definition of addition of cuts is that $A+0^* = A$ even when A contains only negative numbers. Addition is commutative ($A+B = B+A$) as well as associative ($A+[B+C] = [A+B]+C$) and the set of cuts forms what is called an Abelian (commutative) group under the operation of addition. Multiplication of cuts having positive smallest upper numbers is similarly defined. Subtraction follows from the operation of negation. There is a unique B such that $A+B = 0^*$. That $B = -A$. The absolute value of a cut, A, is now defined as A if A's smallest upper number is greater than 0 and $-A$ if A's smallest upper is less than zero. Multiplication of cuts having either positive or negative smallest upper numbers is defined using the absolute value to match the conventional rules of multiplication of rational numbers. The arithmetic of cuts is in fact the same as that of rational numbers. If we consider only those cuts for which a (rational) smallest upper number exists, the set of rational numbers can be identified with the set of (rational) cuts.

1.3 Gaps in the System of Rational Numbers and Real Numbers

Theorem: If A and B are cuts, and $A<B$, then there is a cut r* with rational smallest upper number, r, such that $A<r^*<B$.

The proof is that if $A<B$ there is a rational, p, such that $p \in B$ and $p \notin A$. Choose $r>p$ such that $r \in B$. Since $r \in B$ and $r \notin r^*$, then $r^*<B$. Since $p \in r^*$ and $p \notin A$, $A<r^*$. While cuts contain only rational numbers, those with a rational smallest upper number are termed "rational cuts". We can identify these with the rational numbers since they are associated with their rational smallest upper number. From this we see that between any two rational numbers, another rational number can always be inserted. This means that they are everywhere dense among the reals. We may call the set of cuts the set of real numbers. Those with smallest upper numbers as we have defined we may call the rational numbers. The rest we may call irrational numbers since their smallest upper numbers are not rational. (tricky!) It follows that irrational numbers are also everywhere dense among the reals.

Dedekind's Theorem: Let A and B be sets of real numbers such that these requirements are met:

(1) every real number is either in A or in B;

(2) no real number is in A and B;

(3) neither A nor B are empty sets
(4) if $a \in A$ and $b \in B$, then $a<b$.

Then there is one and only one real number c such that $a \leq c$ for all $a \in A$ and $c \leq b$ for all $b \in B$. The proof of this is fairly long and can be found in standard texts. The point is that the real number system is free of gaps unlike the rational number system.

2. ADDITIONAL MATERIAL ON SETS

In Chapter 3 we reviewed the basic operations of set theory. We continue through backtracking a little in order to fill in gaps. The phrase "Cartesian product" needs some clarification. Usually the term is synonymous with *direct product* and means set of all possible pairs where one member of the pair is taken from each set. Consider the case of the product of the real line with itself. If we for instance have identified two quantities on the real line, x and y, the direct product is made up of: (x,x), (x,y), (y,x) and (y,y). If we define a distance in this space then it becomes a metric space. Described in terms of opens sets, topological spaces can be constructed on it [Kolmogorov and Fomin, 1975] p. 78. Although we postpone defining a linear space, we will use some common vector notation. Boldface type will be used to designate vectors. $\mathbf{R}^n$ is intuitively a *continuous* space as affirmed by Dedekind's Theorem in section A.1.3. By continuity we mean that there are points of $\mathbf{R}^n$ arbitrarily close to any given point, that a line joining two points can be divided in to arbitrarily many pieces each of which joins two points of $\mathbf{R}^n$. (This would not be true of $\mathbf{A}^n$.) These notions contrast with the properties of an integer lattice defined by an n-tuple if integers such as $(i_1, i_2, \ldots i_n)$ where i_i is an integer. Consider the Euclidean distance function,

$$d(\mathbf{x},\mathbf{y}) = \sqrt{(x_1 - y_1)^2 + (x_2 - y_2)^2 + \ldots + (x_n - y_n)^2} \, ,$$

which measures the separation between two points $\mathbf{x}$ and $\mathbf{y}$. We may select a point $\mathbf{x}$ and refer to the set of all points which are a distance d or less from it. This we can call a *neighborhood*, $N_r(\mathbf{x})$, by setting $d = r$. By considering very small neighborhoods, the continuity of space can be examined. A set of points is called *discrete,* if each point has a neighborhood that contains no other points. A set of points, S, is *open* if every point has a neighborhood entirely contained in S. This does not exclude the possibility of also having neighborhood not entirely contained in S. A set of discrete points is not an open set. But, an open interval on the real line is also an open set of points. However, the set $S = \{x \mid a \leq x < b\}$ is not open since a, the *limit point,* does not have a neighborhood entirely contained in S. The capacity for infinite subdivision of $\mathbf{R}^n$ denotes its continuity. More precisely, any two points of $\mathbf{R}^n$ have neighborhoods that do not *intersect.* Intersection means having points in common. Reviewing Dedekind's Theorem, we see that if we have a closed interval along a line, the contiguous *non-intersecting* interval must be at least partially open. It follows that two limit points on a line cannot be contiguous. The set property of having non-intersecting neighborhoods for all points is called the *Hausdorff property* of $\mathbf{R}^n$ and defines the space as *separable.* (A weaker property is that of being a *normal* space, one in which given a pair of points a neighborhood can be constructed for one that does not include the other point.) Not all spaces have this property.

Through using the distance function, or *metric* (A metric is a distance function. Not all metrics are the same as the one just given.), we have induced a *topology* on $\mathbf{R}^n$. The funny thing about this is that distance is of no great significance in the study of topological spaces since it is not a topological invariant. But "openness" and "neighborhood" are and a metric helps define these. Topological spaces themselves are defined as follows. Let A be a set and $\{U\}$ be a system of open subsets. $\{U\}$ is a topology if

(i) The null set and A both belong to $\{U\}$, (This is why $\varnothing$ is defined as open.)

(ii) The intersection of two (hence any finite number of) members U_1 and U_2 belongs to $\{U\}$,

(iii) The union of any number of members of $\{U\}$ belongs to $\{U\}$.

As a consequence of this induced topology, $\mathbf{R}^n$ is defined as being open. We might consider an elliptical rather than a circular distance function, e.g. [Schutz, 1980] p.3 and test to see whether this affects (i), (ii) or (iii). Since an open set under the circular Euclidean metric will still be open under an elliptical metric, the change has no effect on the topology.

Topology is concerned about the subject of mapping that was discussed in Chapter 3. A mapping, f: $M \rightarrow N$, is continuous at $x \in M$ if any open set of N containing $f(x)$ contains the image of an open set $\subset M$ which contains x. We may consider these open sets to be neighborhoods. Making the neighborhood of $f(x)$ in N arbitrarily small requires for continuity that a corresponding neighborhood of x exist in M. This is essentially the epsilon-delta argument used to define the continuity of functions. A mapping is continuous on M only if it is continuous at every point of M. In other words, it is continuous if and only if the inverse image of every open set in N is open in M. Continuous mappings on $\mathbf{R}^n$ may be differentiable in the same fashion as are functions of several variables. If $f(x_1, x_2, x_3 \dots x_n)$ defined on some open region, S, in $\mathbf{R}^n$ has continuous partial derivatives up to order m on S, the mapping is said to be differentiable of class C^m. From this definition C^0 denotes a continuous function and C^∞ denotes a function all of whose derivatives exist. Since a function, which has all its derivatives, may still not be capable of being expanded in a local Taylor's series (e.g. $\exp(-1/x^2)$), a special notation has been developed for those which can. Thus the class of analytic functions is denoted C^{ω}.

If f is a 1-1 map of an open set M of $\mathbf{R}^n$ onto another open set N also of $\mathbf{R}^n$ it is written vectorially as $\mathbf{y} = \mathbf{f}(\mathbf{x}) : y_i = f_i(x_1, x_2,... x_n)$. If the functions, f_i, are all C^k -differentiable then the map, $\mathbf{f}$, is also C^k -differentiable. If the map is C^1 -differentiable then the Jacobean [see section 2.3.5] matrix $J = |\partial(f_1, f_2,...f_n)/\partial(x_1, x_2,...x_n)|$ exists and if it is non-zero at some point, $\mathbf{x}$, the *inverse function theorem* assures us that the map is indeed 1-1 over some neighborhood of $\mathbf{x}$. Extending the result of Section 2.3.5, if a function $g(x_1, x_2, ... x_n)$ is mapped into a function $g^T(y_1, y_2, ... y_n)$ by the rule $g^T(\mathbf{y}) = g(\mathbf{x})$ where $\mathbf{y} = \mathrm{f}(\mathbf{x})$, and f maps region M into N, then we say the integral of g over M equals the integral of $g^T J$ over N. Comparing this result with that of section 2.3.5, we see that the Jacobian is defined differently in the two contexts because the roles of the variables have been switched. The Jacobian shown here looks "customary" because the y-values appear in the numerator and the x-values appear in the denominator.

$$\int_M g(x_1, x_2,...x_n)dx_1 dx_2...dx_n = \int_N g^T(y_1, y_2,...y_n)J\,dy_1 dy_2...dy_n \tag{A.1}$$

Since g and g^T have the same value at corresponding points of the mapping, the volume element abbreviated $\mathbf{dx}$ is $1/J$ times $\mathbf{dy}$. The Jacobian in Eqn. (A.1) could be written $J(y_i/x_i)$. That in section 2.3.5 would then be written $J(x_i/y_i)$. In order for the volume element representation of the two mappings to be consistent,

$$J\left(\frac{\mathbf{x}}{\mathbf{y}}\right)\cdot J\left(\frac{\mathbf{y}}{\mathbf{x}}\right) = 1\ . \tag{A.2}$$

(Reminder: The bold face denotes a vector and the division sign is an abbreviation.)

This is true and yet there is no theorem that states that the determinant of the reciprocal is equal to the reciprocal of the determinant (because it is not so). The proof lies in the properties of partial derivatives and serves as a reminder that algebraic cancellations that are correct for total derivatives may not be so for partial derivatives.

Whether or not a map is 1-1, the inverse image of the union of two sets is the union of the inverse images, and the same holds for the intersections of sets. However, the image of the intersection of two sets may not equal the intersection of the images. This is the case when a projection is made.

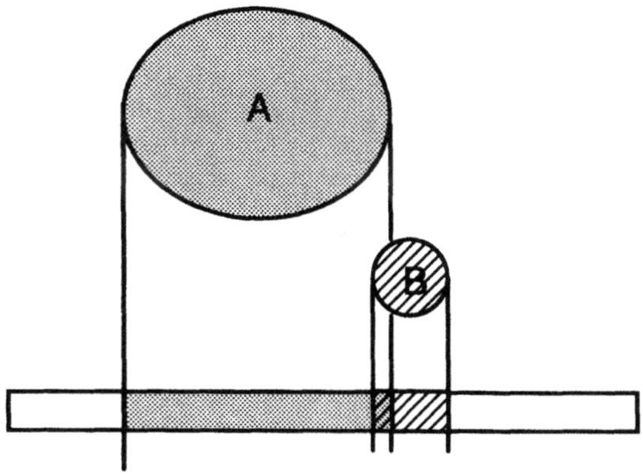

Figure A.1: Many to one projection

In this example two sets, *A* and *B*, are mapped onto the real line. The intersection of the two sets is ∅, but the intersection of their images, the hatched and shaded portion, is not.

2.1 Classes and Equivalence

Arrangement of a set into pairwise disjoint (the intersection of any pair is ∅) subsets plays an important role in geometric measure problems among others. For example a subspace of $\mathbf{R}^2$ can be decomposed into lines parallel to the x_1-axis. The swimmers in a race may be partitioned into different age groups. Any such partition of a group, call it *M*, is *a partition of M into classes*. In order that the classes be logically disjoint, three conditions must be met by any *relation* between members of the same class which acts a class-defining criterion. We call the relation "*R*" and the class member "*a*" (and the class, "*A*" for convenience; we need not refer to it.).

(1) Reflexivity: *aRa* for every $a \in M$.

(2) Symmetry: If *aRb* then *bRa*.

(3) Transivity: If *aRb* and *bRc* then *aRc*.

If a relation meets these three criteria, it is called an *equivalence relation*. This is the only type of relation that can be used for assigning classes. Every partition establishes a binary relation between class members which means "*a* belongs to the same class as *b*". Were one to try to partition the set of real numbers by assigning *b* to the same class as a iff (if and only if) $b > a$ we would run into trouble. If $b>a$, *b* must be assigned to the same class as a but then a cannot be assigned to the same class as *b* since $a<b$.

In the theory of symmetry groups, there exist *conjugacy classes*. The elements of these classes are related by a *similarity transformation* that involves three group members. A similarity transformation relates a third group member to two others as follows: $a^{-1}ba = c$ In this application, these group members are typically rotations and reflections. For the symmetry group of the square, C_4, the division into conjugacy classes is as thus,

$$(E)(C_4, C_4^3)(C_4^2)(m_x, m_y)(\sigma_u, \sigma_v).$$

E is the identity element. C_4 is a quarter rotation, C_4^n is an n-quarter rotation, m_x is mirroring in the *x*-axis and σ_v is a mirroring about a principal diagonal. Although it is not directly apparent, there are equivalence relations for the various classes implied by the similarity transformation.

There is also a different concept of equivalence that can be applied to sets: two sets are equivalent if there is a 1-1 correspondence between their elements. This concept of equivalence is applicable to both finite and infinite sets. It states that two sets are equal if they have the same number of elements. If there is equivalence between sets in this sense, this equivalence is obviously reflexive symmetric and transitive as described for an equivalence relation. Note that two sets may be *equivalent* without being equal. Equivalence of two sets, *M* and *N*, is written *M~N*.

2.2 Countability and Power

The simplest infinite set is the set $\mathbf{Z}_+$ of all positive integers. An infinite set is called *countable* if its elements can be numbered, that is, put in one-to-one correspondence with $\mathbf{Z}_+$. If it is not possible to do this, the set is termed *uncountable.* The set **Z** is countable since for every member of **Z** identified, another member of $\mathbf{Z}_+$ can always be found to be its mate. Doing this in an organized way might be to have the sequence of negative integers mapped onto the sequence of evens and the sequence of positive integers mapped onto the sequence of odds. Obviously there are twice as many integers (positive and negative) as there are positive integers. But, conceptually counting is done one element at a time. There is no possibility of ever running out of integers. On this subject there is a quote from the *Ishavasya Upanishad*, "Taking away this infinity from that infinity, yet infinity remains behind." (This suggests why Eqn.(8.70) takes the form of a ratio rather than a difference to discount the effect of the excess information contained in arbitrary redundancy.) Every subset of a countable set is countable as is the union of two or more countable sets. Also every infinite set has a countable subset.

We may use countability to test the equivalence of sets. Since every infinite set, countable or not, contains a countable subset, we make A be the countable subset of a set M. We can partition A into two more countable subsets, A_1 and A_2. Noting that the word *countable* is only applied to infinite sets, we can, as before, assign the members of a set to one of its proper subsets through countability, $A \to A_1$. Consider next the set $M \backslash A$. It has elements, $e \in (M \backslash A)$. The union, $\{e\} \cup A = M$. The union, $\{e\} \cup A_1$ differs from the former only by a different choice of countable subset. Since a 1-1 correspondence can be made between any two countable sets and $\{e\}$ is obviously in a 1-1 correspondence with itself, there exists a 1-1 correspondence between $\{e\} \cup A$ and $\{e\} \cup A_1$. But the former set is M and the latter is its proper subset $M \backslash A_2$. Hence, every infinite set is equivalent to one of its proper subsets.

In deriving of the above theorem, care was taken to avoid specifying whether M was countable or not. Not all sets are countable. It turns out that the set of real numbers in the closed interval [0,1] is *uncountable.* At this point we have three types of sets: finite, countable and uncountable. If M in the previous derivation were the real interval [0,1], then we could see that a countable set can be a subset of an uncountable one as well as a finite set can be a subset of a countable one. For finite sets, the *power* of the set is the number of elements it has. Georg Cantor has extended this concept to infinite sets by giving the set $\mathbf{Z}_+$ the power of aleph-naught, $\aleph_0$. (Aleph is a letter in the Hebrew alphabet. It is a silent letter.) The power of the real interval [0,1] is that of the *continuum.* It's power is designated $\aleph$. Are there powers between $\aleph_0$ and $\aleph$? Axiomatic set theory, the axiomatized version of the *naive* set theory of Cantor -- the language of which we are now studying -- leaves this question open to definition. When such powers are defined, like with imaginary numbers, some properties and applications should follow. The important thing is that sets that are equivalent have the same power. We have seen that sets can be ordered according to their powers. This ordering is the basis for a well-known theorem: the Cantor-Bernstein theorem.

The Cantor-Bernstein theorem states that given two sets A and B, *if A has a subset equivalent to B and B has a subset equivalent to A,* then the two sets are equivalent. Nothing is said about whether these are proper subsets. Note that they cannot be proper if the sets are finite. The two sets in question must be of the same power if they meet the Cantor-Bernstein test for equivalence, since each set has a subset that is in some sense $\leq$ the other set. Further exploration of ordering is done through defining an ordering relation. Call this relation R. A set, M, is said to be *partially ordered* if there exists a relation acting upon it which meets these three criteria:

(1) R is reflexive. (aRa for every $a \in M$).

(2) R is transitive. (if aRb and bRc, then aRc).

(3) R is *antisymmetric* in the special sense that if aRb and bRa, than $a = b$.

It is not required that R be capable of being applied to all elements of M. Were R to apply to all elements, the set would be *ordered.* (Every ordered set is partially ordered.) The relation defined by the sign $\leq$ fulfills all the requirements for an ordering relation. When it can be applied to two elements, they are said to be *comparable* otherwise they are *incomparable.* Furthermore, an element, a, of a partially ordered set is *maximal,* when $a \leq b$ for all b implies $a = b$. For the element, b, the same relation sets b as *minimal* when it implies $b = a$ for all a. The partial ordering may not result from size as we normally think of it. Thus setting R to mean " is evenly divisible by" and applied to the set of integers, this makes every prime number (greater than one) a minimal element. A mapping that preserves the order of a partially ordered set is called an *isomorphism* and two sets so mapped to each other are *isomorphic.* Isomorphism is an equivalence relation. A set is termed *well-ordered* if every subset not $\varnothing$ has a smallest element. Every set can be well ordered; this is called *Zermello's theorem.*

2.3 The Axiom of Choice

Although this axiom is not often used in stereology, it is so fundamental to set theory that some mention should be made of it. The proof of Zermello's theorem requires it and a student cannot be considered educated without exposure to it. It simply states that *for every subset, A, of a larger set, M, there exists a choosing function, f, such that f denotes an element of A.* The function, f, itself may be dependent upon the choice of A. This function implies some sort of order. We now define a *chain* as a subset of a partially ordered set for which all elements are comparable. A chain is *maximal* when it is not a proper subset of any other chain in the same set.

Hausdorff's Maximal Principle is a variant of the axiom of choice and states that *every chain in a partially ordered set is contained in a maximal chain of that set.* Consider the union of the set of integers between one and one hundred with the set of dust balls in a ventilator. (The latter is a fuzzy set.) Elements of the former set are presumably not comparable with those of the latter yet both sets are capable of being ordered. There are therefore two maximal chains in the union, those of numbers and those of dust. Since every set can be well-ordered, the two maximal chains could be reduced to one. But this would require a decision about where integers rank with respect to dust. Hence we see why it is "a maximal chain" and not "the maximal chain" in Hausdorff's Principle.

2.4 Metric Spaces

We have already commented on metric spaces here and in Chapter 3. We continue with some further details. Note that in the case of an isometry, the metrics used in the from and to spaces need not be the same. Rather if $f : x \mapsto y$ then if $\rho(x_1, x_2) = \rho'(y_1, y_2)$, where ρ is the metric in the from space and ρ' is the metric in the to space, the mapping is an isometry.

Reprise: if every neighborhood of a point contains infinitely many points of M, that point is a *limit point* of M. It need not belong to M. For continua, limit points and contact points are often interchangeable concepts. Every contact point is either a limit point or an isolated point. Sequence and convergence are concepts so similar here to those of elementary calculus that redefining them in this context is not necessary. We note that while a necessary and sufficient condition for a point x to be a limit point of a set M is that there exist a sequence $\{x_n\}$ of distinct points in M converging to x, removing the word "distinct" degrades the type of point from a limit point down to a contact point.

2.5 Open and Closed Sets in Metric Spaces

We now return to this subject discussed earlier with the intention of refining its definitions. This may seem redundant. But, in my experience (L.L.) as an educator, crisscrossing back and forth over the same material in different ways does a lot to clarify its meaning. So, a set that coincides with its own closure is said to be *closed*. A set whose closure contains another set is said to be dense in it. Let A and B be two subsets of a metric space, R, the parent set. A is *dense* in B if the closure of A, $[A]$, contains B. Moreover if $[A]$ contains R, A is *everywhere dense* in R (viz. if the closure of the subset A contains the parent set R.). The set of rational points is dense in the real line, $\mathbf{R}^1$. This is because the closure of the set of rational points includes the set of irrational points and hence the entire line. If a metric space has a countable everywhere dense subset, it is said to be *separable*.

The intersection of an arbitrary number of closed sets is always closed as is the union of a finite number of closed sets. But, the difference of two closed non-disjoint sets will not be closed. Recall that $\varnothing$ is open. A set is defined as *open* if all its points are interior points where an *interior point* of a set A is one which has a neighborhood entirely containing points of A. Clearly, a half-open interval is neither open nor closed. Similar to the case of closed sets, the union of an arbitrary number of open sets is open; the intersection of a finite number of open sets is closed. This is an application

Eqns. (3.1a,b) Thus we note that an infinite union of closed sets may be open while an infinite intersection of open sets may be closed.

Since the real line has the power of the continuum, we may expect that theorems applicable to it may have application, through mapping, to higher dimensional spaces. Consider an open interval (a,z) on the real line. If x is an arbitrary point on (a,z), then there is at least one open interval containing x and contained in (a,z). Let X be the union of all such intervals. Let $a = \inf X$ and $z = \sup X$, where *inf* means infimum or greatest lower bound and *sup* means supremum or least upper bound. It is obvious that X is contained (a,z). Yet, if we choose any point distinct from x on the interval between a and x, call it c, there is another point b which belongs to X which is between a and c because any closed interval in (a,z) containing x belongs to X. Hence X contains (a,b). By the Cantor-Bernstein theorem $X = (a,z)$. With similar reasoning it is possible to show that closed intervals form closed sets and that single-element sets are closed.

The Cantor set is sometimes referred to as a Cantor dust. (Dorothy Parker's proposed epitaph: "Pardon my dust." might apply to this one of Georg Cantor's constructions as well.) The contemplation of a *Cantor dust* is entertaining as well as being a topic embedded in the language of set theory and frequently referenced in contemporary geometrical literature. We begin by defining a closed interval, $F_0 = [0,1]$, which was the continuum of Cantor. Then we commence with a process that begins by deleting the open interval (1/3,2/3). This leaves two closed intervals. The union of these intervals we call F_1. The next step of the process is to remove the open intervals corresponding to the middle third of each interval in F_1. This forms F_2. We continue this process to form a sequence F_0, F_1, F_2, F_3, ...F_n of closed sets. Furthermore since $F_1 \supset F_2 \supset F_3 \supset ... F_n \supset ...$, the sequence is

decreasing. The *Cantor set* is defined as $F = \bigcap_{n=0}^{\infty} F_n$. Since it is the

intersection of a number of closed sets, it is closed. By its construction, F contains 0,1,1/3,2/3,1/9,2/9,7/9,8/9 etc. which are the end points of the deleted intervals. But, it also contains many other points. We may imagine a series to represent a number, $x \in F$ using the pattern of the sequence we have just written. It takes the form

$$x = \sum_{n=1}^{\infty} \left[\frac{a_n}{3^n} \right], \qquad \text{(A.3)}$$

where $a_n \in \{0,1,2\}$ which can be reduced to only two values by taking limits.

The Cantor set has some odd properties. The sum of the lengths of the deleted intervals equals 1, but nevertheless the Cantor set has the power of the continuum. These properties parallel those of the rational number system, $\mathbf{A}^1$. The rational numbers have the power of the continuum but when integrated over any closed interval of the real line, yield zero as the value of the integral. The proof that F has the power of the continuum is interesting to note. Since the values of a_n are limited to two, by substituting 2^n for 3^n in Eqn.(A.3), it is possible to map Eqn.(A.3) term-by-term onto the binary number system. Since any number including irrationals can be represented in binary, F has the power of the continuum.

On a more general level we may consider the completeness of a metric space itself as being analogous to the closure of a set. A metric space, M, is *complete* if every Cauchy sequence in M converges in to a point in M. This is the familiar Cauchy sequence of calculus except that the metric may not necessarily be Euclidean. (For every $\varepsilon > 0$, there exists an integer, I, such that the distance between points in the defining sequence is less than ε for all terms indexed with numbers greater than I.) As expected Euclidean $\mathbf{R}^n$ space is *complete*. This can be proved by looking for voids using what is called the *nested sphere* theorem. It says that a metric space is complete iff every imaginable sequence of nested closed spheres such that their radii $\to 0$ as the index number of the sequence's terms $\to \infty$. has a nonempty intersection. If there were a void of missing points within a sphere, the shrinking down of some particular sequence of nested spheres would put a sphere inside a void giving a $\varnothing$ intersection. The relationship between completion and closure is clarified by saying that a metric space R^* is a *completion* of a metric space R if R is everywhere dense in R^*. If R is complete, $R^* = R$. The space of all real numbers, for example, is the completion of the space of all rational numbers.

A geometrical example of the properties of opens sets introduces the important subject of connectedness. An open set is said to be *connected* if any two of its points can be joined by a curve made up of a finite number of straight line segments. The union of two open disks in $\mathbf{R}^n$,

$$\{x, y \mid x^2 + y^2 < 1\} \cup \{x, y \mid (x-2)^2 + y^2 < 1\},$$

has a contact point at (1,0) since every neighborhood contains at least two points of the set. But the set is not connected because (1,0) does not belong to it. There is a cleavage in the neighborhood of (1,0).

2.6 Topological Spaces

The idea of a topology was introduced at the beginning of Section A.2. We there introduced a metric and defined three properties for open sets, called these a topology and then discarded the metric. It may be clichéd to refer to topology as rubber sheet geometry, but what this colloquial definition offers is the idea that the details of the metric may be ignored; they are outside the definition of topological space. Once again, a topology is a system of open sets including ∅ which is closed under any number of unions and any finite number of intersections. A metric space can be defined as a pair (X,ρ) where X is a set and rho is the metric; a topological space is similarly defined as a pair (X,τ) where X is a set and tau is a topology. This gives us a way of looking at a set X from two different vantage points. The pair (X,τ) may be abbreviated as simply T. In T the null set ∅ is here defined as *closed* in contradistinction to its definition in a metric space. The topology is made up of subsets G of X and the system of which we will call τ. The difference $X\backslash G$ is not ∅ and is closed since G is open..

In spite of there being no metric, the concepts of neighborhood, contact point and limit point, acquired from metric spaces, are retained in a topological space. And, every metric space is a topological space since, as we have seen, in a metric space the union of any number of open sets is still open. Can a set be both open and closed at the same time? In a topological space the answer is yes. It must be both composed entirely of points having neighborhoods entirely contained in the set and be a set containing all its limit points, those which have neighborhoods always containing some points of the set. Apart from the trivial case of ∅, which is defined as both open and closed, special metrics can be contrived to meet these requirements on $\mathbf{R}^n$. An example is

$$\rho(x,y) = \begin{array}{r|l} 1 & x = y \\ 0 & x \neq y \end{array}.$$

You may quite correctly argue that this cooked-up metric does not meet requirement (1) for being a good metric given in Section 3.3.1. And, so it is. Metrics that do not meet requirement (1) are called *pseudometrics* or *écarts*. Also to accomplish this end, the topological space may be composed only of two subsets, its entirety and ∅. (By the way, this is called the *trivial* or *indiscrete* topology.) Specifying the open subsets in a topology is a matter of choice and is the definition of a particular topology.

The problem of choosing which open subsets to specify in a topology, can be approached in an organized way. Intuitively, the choice of subsets must span the space of the topology. A system or family $\mathcal{T}$ of open subsets of a topological space is a *base or basis* for that space if every open set in T can be represented as a union of sets in $\mathcal{J}$ We note at this point that one topology τ on the same set may be *stronger* than another τ', if τ contains τ'. A base need not be the strongest possible topology. In fact a base is not necessarily a topology at all; it induces a topology. Consider the set of all open cubes as a base for $\mathbf{R}^3$ and then consider the set of all open spheres as a topology in $\mathbf{R}^3$. Any sphere can be formed through an infinite union of cubes and visa versa. Yet neither set contains the elements of the other. Either can be a base but neither a topology since neither set is closed under the taking of unions. It can be shown that a system T of open sets, S_i is base for a topological space T iff for any point x belonging to open set $U \subset T$, there is an S_i such that $x \in S_i \subset T$. That is the base must cover all points of T.

2.7 Open and Closed Sets in Topological Spaces

The definitions of open and closed sets in a topological space are consistent with those of a metric space but not defined in the same way. In a topological space a set N is a neighborhood (*τ-neighborhood*) of a point, x, iff N contains an open set to which x belongs. While a neighborhood of a point need not be an open set, an open set is a neighborhood of each of its points. Somewhat tautologically, a topological set is defined as open *iff it contains a neighborhood of each of its points.* On the other hand a subset A of a topological space (X,τ) is closed if its relative complement X-A is open. A set is open if its complement is closed and visa versa. According to the DeMorgan formulas, Eqns. (3.1ab) the union of the complements of a family of sets is the complement of the intersections of that family and the intersection on the complements of the family is the complement of their union. It follows that for consistency the finite unions and intersections of a system of closed sets remains closed. In fact a *family of closed sets* may actually be defined as composed of subfamilies closed under the operations of union and intersection relative to some topology.

Since a set is closed iff it contains its limit points, the concept of limit point needs consideration apart from its definition in a metric space. In a topological sense a point is a limit point in a topological space if every neighborhood of it contains points of the space other than itself. One may say that the neighborhood of a point x intersects a subset A iff x is a point of A or a limit point of A. Hence, as in a metric space, a subset of a topological space is closed iff it contains the set of all its limit points. We note in passing that in a topological space (X,τ) the space itself and $\varnothing$ are both open and closed sets. The concept of closure is essentially the same in both metric and topological spaces and the closure of a topological set is the union of that set with all its limit points. Similar to the rule for a metric space, in a topological space (X,τ) a set A is everywhere dense if the closure of A is X.

2.8 Countable Bases

A topology with a countable base is simply one whose base consists of a countable system of sets. The sets composing the base may themselves be uncountable sets. As with a metric space, a topological space $T = (X,\tau)$ is separable iff there is a countable subset which is everywhere dense in X. It follows that a topology that has a countable base is separable. On the other hand, the space of all bounded infinite sequences

(with the metric $\rho = \sup_{(k)} |x_k - y_k|$)

is not separable because there is no countable everywhere dense subset in that space [Kolmogorov and Fomin, 1975] p.49 . A related and very important concept is that of a cover. A system of sets $\mathcal{S}$ with members S_i is a *cover* of the topological space T

if $X = \bigcup_{(i)} S_i$.

A *subcover* is any subset (or subfamily) of $\mathcal{S}$ which is also a cover of T. That a finite subcover exists is the subject of the *Heine-Borel* Theorem.

2.9 Compactness and Convexity

Compactness is an important property not because it is new and hence unfamiliar but because it isn't. The features analyzed in stereology are for the purpose of derivations assumed to be compact. A torus or a sphere in three-dimensional Euclidean space is contained in a finite portion of that space. A hyperboloid or a paraboloid is not. The former are compact. The latter are not. Topology seeks to define this property of compactness in terms that depend of the set of points comprising the compact object and not upon the space that surrounds it. We begin with the well-known *Heine-Borel theorem*:

If X is a closed bounded subspace of Euclidean n-space, then every open covering U of X contains a countable open covering viz. from the open sets U which cover X, and there can be selected a finite number of these (subcover) which also cover X.

A topological space T is *compact* if every open cover of T has a countable subcover, since that makes it closed and bounded. It follows directly that a compact subspace of a metric space is bounded. Furthermore, if $f:T_1 \rightarrow T_2$ is a continuous mapping of a compact topological space T_1 onto another topological space T_2, then T_2 is also a compact space. Also the product space of topological spaces T_1 and T_2 is compact only if T_1 and T_2 are compact. A topological space T is termed *countably compact* if every infinite subset has at least one limit point in T. This is a relatively weak form of compactness. Every compact set is countably compact, but the converse is not true. See [Kolmogorov and Fomin, 1975] p. 97.

In a metric space compactness is connected to the concept of *total boundedness*. To define total boundedness we must first define an epsilon net. A set N in a metric space R is an ε-net for another set M in the same space if for every $x \in M$ there is an $n \in N$ such that $\rho(x,n) \leq \varepsilon$ where ε is some positive number. Given this definition, if M is a subset of a metric space R and M has a finite ε-net for every $\varepsilon < 0$, M is totally bounded. It follows that every compact metric space (and even every countably compact space) is totally bounded.

Convexity is a topological property of linear subspaces. We will postpone defining what a linear space is and for now just say that linear spaces are composed of lines which obey the customary rules of vector algebra with respect to commutativity, associativity and so forth. The set of all ordered n-tuples obeying the laws of vector algebra is a linear space. So also is the real line with its customary arithmetic. The idea can be generalized to include function spaces. The set of all polynomials on the closed interval [a,b] can also be considered a linear space. It is a subset of all the continuous functions on [a,b] sometimes denoted $C_{[a,b]}$. The elements of either of these sets are an uncountable number of functions and consequently these spaces are infinite-dimensional. However the space of polynomials is contained in that of continuous functions and is therefore a *subspace*. Linear spaces may be real or complex.

A fundamental definition is that for the interior of a set belonging to a real linear space L. The *interior* of set M is the set of points belonging to M such that the point $\mathbf{x} + t\mathbf{y}$ belongs to M for all scalar $|t| < \varepsilon$, all $\mathbf{x} \in$ M and all $\mathbf{y} \in$ L. A set is said to be *convex* if whenever it contains two points, $\mathbf{x},\mathbf{y}$ it also contains all points of the line segment joining $\mathbf{x}$ and $\mathbf{y}$. A *convex linear subspace* is similarly defined in Euclidean space as one where if $\mathbf{x}$ and $\mathbf{y}$ are two points in the space; any point defined by $\mathbf{p} = \lambda\mathbf{x} + (1-\lambda)\mathbf{y}$ with λ being a positive scalar, also belongs to the same subspace. (The term *affine linear subspace* is also used.) And, a convex set is called a *convex body* if its interior is non-empty. Furthermore, if M is a convex set, so is its interior. The intersection of any number of convex sets is a convex set. Just as one may set a winter scene inside a glass ball, there is a convex set to contain any subset of a linear space. (With some needless loss of generality but an eye toward practicality we can restrict our thinking to bounded sets.) Intuition is then correct when we think that for each subset A there is a smallest convex set that contains it. This minimal convex set is called the *convex hull* of A. Geometrically convex solid objects in ordinary space are typical examples of convex sets. Convexity greatly simplifies stereological analysis and many results apply only to convex structures.

2.10 Convex Functionals

A *convex functional* is one which when defined on a real linear space obeys three properties.

(1) $p(\mathbf{x}) \geq 0$, it is non-negative.

(2) $p(\lambda\mathbf{x}) = \lambda p(\mathbf{x})$ for all $\lambda \geq 0$, it is linearly homogeneous.

(3) $p(\mathbf{x}+\mathbf{y}) \leq p(\mathbf{x}) + p(\mathbf{y})$, it obeys the triangle inequality.

Convex functions behave like the length or *norm* of a vector. Different norms can be chosen for vectors just as different metrics can be chosen for metric spaces. It is possible to use a norm to define a body centered on the origin. If p is a convex functional and κ is any positive number, then the set $H = \{\mathbf{x} : p(\mathbf{x}) \leq \kappa\}$ is a convex set. (This is what makes convex functionals "convex". Of particular interest in stereology are specific Minkowski functionals defined on sets.) If κ is finite then H is a convex body. If we set $\kappa = 1$, then every convex functional corresponds to a convex body in this way.

It is possible to apply a convex functional to map all points of a convex body centered on zero onto the real line. In particular we consider the non-integral Minkowsky functional, $p_H(x)$ of the body H.

$$p_H(\mathbf{x}) = \inf\left\{r : \frac{\mathbf{x}}{r} \in H, r > 0\right\} \tag{A.4}$$

The functional of Eqn.(A.4) is finite and convex. Since H is centered on zero (Two sets are *centered* if they have a non-null intersection. Hence, a set is centered on a point if it contains that point.) $\mathbf{x}/r$ must belong to H for some value of r sufficiently large. If H is a sphere of radius R, r is $|\mathbf{x}|/R$ for all $\mathbf{x}$.

2.11 The Hahn-Banach Theorem

This chapter is a kind of quick tour intended to help the reader "reload into memory" the fundamental language of real analysis that is essential to the understanding the language of post-classical stereology and stochastic geometry. The Hahn-Banach theorem is a sort of detour off the main path we have been following. It is a useful tool and a useful piece of terminology.

Let p be a finite convex functional on linear space L of which L' is a subspace. Let p' be another functional defined on L' only. Let $p'(x) \le p(x)$. Then there is a functional p' on all of L, which is bounded by p and equal to p' on L'.

This theorem is useful for proving that a particular functional is capable of being extended over an entire space. It can be extended if it is bounded by a convex functional.

2.12 Normed Linear Spaces

Let us look at what kinds of spaces we have studied. We have examined number systems, the real line and its Cartesian product space $\mathbf{R}^n$. We have added the notion of length and constructed metric spaces. This allowed us to conceptualize open sets, which we used to construct topological spaces. In a sense we were working backward because each space we constructed was a somewhat more abstract and less intuitive version of the one that preceded it. Linear spaces are very similar in concept to the product space $\mathbf{R}^n$ and normed linear spaces are very similar to metric spaces. In the general linear space we think of lines from the origin to points and sometimes call these vectors. In the basic definition of a linear space, which we now develop, we consider only the points themselves.

In an n-dimensional Euclidean space we consider a set $\{P_i\}$ consisting of $m+1$ points where $m<n$. We define the j-th coordinates of the i-th point to be x_{ij}. These points are said to be *linearly independent* if the equations for real a_i

$$\sum_{i=0}^{m} a_i x_{ij} = 0, \quad \sum_{i=0}^{m} a_i = 0 \tag{A.5a,b}$$

imply that all $a_i = 0$. In terms of common experience, two points determine a line in two or three dimensions, and three non-collinear points determine a plane in three dimensions. We can try to generalize this.

The equation of a line passing through (x_1,y_1), (x_2,y_2) in two dimensions can be written in determinant form,

$$\begin{vmatrix} x_1 & y_1 & 1 \\ x_2 & y_2 & 1 \\ x & y & 1 \end{vmatrix} = 0. \tag{A.6}$$

The determinant is partitioned into upper and lower, left and right submatrices. The upper left contains the intercepted points and the lower left contains the independent variables. Similarly the equation of a plane can be written in terms of three non-collinear points,

$$\begin{vmatrix} x_1 & y_1 & z_1 & 1 \\ x_2 & y_2 & z_2 & 1 \\ x_3 & y_3 & z_3 & 1 \\ \hline x & y & z & 1 \end{vmatrix} = 0 \,. \tag{A.7}$$

If we consider just the upper submatrix as a system of homogeneous equations in $n = 4$ unknowns, we see that it is of rank $r = 3$. Hence we have n-r or one solution of the form

$Ax + By + Cz = D$ (where A, B, C and D is the solution),

which passes through (x_1,y_1,z_1), (x_2,y_2,z_2), (x_3,y_3,z_3). If the rank of Eqn. (A.7) were $r = 2$, then there would be linear dependence such that there would be two such solutions corresponding to two planes. The two solutions would correspond to two planes and would be simultaneously satisfied at their intersection, which is a line. This can be generalized to n dimensions. Given that the m+1 points $\{P_i\}$ are linearly independent, then the m+1 by n+1 matrix whose elements are x_{ij} up to the n-th column and whose n-th column is $(1,1,1...)^{\mathrm{T}}$ is of rank m+1. Therefore there are n-m linearly independent solutions for y_j to the set of equations

$$\sum_{j=1}^{n} x_{ij}\, y_j = k \,. \tag{A.8}$$

We see that the number of independent points is equal to the rank of the corresponding set of homogeneous linear equations, Eqn. (A.8), where k is one of the variables.

A *linear subspace*, L is a subset of some space M such that if P and Q belong to L than so does $\lambda P + \mu Q$ where $\lambda + \mu = 1$. If λ and μ are both non-negative, the linear subspace is convex. A set of n linearly independent points is called a basis for an n-dimensional subspace. We can write the coordinates of any point in the subspace in terms of n coefficients of the coordinates of the n points. For example, in a two-dimensional subspace the coordinates of any point can be written $\lambda P + \mu Q$.

The points of a linear space have defined coordinates relative to some origin. In this generalized discussion, we have not payed much attention to the details of the coordinate system. In passing we may note that the origin of reference may be global or may be local with respect to some point otherwise specified. A set operation that varies significantly with the origin is called *reflection*. It is defined, $\breve{A} = -A = \{-\mathbf{x}:\mathbf{x} \in A\}$. Obviously this operation is only meaningful on a linear space or something similar. It is likewise possible under the same circumstances to multiply a set by a constant. The meaning of that operation is to multiply each coordinate of each point by the same constant. If $\breve{A} = A$ for some origin, the set A is *symmetric*.

A *norm* is closely identified with a metric. It is a functional, p, which obeys these rules.

(1) p is finite and convex;

(2) $p(x) = 0$ only if $\mathbf{x} = 0$;

(3) $p(\lambda x) = \lambda p(x)$ for all x in L.

The norm of x is designated $\|x\|$. In common vector space, the basis is orthonormal rather than just any three suitable points. In common vector space the norm is the Euclidean metric, Eqn. (3.4). A complete normed linear space (where completeness is the property that all Cauchy sequences converge to points of the space) is called a *Banach space*.

3. TOPICS CONCERNING MEASURE

Continuing the discussion of Section 3.3.3.2, let us define a sample space U as an interval on the real line (composed of numbers, x) and define a real valued function, $f(x)$. The $f(x) \geq 0$ for all x in U and

$$\int_U f(x)dx = 1.$$

We could define $P(A) = \int_A f(x)dx$

but might have problems with the existence of the integral since A or f may be very ill-behaved. For example, the Cantor dust of Section A.2.5 is not integrable. In order to insure integrability, we require (Lebesgue integration makes this requirement less stringent than it first appears) that f be *absolutely continuous* on some interval over which the integral is taken. A function f is absolutely continuous on [a,b] if given any $\varepsilon > 0$, there is a δ such that

$$\sum_{k=1}^{n} |f(b_k) - f(a_k)| < \varepsilon$$

for every system of pairwise disjoint subintervals $(a_k,b_k) \subset [a,b]$ of total length

$$\sum_{k=1}^{n} (b_k - a_k) < \delta .$$

Every function that is absolutely continuous meets the requirements of uniform continuity and bounded variation as well. This should be reassuring but these additional properties need not be discussed here. When f is absolutely continuous a probability measure can be defined as

$$P(A) = \int_A f(x)dx . \tag{A.9}$$

Probability measures integrate to unity as previously specified. Other measures need not.

3.1 Measures on B^{d}

A *measure* can be thought of as being a generalization of the probability measure previously defined. Measures apply to content as well as probability. A measure meets these criteria, which essentially restate the ones already given:

1) A measure is a real non-negative number associated with a set.
2) The measure of an empty set is zero.
3) Measures are σ-additive.

This list does not exhaust the requirements for defining "measure" as the term is customarily used and additional criteria will be introduced when specific measures are discussed. The σ-additivity criterion is met when the system of sets being measured belongs to a σ-algebra, viz. they can be partitioned into disjoint sets and the measure summed as defined in Section 3.3.3.2. Since our main interest is geometry, the σ-algebra of principal interest to us is that of Borel sets on $\mathbf{R}^{\mathrm{d}}$ where d is the dimensionality of some space. This family of Borel sets is abbreviated B^{d}. For convenience we define a *measurable space* as an ordered double $[X, \Upsilon]$ where X is a set and

Υ is the B-algebra of subsets of X. Thus a sample space is also a measurable space and a probability space is a measurable space with a probability measure specified. If we append a third element, the measure itself, to the measurable space, we form a *measure space*. Probability space is an example of a measure space.

3.2 Lebesgue Measure

This was defined in Chapter 3. In Section 3.3.3.3 we defined ϕ as a set function that could be Lebesgue measure. We now complete the proof. (The reader is encouraged to make a sketch using squares.) We assume that ϕ is additive on a system of elementary subsets. To demonstrate countable additivity, we consider a set A and a system of open

sets S. $A_n \in S, \quad A \subset \bigcup A_n$

which may be infinite. Choose also a family of open elementary sets G_n and a closed subset F of G_n such that $\phi(F) \geq \phi(A) - \varepsilon$. Each $G_n \supset$ each A_n, and define

$$\phi(G_n) \leq \phi(A_n) + \frac{\varepsilon}{2^{n+1}} .$$

For any positive ε, there exists by the Heine-Borel theorem, Section A.2.9, a finite subcover of F composed of elements of $\{G_n\}$ and therefore an integer N such that

$F \subset \bigcup_{i=1}^{N} G_i$. It now follows that

$$\phi(A) - \varepsilon \leq \phi(F) \leq \phi(\bigcup_{i=1}^{N} G_i) \leq \sum_{i=1}^{N} \phi(G_i) \leq \sum_{i=1}^{N} \phi(A_i) + \frac{1-\left(\frac{1}{2}\right)^N}{1-\frac{1}{2}} \cdot \frac{\varepsilon}{2}$$

from which by taking the limit as $N \to \infty$ and recalling that ϕ is additive on disjoint intervals,

$$\sum_{n=1}^{\infty}\phi(A_n) \le \phi(A) \le \sum_{n=1}^{\infty}\phi(A_n) + 2\varepsilon\,,$$

which implies countable additivity since epsilon is arbitrary. If the set function, while meeting the other criteria for measure, is countably additive, then it can be an integral since it is then possible to partition the set to be measured into an infinite number of disjoint subsets as is done in integration. For example, if we want to measure the content of a hyperball, we can divide it into an infinite number if hypercubes and sum their contents. This is important because we can understand the meaning of "content" for hypercubes by analogy with cubes and squares. We cannot visualize a hyperball and therefore can only measure its content (or "volume") by reason alone. The careful definition of Lebesgue measure allows us to do this integration.

3.3 Radon Measures

A generalization of Lebesgue measure called Radon measure is useful as a tool for comparing apparently different measures. The notion of a Radon measure is related to the Radon-Nikodym theorem, which we pursue first. See [Kolmogorov and Fomin, 1975] p.347. To do that we must first define charge. A σ-additive set function defined on a σ-ring or σ-algebra of subsets of a space X which takes both positive and negative signs is termed a *charge* or *signed measure* on X. The word is in analogy with the concept of electrical charge. Up until now in our discussion, measures have been non-negative. The definition of a negative charge on a set is so close to that of a negative value of an integral over some region, that additional definitions are not really needed except that a charge of zero is defined as both negative and positive. It follows from the definition of charge that if there is a charge with respect to some function on some set X then there must be both negative and positive subsets. (They may be null.) The union of these subsets is X and the separation into positive and negative subsets is called the *Hahn decomposition* if it is stated in terms of the subsets of X and the *Jordan decomposition* if it is stated in terms of the function.

If m is a σ-additive measure defined on a σ-algebra Υ_m defined on a space X and Φ is a charge defined on Υ_m, then if Φ is absolutely continuous with respect to m, then there is an m-summable function ϕ on X such that

$$\Phi(A) = \int_{A\subset\Upsilon_m}\phi(x)\,dm \qquad \text{(A.10)}$$

for every subset A of Υ_m. The function ϕ is unique other than for its values on a set of m-measure = 0. This is the *Radon-Nikodym theorem.* The function ϕ is called the *density* with respect to m and is denoted by $d\Phi/dm$. The form of the integral in Eqn. (A.10) is the Lebesgue integral.

The measure m in Eqn. (A.10) when Υ_m is a family of Borel sets is termed a *Radon measure.* Radon measures are finite on bounded sets as are probabilities (but probabilities are not negative). The way the Radon measure is used in Eqn. (A.10) may not be entirely clear. One way of visualizing the integral which is particularly useful in conjunction with Lebesgue measure on continuous Borel sets is to imagine the function ϕ partitioned into values of range $\phi(x)\pm d\phi/2$. The measure dm of the subset in A having the value of $\phi(x)\pm d\phi/2$ is then summed the form the integral. The dummy x is used to sweep ϕ through its entire range on A.

An extreme example of a Radon measure is the Dirac measure,

$$\delta_x(A)=\begin{cases}1 \text{ if } x\in A\\ 0 \text{ if } x\notin A\end{cases}. \tag{A.11}$$

It only takes two values for any union of sets. Since the point x can only fall in one of any number of disjoint sets, the Dirac measure posesses σ-additivity. The Radon-Nikodym theorem gives the conditions for one Radon measure Φ to be absolutely continuous with respect to another Radon measure m, that is, if one measure can be stated as an integral of another as m and Φ in Eqn. (A.10). We know that we can integrate delta functions, so the matter is resolved; the Dirac measure is a Radon measure.

3.4 Point Measures

The Lebesgue measure of a set of isolated points is zero as it is for any countable set. The Dirac measure of a set of isolated points is, however, finite. But, the Dirac measure only takes values of one and zero. Another Radon measure is the *Hausdorff measure* of [Stoyan at al, 1995] p.22. ([Mattila, 1995] states it differently.) It is defined for a set B in $\mathbf{R}^n$ as

$$h_k(B)=$$

$$2^{-k}\nu_k(\mathbf{b}_k)\liminf_{\delta\to 0}\left\{\sum_{i=1}^{\infty}\left(diam(M_j)\right)^k : B\subset\bigcup_{i=1}^{\infty}M_j, diam(M_j)\le\delta\right\}, \tag{A.12}$$

where here $\mathbf{b}_k$ denotes the point set of a k-dimensional ball of unit diameter; the *diam* function is defined as the *diameter* of a set,

$$diam(M) = \sup\left(\rho(\mathbf{x}_a, \mathbf{x}_b) : \mathbf{x}_a, \mathbf{x}_b \in M\right), \tag{A.13}$$

and M_j is any compact set in $\mathbf{R}^n$. This rather cumbersome measure is introduced because when $k = 0$, it counts the number of subsets in B. If B is a set of isolated points, it counts the number of points. But, if $k = n$, the Hausdorff measure is the same as the Lebesgue measure. It is thus possible to define a measure applicable to both countable and uncountable sets. Measures applicable to uncountable sets are sometimes called *diffuse measures*. Measures of countable sets are sometimes called *countable* or *atomic* measures. In practice we are usually concerned about the density of points in a set. If P is a set of points their *number density* in a set B is formally

$$\frac{h_0(P \cap B)}{h_n(B)}.$$

Besides the Dirac function, there is another function that appears in this book and in the literature that is associated with counting things. It is called the *indicator function,* $\boldsymbol{\varepsilon}()$. Relative to a set A, the indicator function is defined,

$$\boldsymbol{\varepsilon}_A(x) = \begin{cases} 1 & x \in A \\ 0 & x \notin A \end{cases}. \tag{A.14}$$

The indicator function is a Radon-measurable or simply a "measurable" function. It is also abbreviated as $\mathbf{1}(\cdot)$ in some texts.

There also exist Lebesgue measure definitions for the variance and covariance that are sometimes cited. Because they are indistinguishable from those given in Chapter 2, there is no need to repeat them here.

4. BIBLIOGRAPHY

These works contributed directly to the Appendix. If not explicitly cited, they are included as readable references for common elementary material included here.

B.V. Gnedenko, *The Theory of Probability*, trans. B. Seckler (Chelsea Publishing Company, New York, 1967).

F. Hohn, *Elementary Matrix Algebra*, The Macmillan Company, New York (1958).

A.N. Kolmogorov and S.V. Fomin, *Introductory Real Analysis,* trans. R.A. Silverstein (Prentice-Hall, Englewood Cliffs, NJ, 1970 reprinted Dover Publications Inc. New York, 1975).

P. Mattila, *Geometry of Sets and Measures in Euclidean Spaces* (Cambridge University Press, Cambridge, UK, 1995).

W. Rudin, *Principles of Mathematical Analysis* (McGraw-Hill, New York, 1953).

B. Schutz, *Geometrical Methods in Mathematical Physics* (Cambridge University Press, Cambridge University Press, Cambridge, UK, 1980).

D. Stoyan and H. Stoyan, *Fractals, Random Shapes and Point Fields* (John Wiley and Sons, NY, 1994).

D. Stoyan, W.S. Kendall and J Mecke, *Stochastic Geometry and Its Applications, 2nd Edition* (John Wiley and Sons, NY, 1995).

M. Woodroofe, *Probability with Applications* (McGraw-Hill, New York, 1975).

Index of terminology

MIX
Papier aus verantwortungsvollen Quellen
Paper from responsible sources
FSC® C105338

If you have any concerns about our products,
you can contact us on
ProductSafety@springernature.com

In case Publisher is established outside the EU,
the EU authorized representative is:
Springer Nature Customer Service Center GmbH
Europaplatz 3, 69115 Heidelberg, Germany

Printed by Libri Plureos GmbH
in Hamburg, Germany